Institute of Climate Change and Sustainable
Development at Tsinghua University

Li Zheng Wang Binbin

C+NbS

GLOBAL PRACTICES OF NATURE-BASED SOLUTIONS

——A Synergistic Exploration Towards the Net-zero Future

China Environment Publishing Group · Beijing

图书在版编目（CIP）数据

基于自然的解决方案全球实践 ：碳中和视角下的协同路径探索 = Global Practices of Nature-based Solutions：A Synergistic Exploration Towards the Net-zero Future ：英文 / 李政 (Li Zheng) ，王彬彬 (Wang Binbin) 主编. -- 北京 ：中国环境出版集团，2022.8

ISBN 978-7-5111-5151-3

Ⅰ. ①基… Ⅱ. ①李… ②王… Ⅲ. ①生态环境保护－案例－世界－英文 Ⅳ. ①X171.4

中国版本图书馆CIP数据核字(2022)第080241号

出 版 人　武德凯
责任编辑　丁莞歆
文字编辑　梅　霞
责任校对　薄军霞
装帧设计　金　山

出版发行　中国环境出版集团
（100062　北京市东城区广渠门内大街16号）
网　　址：http：//www.cesp.com.cn
电子邮箱：bjgl@cesp.com.cn
联系电话：010-67112765（编辑管理部）
010-67147349（第四分社）
发行热线：010-67125803，010-67113405（传真）
印装质量热线：010-67113404

印　　刷　北京中科印刷有限公司
经　　销　各地新华书店
版　　次　2022年8月第1版
印　　次　2022年8月第1次印刷
开　　本　787×960　1/16
印　　张　15.25
字　　数　560千字
定　　价　188.00元（全两册）

Editorial Board

Preface

Nature is not only the "victim" of climate change, but also the "battle ground" for climate change. Nature-based Solutions (NbS) emphasize on respecting law of nature, aiming to enhance the service functions of nature, control greenhouse gas emissions, increase carbon sinks, and improve our capacity to respond to climate risks through afforestation, strengthening farmland management, protecting wetland and ocean and other means of ecological conservation and restoration, as well as improving ecological management. NbS constitute a comprehensive approach to mitigating and adapting to climate change while improving climate resilience.

At the invitation of the United Nations (UN) Secretary-General at the UN Climate Action Summit in New York in September 2019, China and New Zealand jointly took the lead in proposing NbS, which calls for a more systematic understanding of the "harmony between man and nature" and a better understanding of the ecological value of the planet earth where human beings reside on. NbS also advocates for coping with climate risks by relying on the natural forces, building a society of low greenhouse gas emissions and high climate resilience, and fostering a community of life for human and nature. Over the past year or so, China has worked with all related parties to strengthen the implementation of NbS via planning, policies, technologies, investment, etc., and actively promote domestic actions and international cooperation.

In September 2020, President Xi Jinping pledged in his speech at the UN General

Assembly that "China will scale up its Nationally Determined Contributions (NDCs) by adopting more vigorous policies and measures. We aim to have CO_2 emissions peak before 2030, and achieve carbon neutrality before 2060." This is a major strategic decision made by China in its overall consideration of both domestic and international conditions, which further demonstrates China strategic commitment to the thorough implementation of Xi Jinping thought on eco-civilization, and the firm pursuit of green, low-carbon and circular development, as well as its responsibility as a major country in firmly supporting multilateralism, and actively promoting the building of a community with a shared future for humankind. On one hand, it is important to achieve carbon neutrality and emission reduction by accelerating transformation and innovation; on the other hand, it is necessary to increase carbon sinks by virtue of NbS.

Guided by the long-term goal of carbon neutrality, NbS plays an increasingly prominent role in adapting to climate change. In the process of deep emission reduction, the potential of technology in emissions reduction becomes increasingly limited, with increasing marginal cost. NbS can increase carbon sinks by relying on natural ecological functions in agriculture, forestry, ocean, wetland, and other fields, to offset carbon emissions from industry, transportation, and other hard-to-abate sectors, ultimately contributing to achieving carbon neutrality. Therefore, it is greatly important to strengthen NbS and integrate them into current emission reduction technologies and policies, which will play a positive role in renewing and strengthening China's NDCs, and achieving peak carbon emissions and carbon neutrality.

The global outbreak of COVID-19 warns us that we cannot consider development, biodiversity, climate, and the environment in isolation when coping with global challenges, we must integrate policies and actions that conserve biodiversity, address climate changes, and improve ecological and environmental quality into economic development, people's wellbeing improvement, employment promotion, health protection and national ecological security maintenance, thereby achieving synergy and promoting sustainable development. Fundamentally, we need to transform the traditional modes of production, life and consumption, promote economic and social transformation and innovation in technologies, systems and mechanisms, and pursue green, low-carbon and circular development. NbS, as an example of synergy, will produce a series of positive effects, including ecological restoration, biodiversity conservation, carbon sink storage, climate change adaptation, economic development, health improvement, and protection of the rights and interests of the public, especially women and children.

This book includes 28 NbS cases for climate changes from around the world, and it is conducive to learning and understanding the advanced concepts and experience of NbS, learning from excellent practices, and providing innovative ideas for China to promote ecological civilization construction. Meanwhile, multiple practices in China are also included in this report, providing Chinese wisdom and solutions for the global promotion of NbS.

2021 to 2030 is a critical period for global ecological conservation and climate governance where global biodiversity conservation and climate change adaptation will become more synergistic and mutually reinforcing. This book hopes to act as a catalyst for cross-sector dialogue and cooperation, exploit the potential of nature to a greater degree, stimulate momentum in innovation, and make positive contributions to promoting global ecological civilization construction, building a community of life for man and nature, and maintaining a better future for humankind.

China's Special Envoy for Climate Change Affairs

Xie Zhenhua

October 6, 2021

Introduction

NbS were first proposed by the World Bank in its 2008 report entitled *Biodiversity, Climate Change and Adaptation: Nature-based Solutions from the World Bank Portfolio*, highlighting the importance of biodiversity conservation for climate change adaptation and mitigation. In 2009, the International Union for Conservation of Nature (IUCN) submitted a proposal report to the 15th Conference of Parties to the *United Nations Framework Convention on Climate Change* (UNFCCC COP15), emphasizing the important role of NbS in addressing a series of societal challenges such as climate change (IUCN, 2009).

The NbS were applied in depth at the regional level at first in the European Union (EU). In 2014, the EU launched the "Horizon 2020" agenda, which emphasized the role of science and technology in promoting employment and economic growth. In 2015, the EU incorporated NbS into the agenda to scale up the research and pilot projects. In the same year, the EU published a report entitled *Nature-based Solutions and Re-naturing Cities*, in which, NbS were defined as solutions that are "inspired and supported by nature, which are cost-effective, simultaneously provide environmental, social and economic benefits and help build resilience". It aims to emphasize that NbS turn the current challenges into innovative opportunities and change the natural capital into a source of green economic growth (European Union, 2015).

At the World Conservation Congress (WCC) 2016, the IUCN adopted the definition of NbS as "actions to protect, sustainably manage, and restore natural and modified ecosystems that address societal challenges effectively and adaptively, simultaneously providing human well-being and biodiversity benefits". In the same year, the IUCN released a research report in which the concept, connotations, social value, implementation solutions and practice cases of NbS were systematically summarized (Cohen-Shacham E et al., 2016).

The Nature Conservancy (TNC) published an article in October 2017 together with expert teams from 15 research institutions in order to estimate the NbS' potential in climate change mitigation in a more scientific way. They analyzed the potential emission reduction benefits of 20 nature-based climate change mitigation pathways, and quantified the enormous potential of NbS to provide up to 37% of the emission reductions needed by 2030 to keep global temperature increases under 2°C (Griscom B W et al., 2017).

The UN Climate Action Summit held in New York in September 2019 is a milestone in the development of NbS. NbS were listed as one of nine tracks for accelerated action. The summit saw the establishment of a global NbS coalition co-led by China and New Zealand, and the coalition launched the coalition's latest reports, The *Nature-based Solutions for Climate Manifesto(Climate Manifesto* in short*)* and the *Compendium of Contributions Nature-based Solutions(Compendium* in short*)*. The *Climate Manifesto* has won the support of more than 70 governments, private sectors, civil societies and international organizations, laying a solid foundation for further research and practice of NbS. The *Compendium* compiles 196 cases and initiatives in such areas as forestry, agriculture, ocean, water resource, biodiversity conservation and desertification control, providing new ideas for global climate actions.

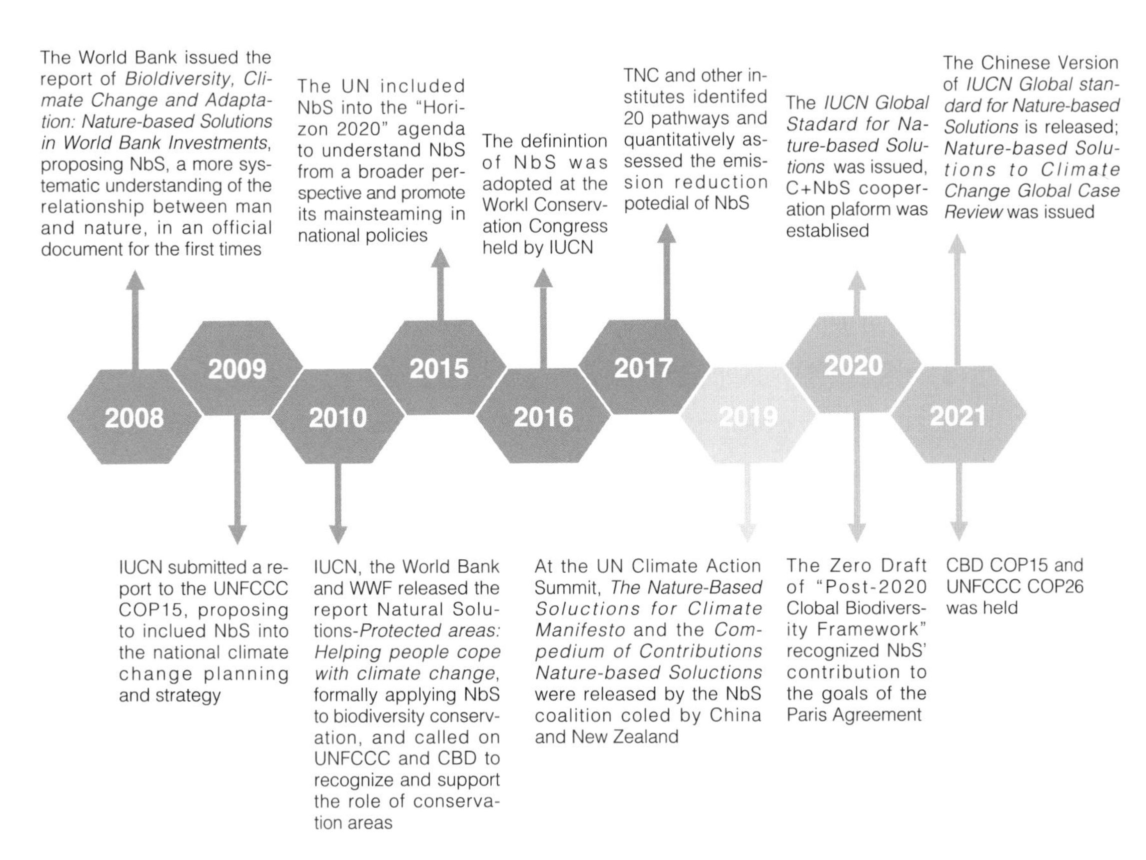

Proposal and Development Milestones of NbS

In 2021, the 15th Conference of Parties to the *Convention on Biological Diversity* (CBD COP15) was held in Kunming, China, and the 26th Conference of Parties to the *United Nations Framework Convention on Climate Change* (UNFCCC COP26) was held in Glasgow, United Kingdom. NbS, as the important link between biodiversity conservation and climate change response, was an important topic at the two conferences. As one of the co-leaders of the global NbS coalition, China is actively learning and practicing NbS-related concepts, and requires access to advanced international experience, especially specific cases that are in line with China's national conditions, have guiding and practical significance, and can be promoted and implemented. Additionally, NbS are in a critical period of development. China, as the initiator of the concept of ecological civilization, has carried out many successful practices in coordinating ecological conservation and climate change mitigation at home, providing innovative and promising new ideas for governments, enterprises and organizations in other countries, especially the developing countries, to address climate challenge and benefit the people by NbS.

Based on the *Compendium of Contributions Nature-based Solutions*, with the NbS case studies conducted by other international organizations, regional governments, universities and think tanks before and after the UN Climate Action Summit in 2019, a case database consisting 300 NbS cases from around the world classified based on characteristics is established in this book. In addition to the 4 major ecosystems involved in NbS, namely forestry, grassland, agriculture and wetland, this book adds four new NbS application areas of city, country, platform & initiative and corporation, aiming to help NbS project designers around the world to understand the latest trends and provide them with innovative ideas.

To further summarize the advantages of cases and explore outstanding cases in line with China's national conditions, a new set of case screening and evaluation criteria is formulated in the book under the guidance of the Sustainable Development Goals of the United Nations and the idea of ecological civilization, with reference to the *IUCN Global Standard for Nature-based Solutions* (*Global Standard* in short), and in combination with China's national conditions and local characteristics. These criteria include six primary screening indicators and three advanced evaluation dimensions, in the hope of providing references for policymakers, research institutions, non-governmental organizations (NGOs), enterprises, communities and individuals. In this book, 28 outstanding cases were selected from 300 cases, including ten Chinese cases and 18 foreign cases. The foreign cases were implemented in Europe, North America, South America, Africa, Asia and Oceania, making them highly representative. This book also analyzes the success factors of each case and sums up the characteristics of cases in the same category, providing targeted references for the world, especially developing countries, in the implementation of NbS projects in related fields, planning of urban NbS development, deployment of national NbS strategies, and encouraging enterprises to participate and play an active role in NbS.

Two observations were made during the preparation of this book.

Firstly, with the evolution of NbS, an increasing number of cases comprehensively highlights the contributions and potential of NbS in ecological conservation, however there remains a lack of review from the perspective of climate change. This book is intended to recommend cases on climate change that are differentiated from other general NbS case reports; thus, greater emphasis is placed on climate factors in the design of screening indicators and dimensions. However, it was found in the actual screening process that few NbS have been truly designed in mind for climate change. Climate change is mentioned in some cases, but information on measures for mitigating or adapting to climate change, benefits and subsequent improvements is still insufficient to support the in-depth analysis of case reports.

Secondly, the definition of NbS, which is now widely accepted internationally, is given by the IUCN as "actions to protect, sustainably manage, and restore natural and modified ecosystems that address societal challenges effectively and adaptively, simultaneously providing human well-being and biodiversity benefits". This definition emphasizes the conservation, management and restoration of natural or artificial ecosystems. From these three perspectives, NbS can provide up to 37% of the emission reductions needed by 2030 to keep global temperature increases under 2°C. At the 9th meeting of the Central Financial and Economic Affairs Commission in March 2021, President Xi Jinping pointed out that "to peak carbon emissions and achieve carbon neutrality is an extensive and profound systemic reform for the economy and society, and should be incorporated into the overall layout of building an ecological civilization." He called for a spirit of perseverance in achieving the goals of peaking carbon emissions by 2030 and achieving carbon neutrality by 2060. Carbon neutrality is achieved by more innovation, which means we should exploit the potential of nature to a greater degree. The new energy is mainly derived from nature with the technological innovation, which embodies "solutions that are inspired and supported by nature, and draw upon nature" (European Union, 2015). This book selects the case of Baofeng Energy Group which produces green hydrogen through water electrolysis using solar energy. Although this case does not meet the IUCN's definition of NbS, the production process of green hydrogen is a technological innovation that utilizes nature without endangering under the approach of exploiting the potential of nature to find innovative solutions.

In view of the above three observations, we propose the following ideas and suggestions.

Firstly, global ecological conservation and climate governance will proceed to a critical period from 2021 to 2030. In which the global biodiversity conservation and climate change response will become more synergistic and mutually reinforcing. NbS are originally intended to tackle climate change and emphasize the importance of biodiversity conservation for climate change adaptation and mitigation. It is suggested that we should further develop and promote NbS-related work from the perspective of climate and biodiversity synergy in the process of achieving carbon neutrality, and contribute additional useful cases of NbS to climate change in the critical period of global governance in the next decade.

Secondly, carbon neutrality must be achieved with systemic changes. It is suggested that we plan the present from the perspective of the future and rethink the connotations and extension of NbS from the perspective of openness and inclusiveness, so as to adapt to the new situation and make NbS more inclusive, leaving ample exploration space for all parties. By tapping the full potential of nature, NbS will contribute more to the overall situation than is currently assessed.

This book was prepared under the careful guidance of Xie Zhenhua, China's Special Envoy for Climate Change Affairs. Professor He Jiankun, Director of the Academic Committee of the Institute of Climate Change and Sustainable Development, Tsinghua University, serves as the General Advisor steering the strategic direction. Liu Ning, Deputy Director-general of the Nature and Ecology Conservation Department of the Ministry of Ecology and Environment, has been encouraging and following the progress of relatea work. The core contributors to this book are from the Institute of Climate Change and Sustainable Development, Tsinghua University. Huo Li from TNC and Chang Jiang, a researcher at the Chinese Research Academy of Environmental Sciences, made their contributions at different stages of compilation. Luo Ming, Deputy Director of the Land Consolidation and Rehabilitation Center of the Ministry of Natural Resources, Zhu Chunquan, Head of the Nature Initiatives and Tropical Forest Alliance at the World Economic Forum in Greater China, and Dr. Li Lin, Director of Global Policy and Advocacy at the World Wide Fund for Nature (WWF), provided comments for this book as external reviewers. We hereby extend our sincere thanks to all the experts and colleagues who participated in and supported the preparation of this book.

As a strategic partner of the C+NbS cooperation platform, The Nature Conservancy (TNC) plays a key role in NbS development and it renders a full range of support to the work of the Institute for Climate Change and Sustainable Development (ICCSD) of

Tsinghua University. Special thanks go to Ms. Ma Jinhong, the chief representative of TNC China, and her team. Thanks for the support for the C+NbS cooperation platform from the Global Climate Change and Green Development Fund of Tsinghua University Education Foundation, Yanbao Charity Foundation and HSBC Bank.

This book has refer to the *Global Standard*. Zhang Yan, director of the International Union for Conservation of Nature China office, has provided us a lot of assistance. On June 23, 2021, IUCN released the Chinese version of *Global Standard*. And this book is honored to be the first one to refer to the standard.

This book is an integral part of the achievements of C+NbS cooperation platform. Since its launch in April 2020, ICCSD has organized nine monthly workshops and established close relationship with relevant NbS institutions and counterparts both at home and abroad, influencing over more than 600,000 people worldwide. In the future, ICCSD will remain engaged with partners in the global theoretical research and follow up the progress of NbS. It will continue to provide advice for domestic development and contribute to the goal of carbon neutrality. At the same time, it aims to build a bridge between China and the world, and offer China's solution to global NbS and the harmonious coexistence between human and nature.

Editors-in-chief
Executive Dean of the Institute of Climate Change and Sustainable Development,
Tsinghua University
Li Zheng

Dean Assistant of the Institute of Climate Change and Sustainable Development,
Tsinghua University
Wang Binbin
July 1, 2021

Contents

Part 1 Methods

Part 2 Excellent Cases

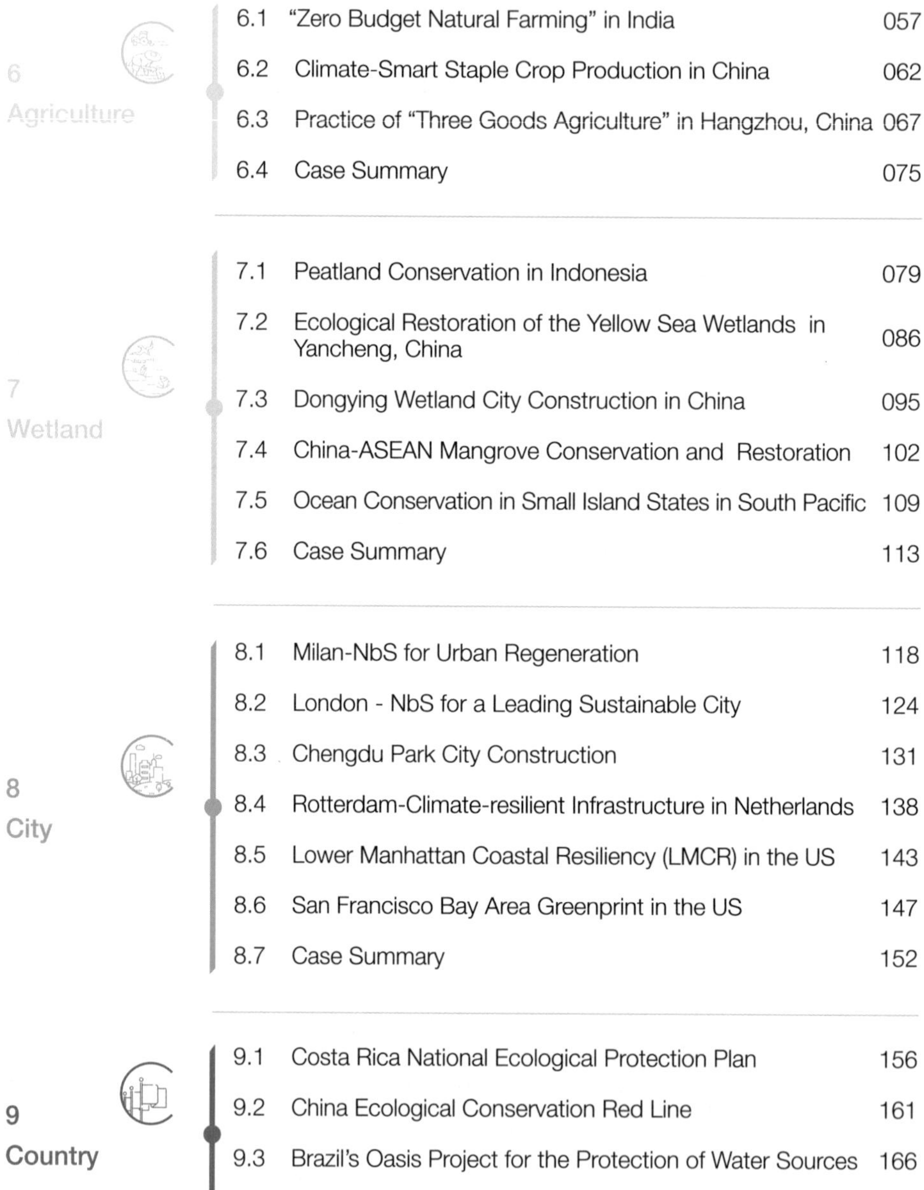

C+NbS

Part 1

Methods

1 Research Schedule

The research of this book first initiated in June 2020, and lasted for 12 months. The Chinese and English version of the *Best Practices of Nature-based Solutions—A Synergistic Exploration Towards the Net-zero Future* was released in July 2021 (Figure 1-1):

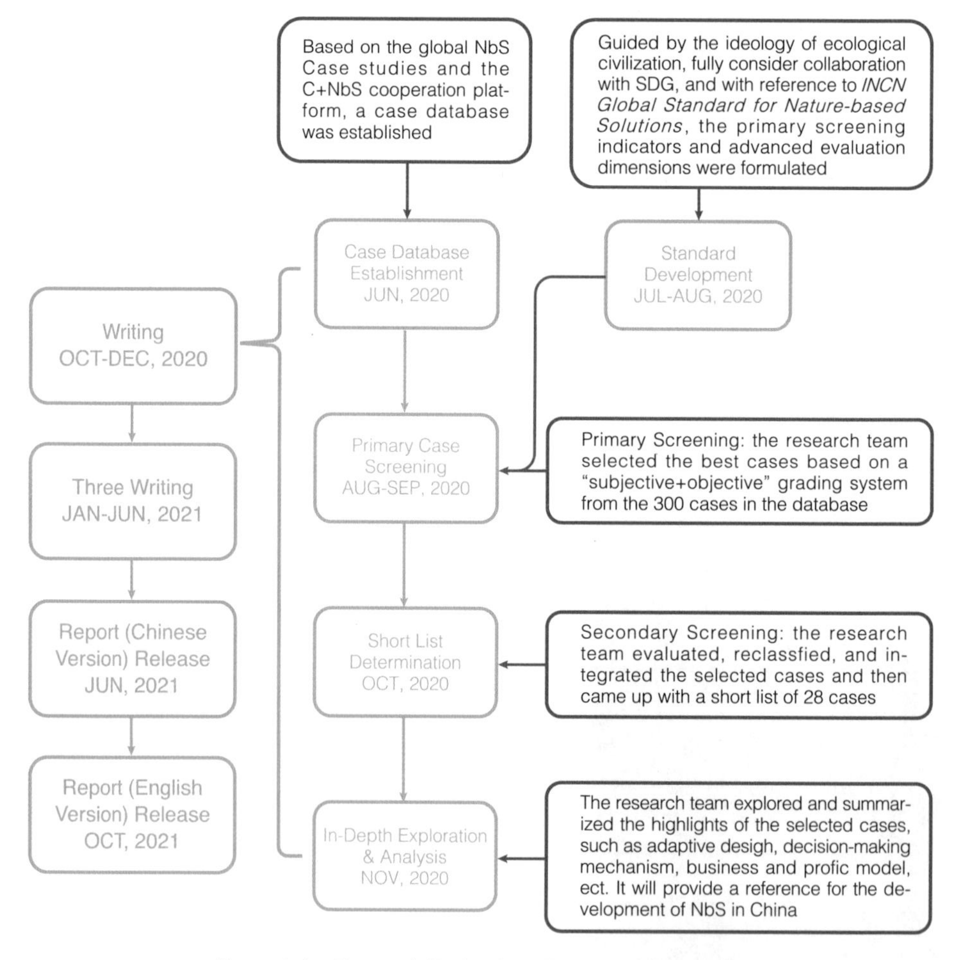

Figure 1-1 Research Technology Route and Schedule

2 Case Database

A case database is established for this book based on the existing NbS case studies around the world and the C+NbS cooperation platform initiated in China. There are 300 cases in the case database, which are selected from three types of case sources categorized by time after the United Nations Climate Action Summit in 2019 (Summit in short), and between the time of the platform launch and report release time (Table 2-1).

The first type of case source mainly includes the NbS case platform launched in the EU's "Horizon 2020" agenda and the NbS-related research reports released by C40 Cities and The Nature Conservancy before the Summit.

The second type of case source includes the report *Compendium of Contributions: Nature-based Solutions* released by the UNEP during the Summit and the platform it established at the Summit.

The third type of case source includes the studies and reports released by WWF and Nature4Climate (N4C) after the Summit, and the NbS case study platform launched by TNC, IUCN, Oxford University, and the Institute of Climate Change and Sustainable Development (ICCSD) of Tsinghua University.

Table 2-1 Reports and Platforms Referenced by the Case Database

No.	Institution	Name	Type	Release Date
1	EU	EU City NbS Case Studies - Naturvation Platform https://naturvation.eu/cities	Platform	"Horizon 2020" Agenda in 2014
2	EU	EU City NbS Case Studies - Oppla Platform https://oppla.eu/nbs/case-studies	Platform	"Horizon 2020" Agenda in 2014
3	C40 Cities	*Sustainable Urban Regeneration Strategies: Through Green Infrastructure for Climate Change Adaptation*	Report	2016
4	TNC	*Beyond the Source: The environmental, economic and community benefits of source water protection*	Report	2018
5	UNEP	UNEP - NbS Contributions Platform www.unenvironment.org/nbs-contributions-platform	Platform	UN Climate Action Summit in September, 2019
6	UNEP	*Compendium of Contributions: Nature-based Solutions*	Report	UN Climate Action Summit in September, 2019
7	WWF	*Climate, Nature and Our 1.5°C Future*	Report	December, 2019
8	N4C	N4C Global NbS Case Studies http://nature4climate.org/nbs-case-studies/	Platform	Launched in 2019
9	IUCN	*Global Standard for Nature-based Solutions*	Report	July, 2020
10	Oxford University	Oxford University NbS Initiative - Case Studies http://www.naturebasedsolutionsinitiative.org/nbs-case-studies/	Platform	Launched in 2020
11	TNC	TNC Global NbS Case Database	Platform	Launched in 2020
12	ICCSD	C+NbS Cooperation Platform	Platform	Launched in 2020

The 300 cases are divided into eight categories, mainly based on the initiator, targets and the types of land use: forestry, grassland, agriculture, wetland, city, country, platform & initiative, and corporation (Figure 2-1). The cases in each category are screened, refined, and summarized with the aim of providing a reference for different parties to carry out similar projects.

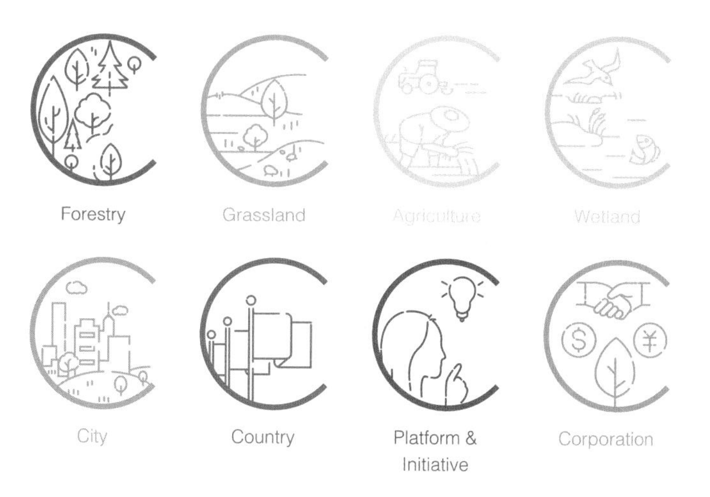

Figure 2-1 Case Classification of this Book

3 Research Criteria

This book is developed under the principal of global climate change mitigation, identifying a series of preliminary evaluation criteria suitable for promoting NbS rollout in China. In addition, comprehensive evaluation on hundreds of excellent global NbS projects to climate change were conducted for in-depth interpretation and analysis.

Guided by the idea of ecological civilization, incorporating the 17 UN Sustainable Development Goals (SDGs), with reference to the *Global Standards* (Appendix 1), this book has developed the research criteria into two parts: the primary screening indicators and the advanced evaluation dimensions.

There are six primary screening indicators which mainly focus on each case's ability to solve problems and protect the interests of stakeholders in order to select cases that can effectively address societal challenges from hundreds in the case database. Furthermore, in this book formulates three advanced evaluation dimensions: governance models, financial and market mechanisms, as well as diversity and inclusion. These dimensions facilitate the secondary screening and in-depth analysis on the selected primary cases, so as to ensure that they have profound reference and learning significance for countries given the different conditions.

By focusing on the adaptability of NbS, especially in developing countries, this comprehensive research criterion is expected to extract worthwhile lessons from good NbS projects for the world. Placing important application in the scientific design, application, and conclusion of the contribution and value of NbS to climate change.

3.1 Guiding Ideology: Concept of Ecological Civilization

The 17 SDGs and 169 targets issued by the UN in 2015 are designed for guiding the global sustainable development and put forward guiding suggestions for different areas of sustainable development (Figure 3-1). With a similar purpose to address societal challenges, the implementation of NbS can bring co-benefits to sustainable development at social, economic, and environmental levels. Christopher M Raymond has developed a framework for assessing and implementing the co-benefits of NbS in urban areas, including four dimensions (climate and environment, biodiversity, ecosystems, and socioeconomics) and ten challenge areas (Raymond et al., 2017). In the book "Nature-based Solutions for Resilient Ecosystems and Societies", the authors have analyzed the pathways through which NbS can help achieve the 2030 Agenda and the 17 SDGs (Acharya P et al., 2020). And in 2020, WWF released the report "Nature in All Goals", presenting 17 NbS cases to reveal their contribution to the achievement of each SDGs

Figure 3-1 UN Sustainable Development Goals (Source: United Nations)

(Osieyo, 2020). Therefore, when developing the criteria, this book focuses on the synergies of the indicators and dimensions with the SDGs, in order that the selected cases can bring co-benefits to the sustainable development of the region.

In addition, the research criteria also reflect the content of the concept of ecological civilization. "Ecological Civilization" is a form of human civilization based on human reflection on traditional industrial civilization following the primitive civilization, agricultural civilization and industrial civilization. This concept first appeared in the 1980s and was put forward by environmentalists in the former Soviet Union in the article *Ways of Training Individual Ecological Civilization under Mature Socialist Conditions* (Ye Q, 1984). In 2007, China marked it as an explicit goal of the country and made the term widely used after that (Liu Jing, 2011). "To respect, conform to, and protect nature" in the theory of ecological civilization is consistent with the basic principle of "protecting the natural ecological bottom line, restoring the natural ecological background, and respecting the laws of nature" of NbS (Liao Maolin, 2020). Both ecological civilization construction and NbS share the same fundamental goal to seek the harmonious coexistence of mankind and nature. Based on the symbiotic relationship between humankind and nature, NbS advocate the sustainable use of ecological service value to create synergy among nature, society and the economy, so as to cope with societal challenges such as climate change (He Qingtang, 2019). The theory of ecological civilization is the top-level design and strategy for guiding human development, while NbS are the methodology of ecological civilization in practice (Lu Feng, 2020). Excellent NbS should be consistent with the theory of ecological civilization in both design and practice. The connotation of ecological civilization is embodied in the following eight "views" (Table 3-1) and they can find their equivalence in the research criteria of this report.

Table 3-1 Connotations of China's Ecological Civilization

No.	Connotation	Interpretation
1	The view of history	Ecological prosperity leads to civilization prosperity, while ecological decline leads to civilization decline
2	The view of natural science	We must endeavor to foster a new relationship where human and nature can both prosper and live in harmony
3	The view of green development	Lucid waters and lush mountains are invaluable assets to society. Thus, improved environment is boosted productivity
4	The view of public well-being	A sound ecological environment is the most inclusive well-being of the people
5	The view of systemic governance	Protecting the ecosystem requires us to follow the innate laws of the ecosystem and properly balance all elements and aspects of nature
6	The view of rule of law	The promotion of ecological civilization from a national perspective must rely on institutions and the rule of law
7	The view of public participation	The construction of ecological civilization is a common cause for all humankind, requiring both top-down institutional design by the government and bottom-up action by its citizens
8	The view of global multilateralism	International law and the principle of equity and justice are the basis of global environment governance

3.2 Primary Screening Indicators

Under the guidance of SDGs and the idea of ecological civilization, in combination with *Global Standard*, six primary indicators were established in this book to screen out cases from the global case database to effectively cope with climate challenges, protect people's rights and interests, and achieve multiple benefits (Table 3-2) .

1. Mitigating and Adapting Climate Change

NbS were first proposed to deal with climate change, which tops the list of the seven societal challenges defined in the *Global Standard*. NbS aiming at coping with climate change should play an active role in either climate mitigation and/or adaptation:

Table 3-2 Primary Screening Indicators

No.	Primary Screening Indicators	UN SDGs	Ecological Civilization	*IUCN Global Standard*
1	Mitigating and Adapting Climate Change	SDG13 Climate Action	Connotation 8: the view of global multilateralism	Criterion 1: NbS effectively address social challenges
2	Preserving Biodiversity	SDG 14 Life Below Water SDG 15 Life on Land	Connotation 1: the view of history Connotation 2: the view of natural science	Criterion 1: NbS effectively address social challenges Criterion 3: NbS result in a net gain to biodiversity and ecosystem integrity
3	Addressing Multiple Societal Challenges	SDG2 Zero Hunger SDG3 Good Health and Well-being SDG6 Clean Water and Sanitation SDG9 Industry, Innovation and Infrastructure SDG11 Sustainable Cities and Communities	Connotation 4: view of public well-being Connotation 8: the view of global multilateralism	Criterion 1: NbS effectively address social challenges Criterion 8: NbS are sustainable and mainstreamed within an appropriate jurisdictional context
4	Achieving the Synergy of Multiple Objectives	—	Connotation 5: the view of systemic governance	Criterion 2: Design of NbS is informed by scale Criterion 6: NbS equitably balance trade-offs between achievement of their primary goal(s) and the continued provision of multiple benefits
5	Promoting Local Economic Development	SDG1 No Poverty SDG8 Decent Work and Economic Growth	Connotation 3: the view of green development	Criterion 4: NbS are economically viable
6	Caring for Vulnerable Populations	SDG 3 Good Health and Well-being SDG10 Reducing Inequalities	Connotation 4: view of public well-being	Criterion 5: NbS are based on inclusive, transparent and empowering governance processes

"mitigation" refers to reducing the concentration of greenhouse gases in the atmosphere by reducing emission sources and increasing carbon sinks; and "adaptation" is mainly reflected in reducing exposure and vulnerability, improving the adaptability of the whole of community, and effectively resisting the impacts and challenges brought by extreme climatic events and changes in average temperature (drought, sea level rise, etc.).

2. Preserving Biodiversity

NbS themselves originate from the goods and services provided by the ecosystem and rely heavily on the integrity and diversity of that ecosystem. The *Global Standard.* proposes that NbS should result in a net-gain to biodiversity and ecosystem integrity (Criterion 3). Therefore, NbS projects should strive to achieve a net-gain in biodiversity and ecosystem integrity on the premise of ecological environment protection, and ecosystem and biodiversity conservation, so as to maintain its sustainable operation.

3. Addressing Multiple Societal Challenges

The IUCN has defined seven important societal challenges facing mankind: climate change mitigation and adaptation, disaster prevention and reduction, economic and social development, human health, food security, water security, and eco-environmental degradation and biodiversity loss (Figure 3-2). If NbS projects can address other societal challenges while addressing climate change and biodiversity loss, the co-benefits can create synergy achieve phenomenon results.

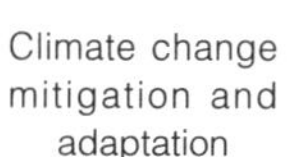

Figure 3-2 Societal Challenges Addressed by NbS
(Source: IUCN)

4. Achieving the Synergy of Multiple Objectives

The ecosystem is an interdependent and interconnected system of elements. Good NbS projects must balance the interests of social development, economic progress, environmental protection, and other aspects, and achieve multiple goals in a synergistic manner. The *Global Standard* requires that the design of NbS should recognize the interactions among the economy, society and ecosystems (Criterion 2), fairly balance the primary goal and other benefits (Criterion 6), take into account all elements of the natural ecology, including those on and under the mountains, above and below the ground, on the land and in the sea, and in the lower and upper reaches of river basins, and carry out overall protection, macro-control, and comprehensive management. Therefore, in the design and implementation of NbS, we should adhere to the view of systematic governance, trade off the interests of all parties, take overall optimization as the goal, and maintain ecological integrity.

5. Promoting Local Economic Development

The *global standard* requires that NbS must be economically viable (Criterion 4), realize profit, increase economic income, contribute to the market and employment while coping with climate change and protecting the ecological environment, and gradually break away from capital investment to achieve their own sustainability, which is key to the long-term effectiveness of NbS. It is an important means of achieving green growth and poverty reduction to implement NbS to protect the local environment and stimulate economic development.

6. Caring for the Vulnerable Populations

"The living conditions of vulnerable groups directly determine the stability and health of a well-ordered society, and they are also an important symbol of social fairness and justice (Zou Haigu, 2012)." The *Global Standard* requires NbS to respect and uphold legal and customary rights to access, use, and manage land and natural resources based on inclusive, transparent, and empowering governance processes (Criterion 5), especially protecting the related rights of vulnerable and marginalized groups. The implementation of NbS projects should ensure social equity, especially by taking into account the special needs and interests of vulnerable groups such as women and children, the disabled, patients with chronic diseases, the elderly, ethnic minorities and indigenous peoples.

In this book, the six primary screening indicators are scored combining objective and subjective criterias, offering objective statistics of the expression and quantitative data of given indicators in case materials and reports, and subjective judgment of the highlights, local adaptability, and replicability of cases. See Table 3-3 for the specific scoring standards of each indicator. By calculating and comparing the total scores of all cases in a certain category (forestry, grassland, agriculture, wetland, city, country, platform & initiative, and corporation), four to six best cases in this category were identified to enter secondary screening and in-depth case study. In Appendix 2 for the scores of the final selected cases.

Table 3-3 Scoring Criteria

Score	Description
0	The case does not reflect the content corresponding to the indicator, or reflects part of the contents of the indicator, and it has no substantive results
1	The case reflects the content corresponding to the indicator, and the description is more detailed and has some substantive results, but the scale is small and replicability is not strong
2	The case reflects the content corresponding to the indicator, the description is detailed with quantitative and substantive results, and has a certain scale effect or policy reference value for other regions and even the whole world
3	The case reflects the content corresponding to the indicator, the description is detailed with outstanding quantitative and substantive results, has a scale effect and positive significance for local, regional, and even global aspects, and has far-reaching policy reference value

3.3 Advanced Evaluation Dimensions

In order to fully explore the highlights and advantages of case management and decision-making mechanism design, marketization, scale, and benefits, make the selected cases better serve domestic and foreign policymakers, and provide stakeholders with ideas and methods worth learning, the in-depth exploration covers the three advanced evaluation dimensions of governance models, financial and market mechanisms, diversity and inclusion (Table 3-4), and conducts the secondary screening, classification, and integration of selected outstanding cases, ultimately obtaining a shortlist of 28 cases for in-depth analysis and potential exploration (Appendix 3).

Table 3-4 Advanced Evaluation Dimensions

No.	Advanced Evaluation Dimensions	Ecological Civilization	UN SDGs	*IUCN Global Standard*
1	Governance Models	Connotation 6: the view of rule of law	SDG16 Peace, Justice, and Strong Institutions	Criterion 2: Design of NbS is informed by scale
			SDG17 Partnerships for the Goals	Criterion 5: NbS are based on inclusive, transparent and empowering governance processes
				Criterion 7: NbS are managed adaptively, based on evidence

(continued Table)

No.	Advanced Evaluation Dimensions	Ecological Civilization	UN SDGs	*IUCN Global Standard*
2	Financial and Market Mechanisms	—	SDG8 Decent Work and Economic Growth	Criterion 2: Design of NbS is informed by scale
				Criterion 4: NbS are economically viable
3	Diversity and Inclusion	Connotation 7: the view of general participation	SDG5 Gender Equality	Criterion 5: NbS are based on inclusive, transparent and empowering governance processes
			SDG17 Partnerships for the Goals	

1. Governance Models

Governance models and decision-making management mechanisms are the key factors to the success of NbS projects. An excellent governance model can realize adaptive management and innovation on the premise of ensuring the effective operation of the project through multi-stakeholder participation. An effective innovation governance model should be based on local geographical and cultural characteristics, including the linkage between top-down and bottom-up models, open supervision model, reliable information communication and dialogue mechanism, innovative performance evaluation system, and so on. A framework for iterative learning that enables adaptive management is required, with which project designers regularly monitor and evaluate projects based on scientific understanding, traditional practices, and local knowledge, adjust the implementation plan in time, minimize the waste of manpower, capital, and other resources, improve the success rate and durability of NbS, and effectively promote the efficient use of resources.

2. Financial and Market Mechanisms

Financing is a necessary process for the implementation of NbS projects, and NbS are funded by governments, enterprises, international organizations, foundations, individuals, and so on. Excellent NbS should use or design reasonable and effective financial instruments and market mechanisms to avoid unnecessary risks and ensure the stable supply of funds, thus increasing the possibility of the long-term success of NbS; Meanwhile, the *Global Standard* requires NbS to actively seek complementarity with other types of projects such as engineering projects, information technology, and financial measures thereby realizing collaborative management between NbS and other departments of social development (criterion 2).

3. Diversity and Inclusion

The basic principles of NbS require projects to produce social benefits in a fair and just manner, and promote the wide participation of all-stakeholders. Excellent NbS should ensure the diversity of behavioral agents, pay attention to urban-rural balance and generational balance, realize the full participation of government, enterprises, international organizations, think tanks, communities and individuals, focus specially on gender balance, and encourage women to participate in NbS(Column 1).

Column 1

As the main force for the design, implementation and further promotion of NbS, central and local governments can provide convenient social conditions for an extensive NbS projects, including formulating relevant policies and facilitating the rule of law, promoting ecosystem payment markets and certification projects, furthering natural capital investment, improving land use policies, etc. The government's attention to and participation in NbS can promote the perfection of the relevant policies, laws and regulations, and is conducive to the mainstream development of NbS.

The private sector is also an important stakeholder of NbS, constituting the backbone force for promoting NbS from theory to practical application. Enterprises' participation in the investment and design of NbS projects is not only a manifestation of them fulfilling their social responsibility and coping with social challenges, but also an important way in which they can improve the sustainability of their own production and operation, expand their businesses, and increase their profits.

International organizations such as the United Nations, World Bank, IUCN, WWF and TNC are the original proposers and the most important early promoters of the concept of NbS, participating in the design, funding and concrete implementation of many NbS projects.

Research institutions, universities and think tanks can provide an important theoretical basis and support for NbS, incorporate NbS into their policy recommendations to advance the formulation of the relevant laws and regulations, as well as promoting the development of NbS in the form of funding specific projects.

Individuals and communities are the most important implementers of projects and the direct beneficiaries of NbS. It is vitally important for NbS to balance the stakeholders by fully considering their interests and bringing them into the implementation and management.

Ensuring the participation of women is an important manifestation of the inclusiveness of NbS. Women are vulnerable to the negative impacts of climate change and ecological damage, but they form an important force for promoting economic and social development and participating in decision-making and management of NbS. As such, it is necessary to pursue mutual respect for gender equality, encourage women to play roles in NbS projects, and improve their enthusiasm and participation in public affairs.

Part 2

Excellent Cases

4 Forestry

Ecosystems such as forests have enormous social and economic value. They are the home of 350 million people, and support the livelihoods of more than 1.5 billion people (20%) around the world, while providing employment for 50 million people (Sunderland, 2013). At the same time, forests are an important resource for humankind to tackle climate change. To mitigate climate change, forests can produce carbon sinks to absorb carbon dioxide from the atmosphere; furthermore, forest management and restoration can avoid or reduce the carbon accumulated in the past decades or even hundreds of years to be decomposed and discharged into the atmosphere in a short time. To adapt to climate change, increased forest vegetation can reduce soil erosion, effectively conserve soil and water, and increase community income to improve its resilience to climate change. Forestry is one of the most typical sectors where NbS play a role, and there are a large number of cases both domestically and abroad dedicated to the conservation, restoration, and sustainable management of forests. In the category of forestry, three cases were selected to provide a reference for the design, planning and implementation of NbS forestry projects: National Forest Conservation Program of Colombia, Community Forestry Campaign in Nepal, and the "Trillion Trees" Initiative.

4.1 National Forest Conservation Program of Colombia

Case Highlights

Long-term sustainable development of protected areas can be promoted through innovative financing plans. The "Heritage Columbia Plan" program attempts to create an extinguishable transition fund that leverages the economy with public and private investments to ensure the possibility of long-term investment in a period of 20 years, and effectively help build partnerships between governments and international development agencies, donors, private sectors and NGOs to increase the efficiency of the use of funds.

4.1.1 Project Background

Located in the northern part of South America, Colombia has a tropical climate that varies with the terrain. Topographically, Colombia is roughly divided into mountainous areas in the west and plains in the east. In the west, there are mainly the Andes Mountains, and many coastal plains; in the east, there are mainly the Orinoco and Amazon plains. Due to geographical conditions, Colombia is rich in landscape types, including the Andes, the Amazon Forest, the Pacific and Caribbean coast, and the Orinoco plain. Colombia has a history of disasters due to strong climatic anomalies (mainly related to El Nino) and extreme hydro-meteorological events. Climate change increases average temperatures, changes precipitation patterns, and exacerbates local droughts, floods, and landslides, posing different threats to different parts of Colombia. In this context, Colombia's forests are particularly valuable. Therefore, Colombia's climate change policies attach great importance to the role of forest resources in improving landform and landscape management, with emphasis on forest management in national protected areas and the rational and efficient use of exploitable land resources in the vicinity of protected areas.

Colombia has about 60 million hectares of natural forests (WWF, 2019) which store about 26 billion tons of carbon dioxide. Yet many of Colombia's protected areas are inefficiently managed due to the lack of resources. From 1990 to 2016, about six million hectares of forests were cut down across the country, one of the important reasons being the encroachment of surrounding residents on protected areas and natural resources. Deforestation accounted for about 16% of Colombia's total annual greenhouse gas emissions in 2012. Colombia also faces severe poverty, with more than half of the population living below the poverty line, and poverty rates as high as 70% in some rural areas. Addressing these challenges requires full recognition of the threats that climate change poses to the poor and vulnerable groups. Improving land use, and achieving a balance between tackling climate change and reducing poverty is the core of the country's climate-friendly sustainable development strategy.

4.1.2 Project Overview

The "Heritage Colombia Plan" program was launched in 2015 during the 21st Conference of Parties to the United Nations Framework Convention on Climate Change (UNFCCC COP21) in Paris, France. Colombia's Ministry of Environment and Sustainable Development, Colombia's National Parks Agency, the Gordon and Betty Moore Foundation, WWF, Wildlife Conservation Society (WCS), and Conservation International (CI) are some of the partners among others involved in the program. The 20-year (2018–2038) long program aims to standardize and improve the management of national protected areas by deforestation reduction and reforestation, while achieving multiple benefits such as conserving biodiversity, improving water resources and adaptability to climate change.

The "Heritage Colombia Plan" program not only directly contributes to Colombia in advancing the *Strategic Plan for Biodiversity 2011-2020* and its goals, but also supports the implementation of many of Colombia's climate and development policies, including national climate change adaptation plans, integrated strategies for deforestation control and forest management, climate change strategies for national parks, and national strategies for ecological restoration of protected areas. In the past, the Colombian Government's limited cooperation with stakeholders (including civil society, communities, private sectors and local government) has restricted its ability to achieve the United Nations Millennium Development Goals (MDGs). With the participation of multiple actors, the "Heritage Colombia Plan" effectively helps build partnerships between the government and international development agencies, donors, private sectors and NGOs to increase the efficiency of the use of funds.

Funded by the World Bank (WB) and the Inter-American Development Bank (IADB), the program aims to promote long-term sustainable development of protected areas through innovative financing schemes. The financing mechanism is based on the establishment of an extinguishable transition fund that leverages the economy over 20 years using investments in the public and private sectors to ensure the possibility of long-term investment. Specifically, the program established the "Project Finance for Permanence (PFP)" model that allows all participants to make financial contributions or policy commitments under the cooperation framework to fill long-term funding gaps for the development and sustainable management of protected areas. Donors need to mobilize as many resources as possible to support the priority actions most in need of support, and the Colombian Government is committed to progressively increasing supporting funds and policy support to ensure long-term funding. There are three main forms of funding: An extinguishable transition fund was established at the beginning of the initiative to fill the funding gap using government and private capital over a specified period of time (Figure 4-1); The Colombian Government developed new sustainable financing models for protected areas, such as payment for environmental services, ecological compensation mechanism and other financing channels. The government also implemented a 5% carbon tax to support the implementation of this initiative; The government provides a financial guarantee for the program. On the basis of a guarantee of 50% supporting funds during the implementation of the program, the government commits to increase its conservation funding every year, and will undertake the whole financing burden by the end of the program.

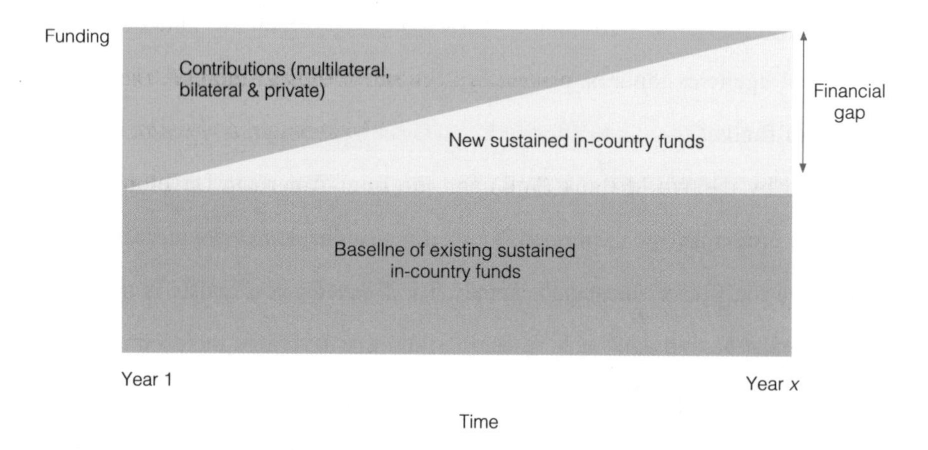

Figure 4-1 The "Heritage Colombia" Program Stakeholder Contributions

The “Heritage Colombia Plan” sets three goals for forest conservation: Establish 3.5 million hectares of protected areas; Increase 14 million hectares of effectively managed areas in the existing protected areas; Establish nine “Landscapes of Intervention” around the protected areas for conservation of biodiversity and sustainable land use. The total area of these landscapes is 35.8 million hectares, which makes up more than 30% of Colombia's land area. The program can help protected areas store large amounts of carbon and adjust the balance of water resources and improve the ability of more than 32 million people to adapt to climate change.

The program involves multiple stakeholders, including national and local governments, NGOs, enterprises, universities and communities. The contributions from each participant are summarized in Table 4-1.

Table 4-1 Stakeholders Contribution in the "Heritage Colombia Plan" Project

Participants	Governments	Donors and Private Sectors	Civil Society and Foundations	Academia and Research Institutes
Contributions	Develop rules, support the expansion of the goals and scope of protected areas through policies; design and plan sustainable financial mechanisms that provide important financial and policy support for the operations of the program	Provide significant replenishment of funding to support priority actions, and provide a new paradigm for the international community to promote mixed financing and government-enterprise cooperation in promoting sustainable development	Promote the engagement and interaction of multiple stakeholders through international and local expertise, network resources, and experiences. Focus on climate equity and gender equality, while supporting the sharing of experiences internationally	Provide the technical and scientific knowledge needed to carry out the program. Promote the development of theoretical basis for synergy with biodiversity conservation and the Sustainable Development Goals. Support the exchange of best practices and lessons learned

4.1.3 Project Results

The program, which is expected to be implemented in a 20-year period (2018-2038), will reduce emissions of about 100 million tons of carbon dioxide equivalent, and provide a carbon sink of about five million tons through reduced deforestation, reforestation, and forest management. In the first decade, the program will focus on five "core landscape areas" covering an area of 18 million hectares, or about 15% of the country's total land mass. Activities targeted at the "core landscape areas" can yield significant early benefits while effectively reducing risks and stresses to ecosystems caused by climate change. In addition to climate change mitigation, the program can benefit local residents by providing a large number of jobs in poor areas. The nine core "landscapes of intervention" planned for the program will directly benefit about 3,800 families. Moreover, more than 540 municipalities, 251 indigenous shelters, 89 Afro-Colombian communities and more than 25 million inhabitants will receive direct or indirect benefits.

4.1.4 Project Information

Case Source: UNEP-NbS Contributions Platform.

Project Participants: Colombia's Ministry of Environment and Sustainable Development, Colombia's National Parks Agency, the Gordon and Betty Moore Foundation, the Natural Heritage Fund, WWF, WCS and CI.

Project Location: Columbia.

Project Duration: 2018–2038.

Project Link: UNEP, https://wedocs.unep.org/bitstream/handle/20.500.11822/28927/Heritage_Colombia.pdf?sequence=1&isAllowed=y.

4.2 Community Forestry Campaign in Nepal

Case Highlights

The *Forest Act* provides a legal basis for the Community Forestry Campaign in Nepal and allows the establishment of Community Forestry User Groups (CFUGs). The law further encourages communities' participation by transferring the use right of state-owned forests to CFUGs; the grassroots electoral system encourages low-income groups, indigenous people and women to participate in community governance which can enhance the ability of vulnerable groups to participate in community governance and development.

4.2.1 Project Background

Nepal is a landlocked mountain country in South Asia, located at the southern foot of the Himalayas. It adjoins China on the north and is surrounded by India to the east, west and south. The area covered by forests (including shrubs) and included in the national park system is about 58,300 square kilometers, accounting for about 40% of the country's total land mass. Nepal has a low level of socio-economic development. It is one of the poor countries in the world and ranks the 157th out of 186 countries on the Human Development Index. Nepal has received a lot of international aid in the past few years. Official development assistance accounted for 5% of Nepal's gross national income from 2009 to 2013 (World Bank, 2014).

4.2.2 Project Overview

The Nepalese Government has extensive management and reform experience in the public sector and natural resource management. Nepal's Forest Management Department has delegated power to all levels of offices, adopted an officer responsibility system and appointed "forest officers". The Community Forestry Campaign in Nepal began in the 1970s, and has grown with the support of government forestry policies and legislation since then, which has covered most of Nepal's forest areas. In 1993, the Nepalese Government passed the *Forest Act*, which allowed people to form forest autonomous groups and provided a legal basis for the Community Forestry Campaign in Nepal. The *Forest Act* proposes to establish CFUGs as a form of organization for communities to participate in forest use and management. In 1995, the Nepalese Government introduced the *Forest Code* and the *Community Forest Operation Guide*, and established the Federation of Community Forestry Users Nepal (FECOFUN). These systems and mechanisms ensure that forest users can protect their rights and have the opportunity to participate in administration and discussion of state affairs (Ministry of Forest and Soil Conservation, 2013).

The existing protected areas, ecological areas and forests in Nepal are shown in Figure 4-2. Community forestry in Nepal transfers the use right of state forests to communities, that is, CFUGs can be established with the approval of regional governments (Pokharel R K et al., 2012). The ownership of forests remains with the government, while the right of land resource management and control is owned by CFUGs. As a result, CFUGs are empowered to use and manage forests. Profit distribution and forest rights generated on this basis are officially recognized (Pardo R, 1993). Key forest management actions taken by CFUGs include wildfire control, open grazing management, illegal encroachment control, and native species regeneration. The Nepalese government has made full use of administrative guidance and mobilization to help CFUGs develop effective governance models and employment channels. Main actions include:

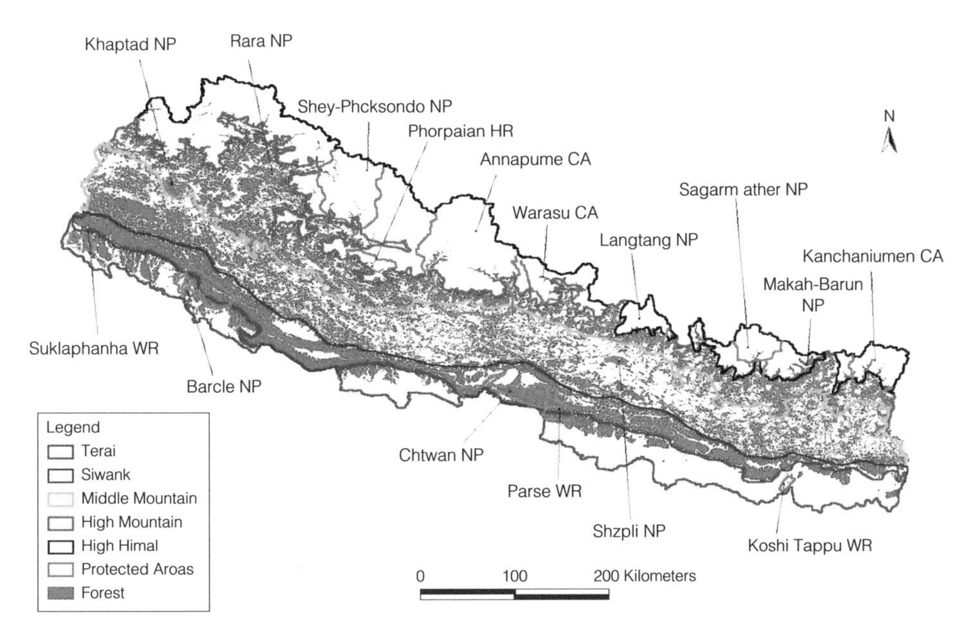

Figure 4-2 Forest Reserves in Nepal

(Source: Bhattarai B 2016)

●Build necessary capacity and provide technical support to help local residents increase their employability and income;

●Further enhance the capacity of women, youth, indigenous people, ethnic minorities and vulnerable groups through administrative legislation, and create the foundation and conditions for their participation in community governance and access to community assistance;

●Increase the area of natural forests, protect native species, and guide CFUGs to protect forest resources in the most effective and sustainable way;

●Innovate in management methods, recognize the value of local communities in climate actions, and encourage community residents to take active actions, so as to form a bottom-up synergy, improve the efficiency of grassroots governance, and integrate capacity building into every grassroots community in a fair and effective way.

The implementation of the Community Forestry Campaign in Nepal is funded by the Nepalese Government's Divisional Forest Office, the Ministry of Forests and Soil Conservation, local governments and NGOs. After acquiring the right to use and manage forests, CFUGs increase their incomes by profiting from the forest resources under their management. Almost all CFUGs have their own funds which are used to implement forest management plans and serve community development. The Nepalese Government has also provided economic incentives through innovation in policy mechanisms: At least 25% of the income of the forestry campaign across Nepal goes to sustainable forest management, about 35% of the income is allocated for poverty alleviation, and the remainder of the income is invested in infrastructure in rural areas, such as establishing solar power and small hydro-power stations, community hospitals and schools, early warning systems in flood-affected zones, rural roads maintenance, and community level basic social security implementation. These actions have attracted more forestry enterprises and related industries, promoted the development of eco-tourism, and created employment opportunities for poor families who live off forests. They also attract more communities to consider forestry as its major source of income, while making the Community Forestry Campaign in Nepal sustainable.

In addition to forest conservation and sustainable development, CFUGs also play a prominent role in social equity and capacity building. As bottom-up, community-centered organizations, the income earned by CFUGs through sustainable forest management and development is consolidated into a specialized fund for unified management; the unified management of the community not only guarantees the community's long-term and stable input in the sustainable development of forests, but also forms a positive incentive for the residents. The Community Forestry Campaign in Nepal has achieved significant results in terms of gender equality, helping indigenous people, creating jobs, and developing community culture. Under the existing forestry legislation in Nepal, the chairperson or secretary-general of the executive committee elected democratically by the community in a bottom-up model must be a woman. As a result, at least 11,000 women took leadership positions in the implementation framework of the Community Forestry Campaign in Nepal after the 2017 election (Figure 4-3). In order to promote social equity, this model not only creatively redistributes profits in the community and guarantees community rights, but also creates conditions for individual community's capacity building, enhancing a community's comprehensive governance capacity, and narrowing the gap between urban and rural areas. CFUGs are a model of community autonomy. Nepal has extended capacity building to every community to ensure efficient and effective governance actions by establishing an effective bottom-up forest management system.

Figure 4-3 Women's Participation in Community Governance in Nepal
(Source: UNEP)

4.2.3 Project Resutls

The Community Forestry Campaign in Nepal is based on communities themselves, and extended throughout the country through the *Forest Act of* 1993. To date, a total of

22,266 CFUGs have participated, covering approximately three million households in rural and urban areas across Nepal. According to statistics of the Nepalese Government, as of 2016, nearly 50% of the country was covered by forests, an increase of more than 5% compared with the end of the last century. Today, almost all of Nepal's forest resources at medium altitudes, more than 22 million hectares, have been built and managed by communities as sustainable forests, which can tackle climate change and produce multiple social benefits. The main goal of the Community Forestry Campaign in Nepal is to reduce deforestation and forest degradation, and increase ground covered by native forests, so as to effectively increase carbon sinks. At the same time, almost all CFUGs are carrying out community-based climate change adaptation actions, and these actions take into account the community environment and vulnerability to climate change, increase the opportunity of communities to gain local food all around the year, effectively protect plants, animals, and healthy ecosystems of water sources, and help improve the country's ability to adapt to climate change. With the active promotion of the Nepalese Government, indigenous people and ethnic minorities have taken the initiative to protect the local ecological environment. The government has also developed a series of ecological products and ecological industries to increase the income of community residents, attract young people to return home for employment, and promote community development. The above achievements have promoted the balanced development of urban and rural areas, and also led to a virtuous circle of economic and social development, demonstrating the long-term effectiveness of the policy.

The Community Forestry Campaign in Nepal not only increases the forest carbon sink with clear ecological and environmental benefits, but also develops sound governance systems over the past 30 years, and community forestry is explicitly recognized in national legislation. At the same time, the grassroots electoral system encourages low-income, indigenous and the female population to participate in community governance, and ensures that activities are carried out in a fair and transparent manner. This measure not only ensures that the Community Forestry Campaign in Nepal can promote the balanced development of the whole country, but also effectively realizes the fairness and justice within the community and among different groups, at the same time, enhances the ability of vulnerable groups to participate in community governance and development. The campaign provides solutions for other countries to learn from for a long time to effectively implement community-based climate change mitigation and adaptation actions.

4.2.4 Project Information

Case Source: UNEP-NbS Contributions Platform.

Project Participants: Nepalese Government and local communities.

Project Location: Nepal.

Project Duration: Since 1978.

Project Link: UNEP, https://wedocs.unep.org/bitstream/handle/20.500.11822/28836/Forestry_Nepal.pdf?sequence=1&isAllowed=y.

4.3 "Trillion Trees" Initiative

Case Highlights

The initiative integrates projects in a more efficient manner, provides a platform for financial and political support, tries to use a global threshold-free platform to solicit solutions to promoting the development of projects from people of insight worldwide, and attracts the participation of partners worldwide to the greatest extent.

4.3.1 Project Background

In 2006, the UNEP first proposed the "Global Billion Trees Campaign" with the goal of planting one billion trees around the world in one year to mitigate global climate change. As of 2011, 12 billion trees had been planted worldwide under the funding of the campaign. In March 2018, driven by international organizations such as the UNEP, Prince Albert Ⅱ of Monaco, Patricia Espinosa, Executive Secretary of the *United Nations Framework Convention on Climate Change* (UNFCCC), and other international celebrities jointly signed the *Trillion Trees Declaration* in Monaco, which pushed "Plant-for-the-Planet" to a new height.

4.3.2 Project Overview

At the 2020 annual meeting of the World Economic Forum (WEF) in Davos, the idea of "one trillion trees" was listed as a formal initiative of the meeting, and submitted to the representatives participating in the conference for discussion (Figure 4-4). Ultimately, representatives of governments from many countries, including China and the United States, and 300 private sector representatives jointly adopted the "Trillion Trees" Initiative. The initiative aims to promote large-scale reforestation investments, and improve the ability of the world to cope with challenges such as climate change and slow down the loss of biodiversity. This initiative which lasts for 10 years is also an important part of "the UN Decade on Ecosystem Restoration 2021–2030", which provides an important opportunity to accelerate cooperation and action in the fields of global climate and nature. The 1t.org platform is launched in sync with the "Trillion Trees" Initiative, aiming to further promote organizations to plant one trillion trees worldwide to cope with climate change.

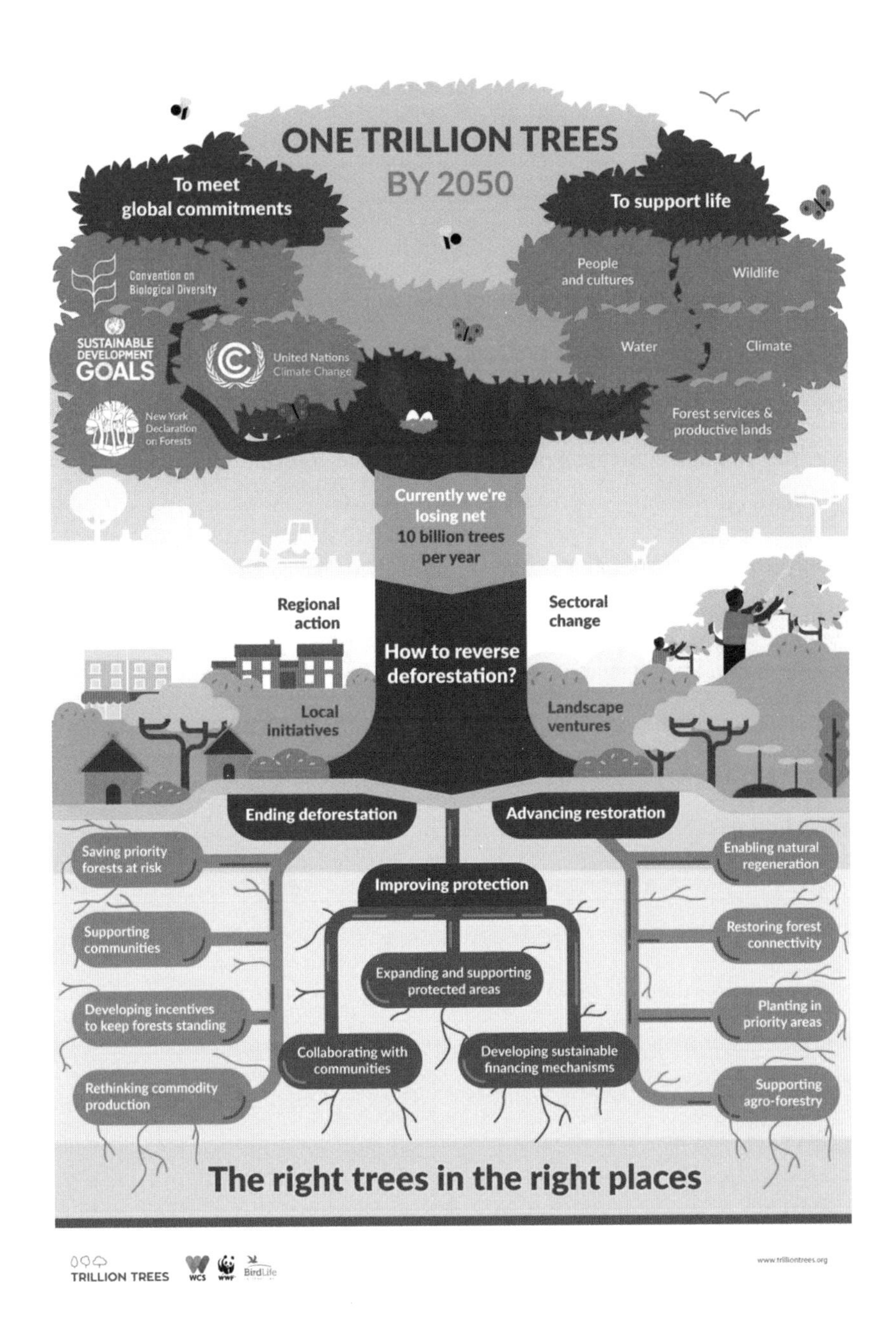

Figure 4-4 "One Trillion Trees" Plan Declaration Launched by Participating Organizations after the World Economic Forum in Davos

To promote the implementation of the initiative, WEF has set up a secretariat to cooperate closely with the World Resources Institute (WRI), Swiss Federal Institute of Technology in Zurich, UNEP, and other institutions to facilitate all stakeholders action and execution. Stakeholders who are willing to publicly demonstrate their achievements and progress in forest conservation, restoration and ecological restoration, and actively promote carbon emission reduction targets in line with scientific afforestation principles and NbS philosophy can apply to join the "Trillion Trees Initiative", become a member of 1t.org, the World Economic Forum will distribute relevant technical and resource support (Li, 2020).

In July 2020, the two major parties in the United States, Democrat and Republican, adopted the *Great American Outdoors Act*, planning to plant 1.2 billion trees in the national forests of the United States while supporting nearly 50,000 jobs. In August 2020, the "Trillion Trees Initiative" established its first national branch in the United States jointly led by the Society of American Foresters and WEF. Twenty-six NGOs, cities and companies have pledged to join the branch, including Detroit in Michigan, Dallas in Texas, MasterCard, Microsoft, Bank of America and American Forest Foundation. The branch has a stakeholder committee, with representatives from senior government and the private sector, representing the two major parties, serving as joint members responsible for providing strategic plans and proposals for the whole branch ensuring the effective completion of the goal of "one trillion trees". The committee members include experts in such fields as forestry, social mobilization, publicity and communication.

Corresponding to the national branch, enterprises have also launched action plans. The president of Salesforce[①] has announced that Salesforce will support and mobilize

① Founded in March 1999, Salesforce is a customer relation management (CRM) software service provider and a key partner of the "Trillion Trees Initiative".

the protection and restoration of more than 100 million trees in the next decade, directly invest in forests around the world, and support the restoration of mangroves in Myanmar, the protection of tropical forests in Brazil and peatlands in Indonesia through the purchase of carbon credits. Hewlett Packard took "zero deforestation" as its corporate strategy, aims to restore and protect more than 80,000 hectares of forests in China and Brazil before 2024 by cooperating with the WWF, so as to achieve the goal of papermaking with zero deforestation, and plans to plant one million trees by the end of 2020 by cooperating with Arbor Day Foundation.

The "Trillion Trees Initiative" established a project on Uplink[①] to maintain innovation and solicit solutions to promoting project development worldwide through the threshold-free platform, from which a case collection is formed to promote project experience. Many participants are also using Uplink to look for innovative solutions and attract more public participation. For example, Ecotree in Europe uses Uplink to attract the general public: an individual can participate by investing ¥150 to adopt a tree at the initial stage. Ecotree plans to plant 100 million trees worldwide within 10 years and will select native tree species with economic benefits. It is estimated that the whole project will produce a profit of about $3.2 billion. After the planted trees complete their life cycle, the tree trunks will be made into recyclable furniture and the economic income gained will be used to ensure the sustainability of the project.

4.3.3 Project Results

The "Trillion Trees Initiative" aims to protect, restore and plant one trillion trees

① As an emerging platform for innovation and creativity, Uplink proposes innovative programs in line with the United Nations Sustainable Development Goals through a multi-stakeholder mode.

worldwide within 10 years (before 2030) by connecting, mobilizing and building the capacity of communities that carry out afforestation projects. If the US branch carries out its work smoothly, 131 million acres of forest land will be used for restoration and reforestation, 430 million acres of forest land will be protected, the density of trees on 55 million acres of forest land will increase and each tree will serve as a carbon sink of 0.62 t in the United States alone.

4.3.4 Project Information

Case Source: UNEP-NbS Contributions Platform.

Project Participants: Led by the World Economic Forum, and joined by governments, organizations and individuals from China, the United States and other countries.

Project Location: Worldwide.

Project Duration: 2020-2030.

Project Link: "Trillion Trees Initiative", https://www.1t.org.

4.4 Case Summary

With the leadership and support of the Colombian Government, Colombia proposed the "Heritage Colombia Plan" to protect its abundant forest resources. The 20-year program aims to standardize and improve the management of national reserves in such ways as reducing deforestation, forest degradation and reforestation while achieving fruitful benefits such as protecting biodiversity, improving water sources, and enhancing the ability to adapt to climate change. In addition, the program uses a financing model of "Project Finance for Permanence (PFP)" to build cooperation between governments and international agencies, donors, the private sector and NGOs to improve the efficiency of the use of funds and promote the long-term sustainable development of protected areas. The program not only directly advances several sustainable development goals in Colombia, but also provides experience in such fields as capital management model and land planning and utilization for the international community to implement NbS.

The Community Forestry Campaign in Nepal is sourced from the national forestry management reform led by the government and will be implemented over a period of more than 30 years. The reform focuses on setting up a bottom-up governance model and issuing corresponding laws and regulations so that the community forestry management team can actively and efficiently cooperate with the government to implement sustainable forestry projects at grassroot levels. This enables the Community Forestry Campaign in Nepal to mobilize the entire country to participate to the greatest extent, forming a long-term, effective and sustainable operation model. At the same time, the campaign protects the local residents' right to use forest land through relatively sound laws, regulations, election and support systems, and gives local residents the motivation and power to participate in forestry projects on a long-term basis. In the process of elections and operation, the campaign focuses on ensuring the participation of multiple actors, especially the participation of vulnerable groups such as people living in poverty, ethnic minorities, indigenous people and females.

Launched at the World Economic Forum in Davos in 2020, the "Trillion Trees Initiative" aims to realize the goal of "planting one trillion trees by 2030" by establishing a platform. Government officials from China, the United States and many European countries, and representatives of more than 300 participating companies jointly adopted the initiative, officially launching the 1t.org platform. The first branch of the program is set up in the US, with joint leadership by the Society of American Foresters and the World Economic Forum. Apart from the public sector, Hewlett Packard and Salesforce also positively participate in the promotion of the initiative and utilize online platforms to promote public participation and innovative solutions.

5 Grassland

Grasslands are a terrestrial ecosystem in which perennial herbaceous plants are the main producers. The grassland ecosystem has ecological functions such as wind protection, sand stabilization, soil conservation, climate regulation, air purification and water conservation. The grassland ecosystem is an important part of the natural ecosystem, playing an irreplaceable role in developing animal husbandry, maintaining biodiversity, improving environmental governance and maintaining ecological balance. Meanwhile, grasslands are the largest terrestrial ecosystem with huge carbon storage capacity. The carbon pool of the grassland ecosystem mainly includes vegetation carbon pool (aboveground and underground biomass carbon pool) and soil organic carbon pool. According to statistics, the carbon storage of global grassland ecosystems is 266.3 billion tons, accounting for 12.7% of the total carbon storage of the terrestrial ecosystem, among which the biomass carbon storage and soil carbon storage account for 6.0% and 15.5%, respectively (Lieth H et al., 2012).

Desertification control, climate change and biodiversity conservation are among the priority areas in *Agenda 21*. In addition to the impact caused by human activities, the increasing temperature and decreasing precipitation due to climate change are important driving factors of desertification. Therefore, there is a significant synergistic result between NbS and desertification control. Climate change mitigation and adaptation can be achieved through collaborative governance to restore the land lost to desertification, increase carbon storage, and conserve water and soil. Because of the great difficulty, high cost and low survival rate of tree planting and management in arid and semi-arid areas, grassland management and restoration in such areas is an important means of desertification control. This chapter selects two ecological restoration cases, namely the "Great Green Wall (GGW)" in Africa, and Desertification Control of the Mu Us Desert in China. Both cases can provide reference for others to implement "forests + grasslands" ecological restoration at the cross-regional and national levels.

5.1 "Great Green Wall" in Africa

Case Highlights

GGW in Africa advocates the organic combination of top-down and bottom-up governance models: at the regional level, through the coordination of the African Union (AU), governments make unified plans and set long-term goals at the national level; at the local level, considering such limitations as weak administrative capacity and difficulty in intervening with individual behaviors through administrative means, the bottom-up community governance model turns passiveness into activeness to find economic tree species suitable for local conditions, or the model of "forests + grasslands" is adopted to stimulate the participation of local people.

5.1.1 Project Background

Sub-Saharan Africa is considered the "hotspot" that's most vulnerable to climate change. The local ecological environment is fragile and highly dependent on rain-fed agriculture with limited adaptability, constantly challenged by drought and desertification. Given that these areas will inevitably be affected by climate change to some extent, climate change adaptation has been a priority area in the climate policy of sub-Saharan Africa.

5.1.2 Project Overview

Inspired by Green Dam in Algeria① and the Great Wall of China, and co-funded by the World Bank, the European Union and the United Nations, 22 countries (Mauritania, Chad, Niger, Ethiopia, Burkina Faso, Nigeria, etc.) participated in the GGW in Africa led by the African Union. GGW in Africa was first proposed in the 1980s and revived around 2000. It has an epic ambition to grow an 8,000 kilometer natural wonder of the world across the entire length of Africa. If successfully implemented, GGW in Africa will be the largest living green building on the Earth, which is three times the area of the Great Barrier Reef off the coast of Queensland, Australia. See Figure 5-1 for the detailed plan of GGW in Africa.

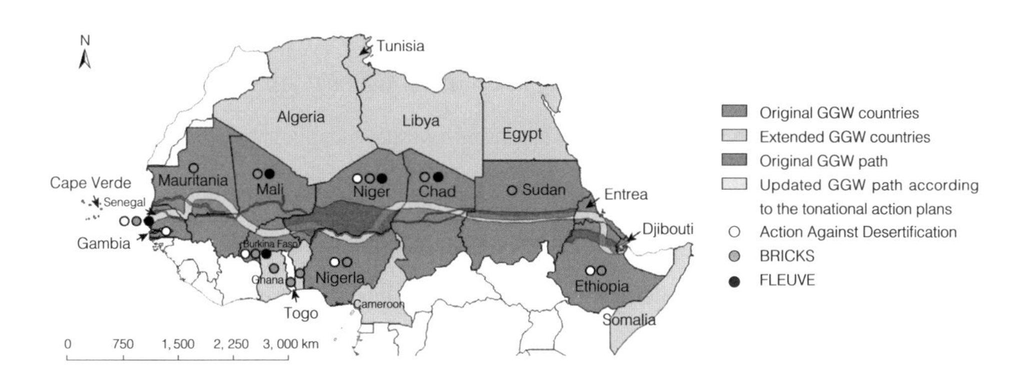

Figure 5-1 The detailed Plan of the "Green Great Wall" in Africa
(Source: Goffner D et al., 2019)

① Green Dam, also known as Green Belt, is a transnational forestry project under construction in North Africa, involving Morocco, Algeria, Tunisia, Libya and Egypt. It aims to create a green plant belt across five North African countries by planting trees and grass, so as to prevent the invasion of the Sahara Desert and desertification.

At first, GGW was designed to create a continuous forest zone across the Sahara region, aiming to slow down desertification and promote agricultural development by adjusting local climate, temperature and wind speed, and reducing soil erosion. After a period of implementation and evaluation, the current GGW calls for a "mosaic" ecological protection barrier by shaping different landscape types instead of building a set of artificial forest barriers that run through from front to back, which will provide a long-term solution for improving the regional environment and increasing community income. GGW is a collection of forestry and non-forestry projects, which are specifically designed and managed by different countries and inter-state organizations located in the Sahara region. GGW advocates the organic combination of top-down and bottom-up governance models: the top-down model at the planning level allows governments to formulate unified plans and set long-term goals at the national level according to local conditions. At the regional level, inter-country cooperation is coordinated through WWF, permanent committees, monitoring stations and national alliances. At the local level, considering such limitations as weak administrative capacity and difficulty in intervening with individual behaviors through administrative means, the bottom-up community governance model turns passiveness into activeness, which stimulates the participation of local people and ensures that the project design and implementation can effectively address local needs.

GGW involves many projects in many countries, which are financed through government and fundraising. In December 2015, the UNFCCC COP21 pledged an additional $4 billion in funding for GGW; the French Government agreed to provide €1 billion in aid by 2020; the President of the World Bank pledged an additional $1.9 billion in funding for GGW and related projects. GGW is supported by the Global Environment Facility (GEF), the European Union, the Food and Agriculture Organization of the

United Nations (FAO), etc., and financed externally through the new multi-donor trust fund PROGREEN established by the World Bank, to coordinate national actions for projects in the Sahel and West Africa, and manage funds to ensure maintenance costs.

GGW includes specific tasks on forestry and grassland, and afforestation (such as desert acacia and baobab) is the main task in forestry. Given the huge amount of personnel and management cost involved in afforestation in desert areas, the promotion of tree species with higher economic value will encourage local residents to carry out follow-up forest management and protection. This practice is now being rolled out on a large scale in Senegal. The afforestation project in Senegal mainly produces Arabic gum for export, and about 18 million trees have been planted on about 100,000 acres. In addition, the improved ecological environment has brought back many animals including gazelles, jackals and songbirds.

However, it should be noted that many countries have chosen to restore degraded grasslands or pastures in GGW due to the high cost of afforestation and the fact that the survival rate of trees planted in arid areas is only around 20% and only 70% in well-managed plantations. In Mali, Burkina Faso and Niger, large tracts of land have been fenced off and local farmers have been instructed to plant around existing trees and sprouts to allow the land to restore from the effects of prolonged overgrazing. In the past 30 years, nearly 12 million acres of grasslands in Niger have been restored in this way.

There are also excellent practice cases of grassland restoration projects. For example, given that the large-scale afforestation is not suitable for the Sahara region, the "ecological fence" project has been carried out in some areas; protection actions have been carried out by identifying priority areas and building ecological fences, so as to achieve the goals of preventing animals from destroying grasslands and restoring the ecosystem. It turns out that there are three to four times as many trees in the ecological fence as outside the fence, and the average tree height is 20 cm higher than that outside the fence. Importantly, the "ecological fence" project is not intended to replace the existing species in the ecosystem, but to increase the speed of tree growth and expansion by reducing the invasion of alien species, improving soil fertility and reducing the risk of fire and other disasters. The "ecological fence" project is in line with the scattered land suitable for afforestation in this region, providing job opportunities for the local community, and helping them in achieving win-win results of targeting poverty alleviation and ecological conservation.

5.1.3 Project Results

By 2030, GGW aims to restore 100 million hectares of currently degraded land, sequester about 250 million tons of carbon dioxide, and create 10 million jobs for local rural areas. GGW is scheduled to be built from 2007 to 2030 and is now in its second planning period (construction progress of 15%). In addition to the aforesaid results, GGW also faces great challenges: On one hand, desertification is rampant in the Sahel region, where every year a large amount of forest land is unregulated or degraded due to agriculture, construction and logging, on the other hand, only half of $8 billion pledged for GGW has been delivered, and the huge cost of tree planting and maintenance has placed a heavy financial burden on the poorer countries in the region, and the survival rate of trees in some areas is only 20% (Aryn Baker et al., 2019). Despite the relatively slow pace of overall construction, the ecological benefits of GGW far exceed expectations. According to the data released by the *United Nations Convention to Combat Desertification* (UNCCD), GGW has made significant progress since its launch in 2007 (Table 5-1).

Table 5-1 Progress of the "Great Green Wall"

Country	Results
Ethiopia	15 million hectares of currently degraded land have been restored and security of tenure on land has been improved
Senegal	11.4 million trees have been planted, and 25,000 hectares of currently degraded land have been restored
Nigeria	5 million hectares of currently degraded land have been restored, and 20,000 job opportunities have been created
Sudan	2,000 hectares of land have been restored
Burkina Faso, Mali, Niger	About 120 communities are involved in GGW. A green belt has been established on more than 2,500 hectares of currently degraded and dry land, and more than 2 million saplings have been planted with 50 local tree species

Source of information: UNCCD.

5.1.4 Project Information

Case Source: UNEP-NbS Contributions Platform.

Project Participants: 22 countries led by the African Union, and sponsors (World Bank, European Union, United Nations, etc.).

Project Location: Sahel region of sub-Saharan Africa.

Project Duration: 2007–2030.

Project Link: "Great Green Wall", https://www.greatgreenwall.org/.

5.2 Desertification Control of the Mu Us Desert in China

Case Highlights

The local government leads and implements a series of ecological governance policies; make precise policies according to different control conditions in the Mu Us Desert; innovate in the mode of desertification control and adopt the method of "prevention and utilization" to promote the development of forestry and sand industry, and achieve win-win results in ecological, economic and social benefits.

5.2.1 Project Background

The Mu Us Desert is located at the junction of the Inner Mongolia Autonomous Region, Shaanxi Province and the Ningxia Hui Autonomous Region, China, the transitional area from Ordos Plateau to Loess Plateau. It is an interlaced area of agriculture, forestry and animal husbandry mainly based on grasslands, as well as an ecotone with special geographical landscape. This region has a typical temperate continental semi-arid climate, with an average annual precipitation of 250-440 mm, and the overall terrain gradually slope downward from northwest to southeast. The main types of aeolian landform include moving dunes, semi-fixed dunes and fixed dunes. The zonation of vegetation in the Mu Us Desert shows a transition of desert grasslands-typical grasslands-meadow grasslands from northwest to southeast. In addition, various intrazonal vegetation[①],

① Intrazonal vegetation, also known as "azonal vegetation", refers to vegetation which is affected by local topography, soil, underground water and surface water and appears in more than two vegetation belts, such as meadow vegetation, swamp vegetation and aquatic vegetation, etc.

mainly including desert vegetation, wetland vegetation and halophytic vegetation, is widely distributed in the Mu Us Desert.

The Mu Us Desert is called "the youngest desert" and "man-made desert". Historically, it used to be a place with plenty of water and lush grass, herds of sheep and cattle, and beautiful scenery of "close to vast lakes and with clear rivers". However, after the mid-Tang Dynasty, due to the expansion of human activities, excessive reclamation and grazing, war and other factors, coupled with climate change, the Mu Us region suffered gradual desertification and eventually formed contiguous tracts of deserts. With the continuous deterioration of the ecological environment in this region, quicksand invaded and buried towns and villages, grasslands were degraded, and crops were not harvested. As a result, the local community suffered, and the Mu Us region was once trapped in a passive situation of "sand in and people out".

5.2.2 Project Overview

The control measures of the Mu Us Desert mainly consist of three parts:

Building of green ecological barriers under policy leadership. Since 1978, China has successively implemented ecological restoration projects such as "Three-North" Shelter Forest, Natural Forest Protection, and Grain for Green. While actively promoting the construction of key forestry projects, local governments have carried out afforestation, implemented ecological governance policies such as ecological migration, banning grazing and enclosure for afforestation, adjusted the structure of agriculture and animal husbandry industries, encouraged the participation of various ownership systems in ecological construction, and increased investment in human and financial resources (Figure 5-2). Thanks to the government leadership, desertification in the Mu Us Desert has been significantly reversed.

(a) New Village for Ecological Migration

(b) No Grazing and Captive Breeding

Figure 5-2 Measures Taken by Local Governments

Precise policies to innovate the models of desertification control. After years of hard exploration and practice, generations of desertification controllers have summed up the desertification control models and experience suitable for the Mu Us Desert and put forward corresponding countermeasures according to different conditions in control areas. For example, establishing a wind-sand prevention system based on the combination of "belt, area and net"; screening out a group of sand-fixing plants suitable for growth, such as Caragana korshinskii, Salix psammophila, Hedysarum scoparium and Pinus sylvestris; adopting the desertification control technique of planting sand-fixing plants in the straw checkerboards for moving or semi-fixed dunes; adopting the method of "aerial seeding + artificial enclosure" in moving dunes with wide area as well as high and dense fluctuation; establishing a shrub-based biological isolation belt for wind prevention and sand fixation on the edge of desert and activated dunes; implementing preferential policies and measures for afforestation, such as anyone who afforests can benefit and allowing inheritance and transfer of the forests (Figure 5-3).

Combination of prevention and utilization to promote the development of forestry

(a) Straw Checkerboard Barriers

(b) Afforestation by Aerial Seeding

Figure 5-3 Models of Desertification Control

and sand industry. Governance drives development, which in turn promotes governance. By means of the unique resources such as light, heat and soil in the sandy area, the Mu Us Desert vigorously develops the local forestry and sand industry, and combines sand control with sand use to achieve win-win results in ecological, economic and social benefits. For example, processing sand industry based on sand plant resources; green industrial sand industry based on light, heat and soil resources in sand areas; sand industry based on sand tourism landscape, etc. Specifically, the biological habit that sandy shrubs need regular stubble tending is used to obtain renewable biomass raw materials (fuels) for biomass power generation; aeolian sand beneficiation is widely used in glass, ceramics, metallurgy, electronics, medicine and chemical industries as raw materials for production; spirulina breeding and deep processing industries are developed (Figure 5-4).

(a) Spirulina Breeding Park

(b) Ecological Park in Desert

Figure 5-4 Forestry and Sand Industry

5.2.3 Project Results

In the past half century, the project of Desertification Control of the Mu Us Desert has seen remarkable results, and the degree of desertification has shown an obvious reversal trend. The control results are as follows.

Both the area of sanded land and the desertification degree have been reduced. The area of China's land loss to and prone to desertification has been reduced for ten consecutive years since 2004, and desertification degree continues to decrease. In recent five years, the area of severely and extremely severely sanded land in the Mu Us Desert has decreased by 6.282 million mu[①]. Relying on national forestry projects and local ecological restoration policies and measures, such as "Three-North" Shelter Forest and Returning Farmland to Forests and Grasslands, the area of sanded land in Uxin Banner, Inner Mongolia Autonomous Region has obviously decreased from 2000 to 2017(Figure 5-5).

① 1 mu=1/15 hectare.

(a) Before treatment

(b) After treatment

Figure 5-5 Historical Changes of the Mu Us Desert

The vegetation coverage has been improved. Vegetation cover is a significant way and one of the main goals of desertification control (Figure 5-6). Taking representative banner counties as an example, the vegetation coverage of Uxin Banner increased from 28% in the 1970s to 80% in 2018; the control rate of sanded land in Yulin reached 93.24%, and the forest coverage rate increased from 0.9% to 34.8%; at the end of 2019, the forest area of Ejin Horo Banner reached 3 million mu, the forest coverage rate reached 36.85%, and the vegetation coverage rate reached 88%.

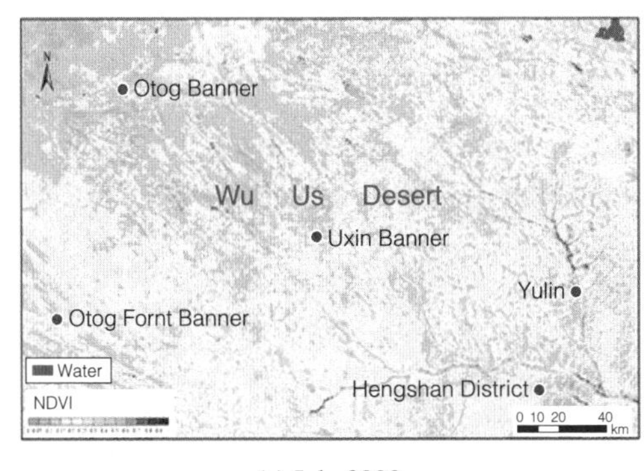

(a) July 2000

(b) July 2019

Figure 5-6 Vegetation Index Comparison of the Mu Us Desert (2000 vs 2019)

Carbon sinks in the ecosystems have increased. According to the long-term measurement of the typical desert ecosystem in the southern edge of the Mu Us Desert by the National Positioning Observation and Research Station of the Mu Us Desert Ecosystem in Yanchi, Ningxia, the shrub ecosystem with Artemisia ordosica, Hedysarum mongolicum and Salix psammophila as typical vegetation plays a vital role as carbon sinks over longer periods of time (absorbing 80 g of carbon per square meter annually).

The living environment and climate have been improved. Climate improvement is an important ecological benefit of desertification control. Taking Yulin City as an example, the frequency of dust storm has decreased from more than 100 days to less than 10 days annually. In 2017, the number of days with excellent air quality in Yulin reached 285 days, and the climate was greatly improved. Sandstorm and dust weather in Ejin Horo Banner has also decreased significantly, and the number of sandstorm days decreased from 23 days a year in 2000 to 13 days a year in 2019, thus effectively curbing the dust weather.

5.2.4 Project Information

Case Source: C+NbS Cooperation Platform.

Project Participants: Central Government of China, regional governments and local community of the Inner Mongolia Autonomous Region, Shaanxi Province and the Ningxia Hui Autonomous Region.

Project Location: Mu Us Desert in China.

Project Duration: 1959-2020.

5.3 Case Summary

The GGW in Africa led by 22 countries contains several sub-projects. It aims to build an 8,000 km long green belt across the entire length of sub-Saharan Africa. Guided by the epic ambition, GGW advocates the organic combination of top-down and bottom-up governance models. At the national level, governments make unified plans and set long-term goals; at the regional level through coordination by the African Union; at the local level, the bottom-up community governance model effectively promotes the participation of communities while ensuring that project design and implementation can effectively address local needs. The strong appeal of GGW comes from its sense of ownership and regional identity that it advocates. At the national level, it has gained more support and response through the concept of "building a new world wonder", and at the community and individual levels, it focuses on engaging the younger generation in the region.

Ecological restoration in the Mu Us Desert is a typical successful case of regional desertification control and an excellent case of practicing the "lucid waters and lush mountains are invaluable assets" theory. The practice of desertification control in China, represented by the Mu Us Desert, first benefited from China's legal system of forestry protection and desertification control, strong management and effective desertification monitoring and reporting systems. In the process of desertification control of the Mu Us Desert, scholars have discovered and innovated many effective sand-fixing measures, including aerial seeding, desertification control by enclosure, mechanical sand-barriers, desertification control by wind, and etc., which provide an important reference for desertification control in China and global. In addition, successful desertification control in the Mu Us Desert is credited to the joint efforts of generations of people in northern Shaanxi and the extensive participation of people at the grassroots level.

6 Agriculture

Agriculture are a major source of greenhouse gas emissions. It is estimated that at least one-fifth of the world's total greenhouse gas emissions come from the agricultural sector. The main sources are land use patterns, gastrointestinal fermentation of ruminants, straw burning, rice production, methane and nitrous oxide released in the production process of organic and chemical fertilizers, forestry, changes in land use patterns, etc. Therefore, there is great potential for energy conservation and emission reduction in the production of food crops and soil carbon sequestration and emission reduction. In addition, agricultural production highly dependent on local climate conditions and is extremely vulnerable to the impact of climate change. Implementing NbS and developing sustainable agriculture can achieve less input and more output, reduce carbon emissions, and improve the adaptability of agriculture to climate change, which is of great significance to sustainable development. Sustainable agriculture can also bring more investment and employment opportunities, thus stimulating regional and global economic growth. According to the research of the Business and Sustainable Development Commission (BSDC), by 2030, the transformation of food and land use patterns can create up to $4.5 trillion of new business opportunities every year (BSDC, 2016). According to the analysis of the Food and Land Use Coalition (FOLU), by 2030, adding $350 billion of new investment in sustainable food and land use system every year can create up to 120 million new jobs (FOLU, 2019). The FAO predicts that the transition to sustainable agriculture and land use types can contribute economic growth of $2.3 trillion to the world and will create 200 million jobs by 2050. In the category of agriculture, this chapter will focus on India and China, two of the largest agricultural producers in the world, three typical cases were selected, "Zero Budget Natural Farming (ZBNF)" in India, Climate-Smart Staple Crop Production in China, and "Three Goods Agriculture" in Hangzhou, China, hoping to provide experience that the agricultural sector can learn from and use as reference in its NbS practices.

6.1 "Zero Budget Natural Farming" in India

Case Highlights

Respecting the natural growth pattern of crops in the production process can bring higher net profit at lower cost; focusing on production problems of smallholder farmers and gender equality, it strives to employ equal numbers of male and female leaders and to provide women with entrepreneurial incentives, etc.

6.1.1 Project Background

India is a large agricultural country, with more than half of its population directly or indirectly dependent on agriculture for employment. Andhra Pradesh (AP) is the agricultural center and one of the largest states[①] in India. Agriculture is the main source of income for 62% of population in AP, accounting for about one third of GDP (Rythu Sadhikara Samstha, 2019). In the past, traditional farming mode was commonly adopted, with large adoption of chemical fertilizers and pesticides, which resulted in high planting costs and serious damage to soil fertility. The agricultural development

① "state" is an administrative unit in India that divides regions and is equivalent to a "province" in China.

was seriously threatened by the poor natural conditions in this region, long-term drought and water shortage, extreme weather events brought about by climate change, uneven rainfall distribution and other phenomena. AP farmlands are rapidly decreasing as soil degradation, climate change, and other issues became increasingly serious with the acceleration of urbanization. In addition, the uncertainty of the agricultural products market makes the situation of farmers with low income worse, and they easily fall into debt crisis. According to statistics, up to 90% of the AP population is under the pressure of repaying debts, with a per capita debt of $1,500. At the same time, this is also a dilemma faced by many countries in the world at present, seriously hindering the world's realization of the Sustainable Development Goals. For this purpose, the FAO began to advocate the application and promotion of agroecology and regenerative agriculture in all countries in 2018 and cooperated with the Indian Government to carry out practice in AP (Figure 6-1).

Figure 6-1 Farmland of ZBNF in India
(Source: Nature4Climate)

6.1.2 Project Overview

In order to help farmers restore the damaged soil and improve the livelihoods of smallholder farmers, the Indian Department of Agriculture and the AP Government launched a regenerative agriculture project named "AP Zero Budget Natural Farming (ZBNF)" in active response to the call of FAO. "Zero Budget Natural Farming" mainly relies on low-cost natural inputs to grow food without using expensive and polluting chemicals. The project aims to promote regenerative agricultural measures on eight million hectares of land in AP from 2018 to 2024 and transform the traditional agricultural mode into "Zero Budget Natural Farming (ZBNF)". It is estimated that the project results will benefit about six million farmers. The project is carried out with the support of farmer associations established by the AP Government and many other NGOs. The ZBNF technology, supported by World Agroforestry Centre, University of Reading, Indian Institute of Science, and Indian Institute of Science Education and Research, focuses on capacity building for farmers and promotion of sustainable community development. The main sources of funding for the project are the "Sustainable India Finance Facility (SIFF)" promoted by the UNEP (totaling $2.3 billion over six years), charitable donations and the FAO.

In the design process, AP ZBNF adheres to the concept of respecting nature, respecting the natural growth mode of crops in the production process, and avoids using any artificially synthesized chemical fertilizers, pesticides and other substances. This farming practice can bring higher net profit at lower cost, thus it is widely adopted by local farmers. As one of the increasingly popular strategies for regenerative agriculture, ZBNF can not only improve the adaptability of agriculture, but also effectively reduce carbon emissions. According to the existing local farming practices, the project makes full use of local wisdom and promotes Climate-Smart Staple Crop Production; the organic content in soil was effectively increased by using local materials and natural fertilizers. The improved soil enhances agricultural productivity, biodiversity, carbon storage, and use efficiency of water resources. Four treatment measures based on local experience in AP ZBNF are shown in Table 6-1.

Table 6-1 Four Measures based on Local Experience in AP ZBNF

Measure No.	Measure Name	Measure Introduction
1	Seed Treatment (Bijamrita)	Treat seeds with cow dung and cow urine and coat them with micro-organisms to protect young roots from fungi, seed-borne diseases or soil-borne diseases
2	Zero Chemicals (Jiwamrita)	Cow dung, cow urine, local palm sugar and bean flour, unpolluted soil and other things are used to stimulate microbial fermentation. Self-made organic fertilizer can make nutrients used by plants, prevent bacteria and increase carbon sequestration capacity of soil
3	Soil Microclimate (Mulching)	The topsoil is covered with nurse crops and crop residues to ensure beneficial microclimate in the soil. Its benefits include the production of decaying matter, protection of topsoil, improvement of water-retaining property in soil, control of weeds, creation of edaphon and provision of essential nutrients for soil
4	Soil Aeration (Waaphasa)	After the improvement of soil structure and change in water content by measures 2 and 3, this measure is taken to promote soil aeration, increase soil water content, improve water use efficiency and enhance the cold resistance of crops

Source of information: AP ZBNF.

Dung, urine and dairy products from locally-bred cows can be used as raw materials invested in "Zero Budget Natural Farming". Therefore, the quantity of local cows is of great importance to the project and a cow is enough to meet the demand of more than 30 acre of land. Besides using local materials, training for farmers is a crucial method to improve the degree of participation in the project. In the implementation of the project, about 6,000 influential major farmers were selected for training and capacity building. As they earned more, they brought in other villagers; project managers sometimes found that these major farmers even showed videos to other villagers at night through small projectors, explaining and spreading knowledge. Moreover, the workload needed to assess the project progress was reduced to a certain extent by training major farmers to use smartphones to track the project progress in time. Such training takes place in different villages every day, significantly increasing the implementation speed and penetration degree of the project.

The implementation of AP ZBNF helps address the high level of debt faced by almost all local farmers, reduces the cost of agricultural transition by up to 50%, and has strong reference value in poverty reduction and gender equality. For example, the project introduces specific crop and livestock models to eligible farmers (farmers with less than 2.5 acres of dry land and less than 1.25 acres of wetland or single female farmers) to improve their income and food security. About 86,000 women's self-help groups participated in the planning, management, and supervision of project implementation with the help of non-profit organizations. The project strives to employ an equal number of male and female leaders and to provide women with entrepreneurial incentives, such as supporting them to build shops in villages and selling organic fertilizers such as natural fertilizers to farmers.

6.1.3 Project Results

In 2017, 580,000 farmers, more than 3,000 villages and 260,000 hectares of land participated in the transition project. More than 90% of the farmers participating in the project saw an increase in income and yield while 10% of farmers reported a decline in yield with increase income. From 2018 to 2019, Indian Centre for Economic and Social Studies conducted a study on AP ZBNF and the results showed that for all crops, the biological costs and planting costs per mu invested in ZBNF are both lower than the chemical input costs under traditional agriculture (Table 6-2).

Table 6-2 Comparison of Net Profit and Planting Cost between ZBNF and General Agriculture

Crops	Net Profit of ZBNF /(USD/hm²)	Net Profit of General Agriculture /(USD/hm²)	Percentage Increase in Net Profit /%	Planting Cost of ZBNF / (USD/hm²)	Planting Cost of General Agriculture / (USD/hm²)	Percentage Decrease in Planting Cost /%
Paddy	667	603	11	526	604	13
Peanut	129	116	11	313	376	17
Corn	639	302	112	454	457	1
Cotton	1,003	573	75	518	567	9
Chickpea	769	655	17	398	464	14

Source of information: Galab S et al., 2019.

6.1.4 Project Information

Case Source: Nature4Climate Global NbS Case Study.

Project Participants: Indian Department of Agriculture and AP Government.

Project Location: Andhra Pradesh, India.

Project Duration: 2018-2024.

Project Link: AP ZBNF, http://apzbnf.in/.

6.2 Climate-Smart Staple Crop Production in China

Case Highlights

The top-down organization and operation mechanism not only strengthens the organization and leadership of the project, but also provides a reliable system guarantee for the normal operation of the project and ensures its sustainable development.

6.2.1 Project Background

Huaiyuan County in Anhui Province, China, is one of the top 100 food producing counties in China. Wheat can be seeded in winter and rice in summer. In 2013, the rice planting area of Huaiyuan County accounted for 84.3% of the total area under cultivation and the wheat planting area accounted for 98.8% of the total area under cultivation. Huaiyuan County is also an important source of labor migration county, very common for rural residents to go out to work. In 2013, the total labor force amounted to 26,140 people, while migrant workers accounted for 49.4% of the total labor force. During the busy farming seasons, most migrated labor cannot return for harvest. In Huaiyuan County, the development of village-level cooperatives is in its infancy stages. There are totally nine farmers' professional cooperatives in the county which are generally small in size with some are no longer operating. There are only two cooperative initiatives related to crop production that serves agricultural companies that transfer land locally. Agricultural machinery services are relatively common in Huaiyuan County, mainly involving the mechanical harvesting of rice and wheat. The basic form of organization is self-initiated and spontaneous, that is, it is provided by farmers who have harvesting machinery and service ability in their own village or neighboring village. However, Huaiyuan County lacks unified plant protection service, and its formula fertilization is still in the demonstration stage.

Ye County in Henan Province, China, is an important agriculture county. The corn plantation area accounts for 88.5% of the total area under cultivation, wheat plantation area accounts for 90.2% of the total area under cultivation. Ye County is also a typical labor migrant intensive county. The total labor force accounts for 56.7% of the total population, with migrant workers accounts for 40.6% of the total labor force. They work all over the country but majority within Henan Province. Close to their home, these migrant workers can return home to participate in agricultural production during the busy farming seasons (sowing and harvesting). In Ye County, cooperatives are also in the initial stage of development. There are totally 14 farmers' professional cooperatives, the earliest one was established in 2008, they are generally small in size and some are no longer in operation. Most farmers' professional cooperatives focus on crop farming and provide agricultural technology services. Agricultural machinery services are available across Ye County mainly for three stages, including mechanical sowing and harvesting of corn and wheat, returning crop stalks to the field. The basic form of organization is also spontaneous, similar to Huaiyu County, service in plant protection and formula fertilization are still in the demonstration stage. ①

As two demonstration sites of the "Climate-Smart Staple Crop Production" project, Huaiyuan County in Anhui Province and Ye County in Henan Province face similar challenges although producing different types of staple crops, different environments and different emphases in production modes: small cultivated land area per capita, scattered land, backward farmland infrastructure, unreasonable production management mode of crop farming, large greenhouse gas emissions, small application range of returning crop stalks to the field and conservation tillage, poor stability of the production system affected by climatic conditions, large input of chemical fertilizers and pesticides, high energy consumption, waste-intensive, low carbon sequestration capacity of soil, and limited fertility.

① Main data source: *Social Impact Assessment Repo*rt on Climate Smart Agriculture Project, issued by College of Humanities and Development Studies, China Agricultural University in 2014, http://www.reea.agri.cn/sttzgg/201906/P020190702531978504130.pdf.

6.2.2 Project Overview

The concept of "climate-smart agriculture" was put forward by the FAO in 2010, aiming at transforming and reorienting the agricultural system, helping and supporting its effective development and ensuring food security in the backdrop of climate change. Three main objectives of climate-smart agriculture are to continuously increase agricultural production and income, to build and improve adaptability to climate change, and to reduce or avoid greenhouse gas emissions where possible. In 2014, the Ministry of Agriculture and Rural Affairs of China and the World Bank jointly launched "Climate-Smart Staple Crop Production Project" at the Global Environment Facility. The project is supported by a grant of $5.1 million from the Global Environment Facility and $25 million from participating provinces and counties, implemented over a period of five years. In two demonstration sites, Huaiyuan County in Anhui Province and Ye County in Henan Province, the project focuses on three staple crops: rice, wheat and corn. The project reduces greenhouse gas emissions from agricultural production, increases soil organic carbon content, and improves ability to adapt to climate change in crop production through such activities as demonstration and application of climate-smart agricultural production technology, policy application and innovation, knowledge management, and expansion and promotion of public knowledge. China's climate-smart agriculture development model can be summarized as "carbon sequestration, emission reduction, grain stabilization, and income increase".

6.2.3 Project Results

In the implementation of the "Climate-Smart Staple Crop Production Project", after five years' collective efforts, domestic and international experts have introduced a number of new internationally frontier technologies and concepts, localized the technologies and concepts, and achieved fruitful results. By the end of 2019, the project had established 100,000 mu demonstration zones in both Anhui Province and Henan Province, the grain production from the demonstration zones saw increase in output, at an increasing rate of by more than 5% annually: the average yield of wheat increased by 5.5%-6.9%, the average yield of summer corn increased by 2.1%-17.2%, and the average yield of rice increased by 4.5%-8.5%; the amount of nitrogen fertilizer per unit area in the project area was reduced by about 10%, the amount of pesticides was reduced by more than 15%, and the soil organic carbon content was increased by 10%. Over the course of the past five years, the project had achieved carbon sequestration and emission reduction by 132,200 tons of carbon dioxide equivalent, exceeding its expected target of 65,000 tons. The project monitoring and evaluation report concludes that technical activities such as testing soil for formulated fertilization, conservation tillage, straw mulching and water-saving irrigation carried out in the project area have played an important role in increase of crop yields, carbon sequestration and emission reduction. At the same time, the project has improved farmers' consciousness of science and technology and farmland management, formulated relevant technical regulations and subsidy systems, established a production mode with high resource efficiency, economic rationality and carbon sequestration and emission reduction capability, enhanced the adaptability of crop production to climate change, and promoted energy conservation and emission reduction of agricultural production in China. In 2019, the utilization rate of chemical fertilizers and pesticides for the top three grains of rice, corn and wheat in China was 39.2% and 39.8% respectively. The application amounts of chemical fertilizers and pesticides showed negative growth for three consecutive years; the pollution prevention and control of large-scale farming had been promoted in an orderly manner and the resource utilization industry with rural energy and organic fertilizer as the main direction was growing; the utilization pattern of straw mainly used for agriculture and diversified development had basically taken shape.

6.2.4 Project Information

Case Source: C+NbS Cooperation Platform.

Project Participants: Ministry of Agriculture and Rural Affairs of China, World Bank, and Global Environment Facility.

Project Location: Huaiyuan County, Anhui Province and Ye County, Henan Province, China.

Project Duration: 2014–2019.

Project Link: Ministry of Agriculture and Rural Affairs of China, http://www.moa.gov.cn/xw/bmdt/202009/t20200922_6352773.htm.

6.3 Practice of "Three Goods Agriculture" in Hangzhou, China

Case Highlights

The project innovatively introduces a charity trust and adds more diversified and influential investment channels to traditional donation methods by relying on the advantages of the trust system; explores the balance between increasing farmers' income and environmental protection, and promotes farmers' sustainable participation through the development of "Three Goods Agriculture"; and develops environmental education courses to promote social participation in environmental protection.

6.3.1 Project Background

Since 2020, Qiandao Lake has been a source of drinking water for about 10 million people in the region of Hangzhou and surrounding cities in China, accounting for at least 50% of drinking water supply in Hangzhou. The overall water quality of Qiandao Lake Basin is good, ranking high in the national water environment quality, but it still face pressure from agricultural diffuse pollution, ecological fragility and other challenges (Figure 6-2). According to the analysis from The Nature Conservancy and the World Bank, the main threat to Qiandao Lake Basin comes from rural areas, which is mainly related to unsustainable agricultural practices such as the overuse of pesticides and chemical fertilizers, and high density livestock and poultry breeding. Diffuse pollution, also called non-point source pollution, mainly involves sedimentation of soil sediment particles, nitrogen, phosphorus and other nutrients, rural domestic refuse and various atmospheric particulates, which enter the water environment through the surface runoff,

Figure 6-2 Real View of Qiandao Lake
(Source: TNC)

farmland drainage and other forms. Diffuse pollution is disperse, concealed, latent, fuzzy and so on. It is difficult to monitor, quantify, study and control, and there is a lack of effective long-term governance mechanisms.

Climate change also has an important impact on the ecology of surrounding waters, agricultural industry and local resident's life. In terms of climate, in recent years, due to the increase of high temperature and dry weather, water shortage and water cut-off have been caused in the water source, which has seriously affected the safety of domestic drinking water for local residents. In recent years, China has focused on the development of industry and manufacturing in various regions, which has sacrificed many natural resources while improving people's wellbeing, and the destruction of water sources has gradually become an obvious social problem. Earnings from industrial development have gradually made people forget the importance of natural resources. The abuse of resources can only worsen the environment and threaten local resident's production and life. Taking the Longwu Reservoir in Hangzhou, Zhejiang as an example. In 2014, before The Nature Conservancy governed the reservoir, substandard amount of nitrogen, phosphorus and dissolved oxygen were detected during quality tests, mainly caused by long-term overfertilization of moso bamboo forests planted in the hills around the reservoir. Since the 1980s and 1990s, local villagers have begun to plant a large number of moso bamboos after profiting from its economic value, using large amounts of fertilizer to increase yields. The government needs to vigorously appeal to farmers near the reservoir to improve their awareness of water protection and change their conventional thinking. In addition, the Qiandao Lake Basin is located at the junction of Zhejiang Province and Anhui Province. Due to its natural geographical conditions, it is extremely vulnerable to the extreme weather caused by climate change. Therefore, mountain torrents, landslides, debris flow and other natural disasters are likely to occur.

From an economic perspective, since most of the farmers around Qiandao Lake grow tea for a living, the quality of these agricultural products is also threatened by the damage to the water source. Therefore, water source protection should be combined with ecological sustainable development. Building long-term mechanisms for water source protection, changing the current production mode and improve the environmental awareness among farmers in the Qiandao Lake Basin have become a top priority.

6.3.2 Project Overview

Ecological protection cannot rely on one's own efforts, but requires the participation of the whole society. In order to explore and effectively implement the long-term mechanism for protecting the water source basin, Alibaba Foundation and Min Sheng Tong Hui Foundation jointly established "China Water Source Protection Charity Trust" with Wanxiang Trust in 2017. In 2018, as the first project funded by "China Water Source Protection Charity Trust", the "Qiandao Lake Water Fund" program was launched in Hangzhou where The Nature Conservancy acted as the scientific advisor.

The project is committed to solving agricultural diffuse pollution in a scientific way and helping the development of ecological industry in the basin through green consumption. Ultimately realize the long-term effect protection mechanism for the Qiandao Lake Water Source Area, change the production mode, and improve the environmental awareness of farmers in the Qiandao Lake Basin. "Qiandao Lake Water Fund" innovatively introduces a charitable trust and stimulates the potential of environmental protection relying on the advantages of the trust system. In the design and construction of the mechanism, the primary goal of the trust is to reduce environmental pollution and ensure economic benefits. Through flexible and ingenious financial design, it adds more diversified and influential investment channels to traditional donation methods, expands the fundraising scope of public welfare and charity projects as much as possible, meets the needs of various charity groups, promotes the growth of raised funds, forms a virtuous circle of environmental protection projects themselves, guarantees the openness and transparency of fund management and operation from such aspects as decision-making mechanisms and information disclosure, and realizes the ecosystem payment mode.

At present, agricultural diffuse pollution in the Qiandao Lake Water Source Area is directly caused by soil erosion, improper use of fertilizers and pesticides among others, fundamentally due to the natural conditions of sloping farmlands which are prone to soil erosion and the unsustainable agricultural production mode. In order to control diffuse pollution in the basin and reform the agricultural production mode, the "Qiandao Lake Water Fund" program adopts the protective agricultural practice of combining green manure mulching, straw covering, natural fruit dropping and economic interplanting. Carrying out pilot projects in Anyang and other locations to promote the comprehensive demonstration of diffuse pollution control in the basin featured "source reduction + process interception + end treatment". In order to give full play to the economic value of "clear waters and green mountains", the "Qiandao Lake Water Fund" program puts forward the business model of "Three Goods Agriculture". "Three goods" means "good products", "good marketing" and "good tourism potential". Good products aims to publicize the ecological methods and growth conditions of agricultural products in the process of governance as well as the embodiment of water source protection to the public. Ant Group's blockchain technology can seamlessly and dynamically trace agricultural products, and publicize the health concepts such as going green, environmental protection and choosing organic, so that the consumers feel safe eating products and thus enhance market competitiveness. Good marketing means establishing a complete sustainable marketing channel based on good products. Brands are built for high-quality crops in the Qiandao Lake Water Source Area through interactive platforms such as e-commerce livestreaming and online stores, helping local villagers expand the market and ensuring economic income sources. Good tourism potential means promoting the popularity of local villages and striving to develop characteristic tourism products. Promote ecotourism and develop experience activities such as parent-child activities, music festivals, Qiandao Lake marathons, homestay, tea culture and so on. Under the premise of environmental protection, the realization of tourist industry and service industry is one of the ways to guarantee the sustainable development of the ecosystem.

While focusing on water resource protection, the project team is also devoted to research and development and practice of environment protection education courses. The project team developed a research course themed "ecological protection", attracting primary and secondary school students as well as executives from Beijing, Shanghai and other regions to practice and learn. "Thinking of the Source while Drinking: Research and Study Route of Qiandao Lake Water Source Protection Program" was rated "Excellent Course of Sustainable Development Education" by the Regional Centre of Expertise on Education for Sustainable Development. Environmental Protection Research and Research Base was awarded as "Ecological and Environmental Protection Volunteer Service Base of Zhejiang A&F University" in 2019.

As the largest water fund program in China at that time, the "Qiandao Lake Water Fund" program cooperates with Anyang City, Henan Province, to build the first comprehensive treatment demonstration base of "source reduction, process interception and end treatment" in the Qiandao Lake Basin. The program has successfully integrated public welfare, scientific research, finance and commerce organically, and created a platform which attracts the joint participation of farmers, governments, enterprises, NGOs and other social subjects. This project makes environmental governance, industrial investment and business cooperation work together on the long-term sustainable development of ecological water.

6.3.3 Project Results

Hundreds of small watersheds are scattered around Qiandao Lake. After estimation, the World Bank finally decided to invest in eight watersheds. In 2020, the ecological water conservation agriculture in Qiandao Lake was extended to 5,157 mu, covering rice, tea, fruits, pecans and other major agricultural products in the typical basin through recommendation of farmers by government, as well as farmers referrals, exceeding the Phase Ⅰ goal of 5,000 mu. In 2024, after the Phase Ⅱ construction of the "Qiandao Lake Water Fund" program, ecological water conservation agriculture is expected to extend to 75,000 mu in the whole Qiandao Lake Basin, which is equivalent to 1% of key land protection can bring about 10% reduction of non-point source pollution in the whole basin.

The project team selected Shangwu Stream in Chun'an and Wulong Stream in Jiande, Hangzhou, as pilot sites to explore the long-term mechanism. Research found governance of local tea planting with fertilizer on ground has a loss rate around 70%, with higher rate if the fertilizer is applied on slope. Excessive use of pesticides will aggravate the pollution of surrounding water flow, and pesticides will eventually enter Qiandao Lake. Therefore, the project team started from technologies and mechanisms, carrying out micro-management on the pilot farmland around the basin, unified prevention and control of pests and diseases, and unified spraying of biological agents and pesticides by drones. As a result, the amount of pesticides is greatly reduced, also reducing the application cost. At the same time, ecological tea gardens have brought 30%-40% of income increase per mu to tea growers. In rice planting, agricultural non-point source pollution prevention and control also achieved breakthrough. The "Qiandao Lake Water Fund" project team selected paddy fields with an area of about 30 mu as experimental fields and increased rice yields and farmers' income through ecological basin protection measures such as soil testing for formula fertilization, precise pesticide application, green manure mulching and rice-fish co-cultivation. The construction of protective "Three Goods Agriculture" achieved greenhouse gas emissions reduction through reducing the use of chemical fertilizers and implementing optimized fertilization and green manure projects. According to the estimate of the team of The Nature Conservancy, the Qiandao Lake Optimized Fertilization Project reduces greenhouse gas emissions by about 39 kg of carbon dioxide equivalent per mu, which can achieve an emission reduction of 195 tons of carbon dioxide equivalent based on the Phase Ⅰ target of 5,000 mu and finally an emission reduction of 2,925 tons of carbon dioxide equivalent based on the Phase Ⅱ target of 75,000 mu; the Qiandao Lake Green Manure Project reduces greenhouse gas emissions by about 47 kg of carbon dioxide equivalent per mu, which can achieve an emission reduction of 235 tons of carbon dioxide equivalent based on the Phase Ⅰ target of 5,000 mu and finally achieve an emission reduction of 3,525 tons of carbon dioxide equivalent based on the Phase Ⅱ target of 75,000 mu.

In addition, taking the ecological characteristics of Qiandao Lake into account, the project team developed brands such as "Qiandao Clear Spring Tea" and "Qiandao Clear Spring Rice" and obtained the qualification to use the regional brand of "Qiandao Lake Tea", realizing the innovation of agricultural added value and improving the living standards of local residents.

6.3.4 Project Information

Case Source: TNC Global NbS Case Database.

Project Participants: Alibaba Foundation, Min Sheng Tong Hui Foundation, Wanxiang Trust, and The Nature Conservancy.

Project Location: Hangzhou, Zhejiang Province, China.

Project Duration: Phase Ⅰ (2017-2020); Phase Ⅱ (2021-2024); Phase Ⅲ (since 2025).

6.4 Case Summary

Due to the long-term abuse of chemical fertilizers and pesticides, the soil in Andhra Pradesh, India, has been seriously damaged and the frequent occurrence of extreme weather events makes the farmland very fragile. Against this backdrop, the FAO, together with the Government of Andhra Pradesh and the Indian Department of Agriculture, actively carried out “Andhra Pradesh Zero Budget Natural Farming (AP ZBNF)” in which the natural growth pattern of crops is respected and the use of artificial fertilizers and pesticides is avoided. The results of “Zero Budget Natural Farming” show that among all crops, the cost of biological fertilizer and planting cost per mu in ZBNF are both lower than the cost of chemical fertilizer in the traditional methods. “Zero Budget Natural Farming” respects local knowledge, uses local materials, and engages communities and local farmers to take part in capacity building. It carries out training for some influential “major farmers”. “To drive more with one”, driving other villagers to implement “Zero Budget Natural Farming” by increasing the income of major farmers and promotes the implementation speed and penetration degree of the project.

Jointly initiated by the Ministry of Agriculture and Rural Affairs of China, the World Bank, and the Global Environment Facility, the Climate-Smart Staple Crop Production Program in China was implemented in Huaiyuan County of Anhui Province and Ye County of Henan Province. Under the leadership of the Ministry of Agriculture and Rural Affairs of China, the project management office established a top-down project steering committee and a project management team involving national, provincial and county representatives. The establishment of top-down institutional framework and operation mechanism not only strengthens the organization and leadership of the project, but also provides a reliable system guarantee for the normal operation of the project and ensures the sustainable development of the project. After five years of implementation, fruitful "triple-win" results have been achieved. 100,000 mu demonstration areas have been established in Anhui Province and Henan Province, the main grain producing areas. The amount of nitrogen fertilizer per unit area in the project area has been reduced by about 10%, the amount of pesticides has been reduced by more than 15%, and the soil organic carbon content has increased by 10%. During the five years of practice, the cumulative carbon sequestration and emission reduction has hit 130,000 tons of carbon dioxide equivalent and the output has increased by more than 5% annually.

Qiandao Lake is a drinking water source for about ten million people in Hangzhou and other cities, with a fragile ecological environment and facing threats and challenges brought about by severe agricultural non-point source pollution and climate change. In view of this, the "Qiandao Lake Water Fund" program was officially launched in 2018, initiated jointly by Alibaba Foundation and Min Sheng Tong Hui Foundation, with Wanxiang Trust as its trustee and The Nature Conservancy as its science advisor. The "Qiandao Lake Water Fund" program is committed to reducing agricultural diffuse pollution through scientific methods, helping the development of ecological industry in the basin through green consumption and exploring the long-term governance mechanism of diffuse pollution in the basin. The "Qiandao Lake Water Fund" program innovatively introduced a charity trust relying on the advantages of the trust system, so as to stimulate the potential of environmental protection cause, which added more diversified and influential investment channels for traditional donation ways; in addition, realizing that the realization of ecological value is the "source" to ensure the sustainable development of ecosystem, the Qiandao Lake program is promoting the market competitiveness of agricultural products, broadening market channels and increasing farmers' income through the business model of "Three Goods Agriculture", and striving to explore various ways to give full play to the economic value of "clear waters and green mountains".

7 Wetland

Wetlands are critical in tackling climate change. With excellent carbon sink capacity, wetlands are important "carbon reservoirs", and "carbon absorbers", as well as "buffers" for climate change. It is known as the "Earth's kidney" because of its powerful function of water storage and purification. In addition, wetlands provide indispensable habitats for a wide variety of aquatic animals, amphibians, birds and other wildlife. Thus, wetland conservation is of critical importance to NbS and enables synergistic responses to multiple societal challenges including climate change, biodiversity loss, water scarcity and extreme weather events. According to the definition in the *Convention on Wetlands of International Importance Especially as Waterfowl Habitat* (RAMSAR) of 1971, "Wetlands are areas of marsh, fen, peatland or water, whether natural or artificial, permanent or temporary, with water that is static or flowing, fresh, brackish or salt, including areas of marine water the dept of which at low tide does not exceed six metres". In the broad sense, there are two categories of wetlands: natural wetlands and artificial wetlands. Natural wetlands include peatlands, beaches, estuaries, marshes, lakes, rivers and salt marshes. Artificial wetlands mainly include paddy fields, water reservoirs and ponds. According to statistics, there are 8.558 million square kilometers of natural wetlands in the world, accounting for 6.4% of the world land mass.

This chapter focuses on the important role of natural wetlands in climate change mitigation and adaptation, selects four typical wetland ecosystems: peatlands, coastal mudflats, estuarine wetlands and mangrove wetlands, five wetland NbS projects are presented. They are designed and implemented in different levels of governing body, including state, cities, and regions. These cases are the Peatland Restoration in Indonesia, ecological restoration of Yellow Sea Wetlands in Yancheng, China, Dongying Wetland City Construction in China, China-Association of Southeast Asian Nations (ASEAN) Mangrove Conservation and Restoration, and Ocean conservation in Small Island States in South Pacific.

7.1 Peatland Conservation in Indonesia

Case Highlights

After the massive peatland fire, Indonesia established National Peatland Restoration Agency (BRG) to manage and promote peatland conservation and restoration; with international and multi-institutional funding and support, Indonesia developed an online monitoring and management platform to achieve scientific management and enhance early warning and monitoring capabilities; in addition, the management trained staff in a coordinated manner so that scientific peatland conservation and restoration in Indonesia could be carried out in an orderly manner.

7.1.1 Project Background

Indonesia has about 24 million hectares of peatlands, about 36% of the world's total tropical peatland area. Peatlands are a major ecological and climate protection barrier from Indonesia to the globe. Conserving and restoring peatlands plays a crucial role in climate change mitigation and adaptation, coupled with the development of a circular economy. Well-preserved peatlands contribute to an effort to protect the largest natural terrestrial carbon pool and maintain biodiversity and the water cycle. However, human activities such as agriculture development, logging and peatland drainage[①] are threatening these fragile ecosystems, and only about 25% of the peatlands in Sumatra, Indonesia remain in good condition.[②] In 2015, a massive fire in Indonesia turned 2.6 million hectares of land in the archipelago into scorched earth. Burning emitted massive amounts of smoke, causing thousands of illnesses and exposing 69 million people to unhealthy level of pollutants. In addition, peat burning releases ten times more carbon than forest fires, so this fire has released about 16 billion tons of carbon dioxide from peatlands in Kalimantan and Sumatra.[③]

7.1.2 Project Overview

The 2015 fires brought unprecedented international attention to protect peatlands. To reduce the impact of fires and protect the vulnerable peatland ecosystem, the President of Indonesia has declared a moratorium on drainage and conversion of uncultivated peatlands, established a special government department to implement the decree and restore about 2.6 million hectares of degraded peatlands in Indonesia, and developed a special information disclosure and monitoring system working with several international organizations to track the real-time status on peatlands, as well as the progress and impact of restoration activities. At the same time, the Indonesian Government has raised funds through multiple channels to carry out "Preventive Measures for Improving Tropical Peatland Use" project and Peatland Restoration Project in Sebangau National Park, Kalimantan Tengah, turning to a proactive approach,

---

① Unlike forests, peatlands have no mechanism to protect themselves from moisture loss to the atmosphere. Draining water accelerates the decomposition of the peat's dense carbon storage and transforms them from firebreaks to fire propagation factors.

② Data Source: WWF Forest Solutions Platform, https://www.unops.org/news–and–stories/stories/restoring–indonesian–peatlands–protecting–our–planet.

③ Data Source: UNEP, https://www.unenvironment.org/news–and–stories/story/fighting–fires–indonesias–peatlands.

enhancing the safety management of peatlands by strengthening the monitoring and evaluation system.

Responding to the 2015 fires and protecting the fragile ecosystem of peatlands, National Peatland Restoration Agency (BRG) was established in 2016 with the task of restoring 2.6 million hectares of peatlands degraded by fire. BRG is mainly responsible for organizing, managing, planning and coordinating measures across sectors for peatland protection and restoration, including the supervision of the construction, operation and maintenance of infrastructure on privileged lands, encouraging the development of a circular economy and conducting publicity and education activities (Dohong A, 2017).

With the help of partners such as the World Resources Institute (WRI) Indonesia Office, the Agency for the Assessment and Application of Technology (BPPT) of Indonesia, the Food and Agriculture Organization of the United Nations (FAO), the United Nations Development Programme (UNDP) and the United Nations Office for Project Services (UNOPS), BRG has established the Peatland Restoration Information and Monitoring System. The system is an online platform tracking ongoing condition of peatlands and monitoring the progress of restoration activities and associated impacts. The data displayed on the platform is publically-accessible, sourced from relevant ministries and agencies, containing annual peatland restoration plans and implementation for each year since 2017, fire hotspot data and comparative charts, the number of villages involved in restoration efforts, and the number of peatland restoration research activities. In addition, the interactive map on the platform helps users to select map layers for peatlands distribution, restoration activities distribution, fire hotspot locations distribution, and restoration-related indicators (Figure 7-1).

Water loss in peatlands is associated with numerous negative impacts, including

Restoration

Planned Planned PHU inerventions by BRG in 7 provinces ▼ in 2019 ▼ 54 PHUs Hide details ▲		Implemented PHUs intervened by BRG in 7 provinces ▼ completed in 2019 ▼ 56 PHUs Hide details ▲	
Acivity	Total	Acivity	Total
1. Canal Blocking	27,375 unit	1. Canal Blocking	1,161 unit
2. Deep Well	24,056 unit	2. Deep Well	2,026 unit
3. Canal backfilling	10 unit	3. Canal backfilling	0 unit
4. Revegetation Area	97,684 unit	4. Revegetation Area	0 unit
5. Revitalization	320 unit	5. Revitalization	0 unit

(a) Example of Peatland Restoration Information

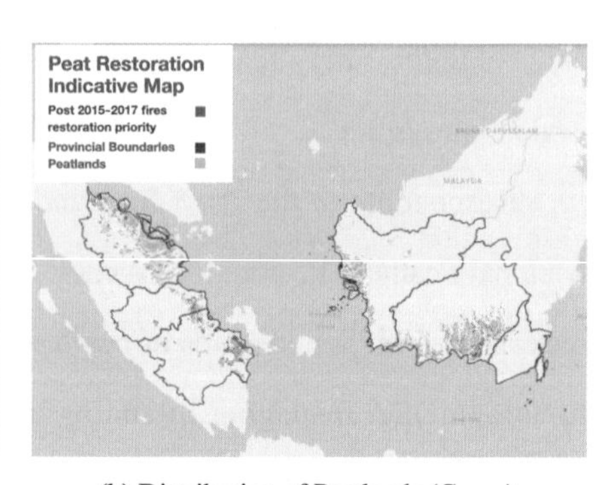

(b) Distribution of Peatlands (Green)
and Distribution of Restoration Areas (Red)

Figure 7-1 Monitoring System Peat Restoration Data Panel

lower water tables, reduced peat surfaces, CO_2 emissions, fires and droughts. To prevent unrestrained water loss from peatlands, BRG has developed a peatland water monitoring system employing information technology innovations, monitoring in-ground water levels, and reporting real-time data transmitted by equipment measuring soil moisture, rainfall, air temperature and humidity, and wind direction and speed. As of December 2018, BRG had installed 142 water level monitoring devices in seven priority restoration provinces, which is of significance to prevent catastrophic forest and peatland fires as well as greenhouse gas emissions.

In 2018, the Indonesian Government launched a two-year peatland project jointly with UNOPS, funded by the Norwegian Government, to support the BRG for developing an efficient and integrated peatland restoration model, starting with restoration actions in seven priority provinces. The project focuses on awareness raising and capacity building, encouraging local residents to make rational use of the natural resources around the peatlands, and develop locally produced environmentally friendly products to provide economic incentives for local residents. The project concentrated on strengthening the connection between villages and preventing fires through different approaches such as capacity building and community prevention.

The Indonesian Government, previously, has accumulated plenty of experience in peatland conservation and scientific testing through international cooperation. For example, WWF has been working with local communities and governments on peatland restoration projects in Sebangau National Park since 2004. In addition, in 2015, UNOPS Indonesia, United States Agency for International Development (USAID) and UNEP jointly launched a project entitled "Preventive Measures for Improving Tropical Peatland Use" to effectively address peatland degradation, with key activities including the establishment of a visual fire risk management system that can provide early warning of fires in high-risk areas for one to three months in advance and help improve local fire risk management capacity. The fire risk management system integrates information from different dimensions, such as weather, socio-economic activities, vulnerability and risk assessment, El Niño phenomenon and climate change, to generate fire risk prediction maps with a computerized system. From there, the project team carried out activities at various levels, including promoting inter-ministerial development of linked fire prevention planning, peatland restoration, etc.

7.1.3 Project Results

Three years since establishment, from 2015 to 2018, BRG has restored about 700,000 hectares of peatlands, equivalent to an area of one million soccer fields. And it committed to restoring an additional 4.5 million soccer-field-equivalent peatlands by 2020. The environment has benefited significantly from the moratorium and restoration task. In protected peatland areas, the forest reduction rate plummeted by 88% in one year, hitting a record low, with the potential to prevent 5.5 to 7.8 billion tons of CO_2

emissions over 15 years.

In 2016, one year after the launch of the "Preventive Measures for Improving Tropical Peatland Use" project, the fire risk assessment system was launched, hundreds of government employees were trained on the system, and fire prevention forums were established in all project coverage areas. Communities and local governments are now realizing the importance of fire management through prevention, and the shift in management paradigm has allowed communities to take precautions and prepare in advance of fire season. Through vulnerability and risk assessments using the platform management system [Figure 7-1(b)], the project has undertaken community-based peatland restoration projects in Kalimantan Tengah, such as the establishment of a peatland restoration project in Buntok covering an area of approximately 20,000 hectares, where peatland fire control is carried out through a model of community mobilization and participation. In addition, sustainable livelihood alternatives are actively developed, for example, through small grants, development of horticulture, fisheries, and other measures to improve incomes, and guiding communities away from illegal logging and burning practices through livelihood improvements so that villagers can become more engaged in peat restoration activities.

As of 2017, a total of 1,400 dams were built, more than 300,000 hectares were irrigated, 70 community nurseries were established, and more than 10,000 hectares of land were revegetated in the most degraded areas of peatlands in Sebangau National Park, Kalimantan Tengah by local communities and governments. The project has also established six local community fire patrols (involving upwards of 28,000 people) to maintain, manage and monitor the dams, as well as carried out reforestation activities. In the future, the project plans to build 850 additional dams, 100 community nurseries, community partnerships involving 30,000 local residents, 15 community fire patrols, including joint law enforcement and community education, while watering 100,000 hectares of land and revegetating 19,000 hectares of land, with the expectation that approximately 1.5 million tons of greenhouse gas emissions will be prevented annually in the 30 years of the project.

7.1.4 Project Information

Case Source: N4C Global NbS Case Studies, TNC Global NbS Case Studies.

Project Participants: Indonesian Government, regional governments and communities, international organizations and donors.

Project Location: Indonesia.

Project Duration: National Peatland Restoration in Indonesia (2016-2020); "Preventive Measures for Improving Tropical Peatland Use" Project (since 2015); Peatland Restoration Project in Sebangau National Park, Kalimantan Tengah (2004-2033).

Project Links: Nature4Climate, https://4fqbik2blqkb1nrebde8yxqj-wpengine.netdna-ssl.com/wp-content/uploads/2019/09/Peatland-restoration-in-Borneo-Indonesia_Sometimes-it-takes-a-disaster.pdf; UNOPS: https://www.unops.org/news-and-stories/stories/restoring-indonesian-peatlands-protecting-our-planet; WWF: https://www.wwf.org.au/ArticleDocuments/360/pub-briefing-borneo-peat-restoration-sebangau-central-kalimantan-15jun17.pdf; Peatland Restoration Information and Monitoring System: https://en.prims.brg.go.id/platform.

7.2 Ecological Restoration of the Yellow Sea Wetlands in Yancheng, China

Case Highlights

Through the top-level design, the relevant policies and regulations are formulated and issued, research bases are set up, scientific conservation approaches are experimented and analyzed, and meanwhile placing attention on communication and exchange with international agencies and foreign institutions to summarize and practice the NbS model of scientific wetland conservation.

7.2.1 Project Background

Yancheng is a city in Jiangsu Province, China. With an area of 455,300 hectares, the Yellow Sea Wetlands in Yancheng, China, is the largest and relatively well-preserved coastal wetland remaining on the west coast of the Pacific Ocean and the edge of the Asian continent that is the largest in size, and relatively well-preserved, it is also the largest coastal mudflat in China (Figure 7-2). Furthermore, it is a declared project of the Migratory Bird Sanctuaries along the Coast of Yellow Sea-Bohai Gulf of China (Phase Ⅰ). It covers a total of 18.86 hectares, mainly covering part of Jiangsu Yancheng Wetland National Nature Reserve for Rare Birds (Short in Yancheng Nature Reserve), the whole area of Jiangsu Dafeng Elk National Nature Reserve, Yancheng Tiaozini Wetland Park and wetland protection areas. The site is home to more than 680 vertebrate species, including 415 species of birds, 26 species of mammals, 9 species of amphibians, 14 species of reptiles, 216 species of fish and 165 species of zoobenthos, serving as a central node along the most endangered and threatened East Asian-Australian migratory bird route and as a resting, molting, and wintering ground for millions of migratory birds worldwide. It provides habitats for 23 internationally important bird species and supports the survival of 17 species on the IUCN Red List, including one critically endangered species: Spoon-billed Sandpiper, and five endangered species: Black-faced Spoonbill, Oriental White Stork, Red-crowned Crane, Spotted Greenshank and Great Knot.

(a) Location Map (b) Natural Landscape

Figure 7-2 The World Natural Heritage Site of Yellow Sea Wetlands in Yancheng, China

7.2.2 Project Overview

Over the years, the Yellow Sea Wetland adheres the motto "from nature, to nature", restoring the wetland ecosystem, exerting the ecological function of the ecosystem, and building a new model of harmonious coexistence between human and nature, wetland protection and operation of sustainable development, providing technical support for the protection of heritage site ecology. In policy making, Yancheng has promoted the establishment of the Yancheng Wetland and World Natural Heritage Conservation and Management Center, and introduced the *Regulations on the Conservation of Yellow Sea Wetlands in Yancheng* and *the Three-Year Action Plan for the Conservation, and the Management and Sustainable Development of Yellow Sea Natural Heritage Site in Yancheng (2019—2021)* to provide a basis for legal and scientific protection.

In terms of academic research, the Land Consolidation and Rehabilitation Center of the Ministry of Natural Resources approved the set-up of the “Nature-based Research Station for Ecological Conservation and Restoration of Yellow Sea Wetlands” at the Yellow Sea Wetland Institute in September 2020. It mainly involves the promotion, practice and application of NbS, highlighting problems such as wetland protection, ecological restoration, and control of invasive alien species (e.g., Spartina alterniflora Loisel), and conducting related training activities.

To promote the conservation, management and sustainable development of the Yellow Sea Wetland World Heritage Site, Yancheng has established the Yellow Sea Wetland Institute in 2017, strengthening international exchanges with and recruiting a team of special experts and scholars renowned at home and abroad from relevant international organizations such as the IUCN, the United Nations Convention on Wetlands, and universities including the University of Cambridge in the UK, the University of Queensland in Australia and Kyungpook National University in South Korea. The team has carried out research on coastal wetland conservation, migratory bird conservation, ecological restoration, alien species control and other topics, striving to build a community of life in the mountains, waters, forests, farmlands, lakes and grasses.

A series of successful pilot projects in the ecological restoration of the Yellow Sea Wetlands have been carried out in Yancheng, China.

1. Ecological Restoration Demonstration Project of Coastal Wetlands in Chuanshui Port

Chuanshui Port Plot is located in the heritage site of Migratory Bird Sanctuaries along the Coast of Yellow Sea-Bohai Gulf of China (Phase Ⅰ) and the heritage buffer zone, which is the second experimental area in the south of Yancheng Nature Reserve. Located in the middle of the entire coastal wetland ecosystem in Jiangsu Province, it is an important node along the migration path from Chongming Island in Shanghai to the core area of Jiangsu Yancheng Wetland National Nature Reserve for Rare Birds, providing essential resting and feeding places for migratory birds. Chuanshui Port Plot is adjacent to the core area of the Dongsha Island Heritage Site, where plovers and snipes mainly reside. Following the ecological restoration, the area will become an extensive high tide habitat and feeding source for them. In addition, Chuanshui Port used to be inhabited by red-crowned cranes. However, there are fewer red-crowned cranes in the southern part of the heritage site due to the reduction of natural wetlands and lack of food in winter. Once restored, it will become an important resident spot for red-crowned cranes in the southern part of the heritage site (Figure 7-3).

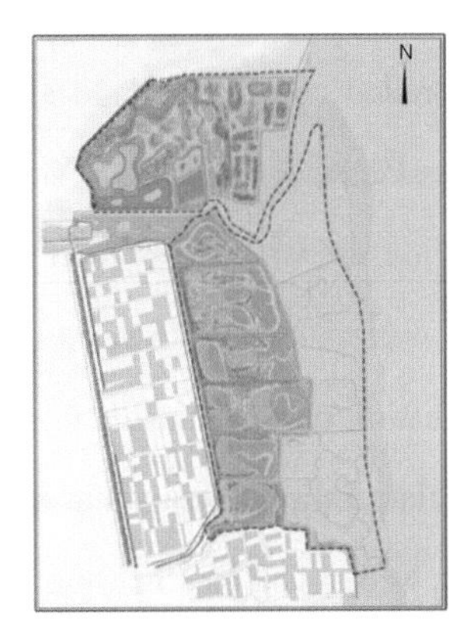

(a) Location Relationship

(b) Plot Master Plan

Figure 7-3 Chuanshui Port Plot

In recent decades, the coastal wetland ecosystem in Chuanshui Port Plot has suffered serious damage caused by natural and human factors, such as coastal reclamation, unreasonable land development, elevated terrain by siltation, wetland aridification and alien species invasion, resulting in severe fragmentation of coastal wetland habitats, gradual decline of biodiversity, increasing fragility of ecosystem and continuous weakening of ecosystem functions; the total area of coastal wetlands in Chuanshui Port Plot has seen decrease, accompanied by the sharp decrease of natural wetlands land mass.

In the context of natural aridification of the coastal wetlands, and combined with the main protection objectives of nature reserves and heritage sites, the demonstration project of coastal wetland ecological restoration in Chuanshui Port Plot restores the wetland ecosystem with NbS, gives full play to the ecological function of the ecosystem, provides technical support for the performance of the eco-conservation function in the heritage site so as to promote wetland conservation and sustainable development. An integrated model is developed for the comprehensive management and restoration of degraded coastal wetland ecosystems based on the creation of multiple types of habitats (seawater habitat, freshwater habitat, seawater-freshwater habitat), water level regulation and multi-level topographic reshaping, and a model for the restoration of salt marsh wetlands (Suaeda salsa mudflats) and seasonal seawater mudflats based on the combination of coastal tidal introduction and regulation, ecological control of Spartina alterniflora Loisel. and natural evolution of coastal ecosystems.

2. Ecological Restoration Program in Doulong Port Plot

Doulong Port Plot is located in the buffer zone of Migratory Bird Sanctuaries along the Coast of Yellow Sea-Bohai Gulf of China (Phase Ⅰ) heritage site, which belongs to the southern buffer zone of the Yancheng Nature Reserve within Dafeng District, south of Doulong Port, with a total area of 1,369 hectares. The northern part of the plot is separated from the Yancheng Nature Reserve by a river, the western part is adjacent to the Dafeng Gan River, the southern part is an artificial water transmission channel for existing pond fish breeding, and a river passes between the eastern part and the core area. The area was formerly a high-density artificial pond fish farming area, which was listed as a mandatory zone for renovation and restoration during the national marine inspection in 2018.

Most areas of Doulong Port Plot are abandoned freshwater aquaculture ponds, with a relatively homogeneous ecological environment that inadequately protects migratory birds (Figure 7-4). However, the area accessible to seawater at Doulong Port Plot is limited because it is affected by long term mudflat evolution; the general water level elevation of the freshwater riverway is around 0 m, making it difficult for substantial amounts of freshwater to enter the project area without passing through the pumping station. Whereas regional water resources are difficult to obtain naturally, the source of freshwater used for aquaculture relies on high water ditches for irrigation, so drought in the wetlands is particularly severe during the dry season. Doulong Port was/has been overloaded with wetland ecosystems due to long-term aquaculture, which leads to water eutrophication, following the abandonment of the aquaculture ponds, the bottom of the fish ponds dried up due to lack of management, leaving it with less biomass, which is not conducive to bird foraging, a single habitat occupied by reed communities was formed. In this project, the topography transformation includes ecological island shaping, shallows shaping, shallows revetment shaping, open water topography shaping, deep water topography shaping and embankment topography shaping. It takes into account the diversification of topography to fully meet the needs of swimming birds and wading birds for habitat. The main plant communities in the project area are submerged plant communities, emerged plant communities, saline rangeland communities and saline shrub communities. Doulong Port Plot can be divided into several functional areas, such as lagoon salt marsh ecological restoration area, salt water dish-shaped lake ecological restoration area, arboreal bird habitat restoration area, freshwater dish-shaped lake ecological restoration area, and bird-friendly ecological planting area. Under the technical aspect of NbS, the policy of "focusing on natural ecological restoration" supplemented by moderate artificial intervention is adopted to create, protect, or restore pure ecosystem elements as much as possible, combined with ecosystem elements and hard engineering interventions.

Figure 7-4 Abandoned Aquaculture Ponds in Doulong Port Plot

3. Wildlife Habitat Conservation Project in Yeludang

Located in the buffer zone of Migratory Bird Sanctuaries along the Coast of the Yellow Sea-Bohai Gulf of China (Phase Ⅰ), Yeludang is on the west coast of the South Yellow Sea, north of the Yangtze River, between the National Elk Sanctuary and the Red-crowned Crane Sanctuary, covering an area of 3,000 mu, embraced by two coastal strips and directly facing the Yellow Sea. In the nearly 50,000 mu primary ecological environment of tidal flat and wilderness, no towns or villages, no lights at night. But every night there are herds of pure wild elk, and other small beasts such as river muntjac and fireflies everywhere in summer. Since 2009, Yancheng Dafeng District Government and NGOs have initiated campaigns to protect Yeludang, specializing in scientific research, culture and nature conservation. At the same time, a coastal wild plant germplasm bank was jointly built with Jiangsu Provincial Key Laboratory of Coastal Wetland Bioresources & Environmental Protection (JLCBE), which now possesses nearly 300 specimens and seeds of the only surviving wild plants of 485 species in the coastal area and carries out solid groundwork for the conservation of wild plants in the coastal area (Figure 7-5).

Figure 7-5 Wild Elk in Yeludang

4. Invasion Prevention Project of Spartina Alterniflora Loisel

Introduced to China in the 1980s, Spartina alterniflora Loisel. (commonly known as smooth cordgrass) has been spreading rapidly in the Chinese coastal belt, posing a threat to the biodiversity of most coastal mudflat wetlands due to its well-developed root system and dispersal ability. It is one of the main reasons for the degradation of Yancheng coastal wetlands (Figure 7-6). Seeking a "Yancheng solution" to the prevention and treatment of Spartina alterniflora Loisel., the government has carried out Spartina alterniflora Loisel. controlling treatment in the Yellow Sea Wetlands with a view to strengthen the containment of the grass, accelerate the protection and restoration of the coastal zone ecosystem, and provide experience and suggestions for the control of Spartina alterniflora Loisel. on a larger scale in China.

Figure 7-6 Status of Spartina Alterniflora Loisel Invasion in the Heritage Site

7.2.3 Project Results

Taking biodiversity protection as an example, Jiangsu Yancheng Wetland National Nature Reserve for Rare Birds has enlarged the buffer zone to return fishery to wetland, having retired nearly 100,000 mu of fish ponds to expand the living space for birds. The elk reserve has taken measures such as internal regulation, silage supplementation, grid management, seine rotation, water system renovation and vegetation restoration to scientifically regulate and manage the elk population as well as to effectively restore and improve the habitat of the elk. According to the recorded data, the number and species of birds in Yellow Sea Wetlands have increased in recent years, with 5,681 elk, accounting for more than 60% of the world total, including 1,820 wild elk.

On July 5, 2019, Yellow Sea Wetlands in Yancheng, China, as the representative of Migratory Bird Sanctuaries along the Coast of Yellow Sea-Bohai Gulf of China (Phase Ⅰ), was inscribed on the World Heritage List at the 43rd session of the World Heritage Committee, becoming the first intertidal wetland world heritage site in China and the second in the world, which filled the blank that there is no coastal wetland type world natural heritage site in China. Yancheng has divided its 2035 urban planning into three stages: In the first stage, by 2021, Yancheng will become an international wetland city, the Yellow Sea Wetland World Heritage Site will be fully protected, a migratory bird migration route city alliance will be established with Yancheng as the core, and a preliminary framework for cooperation will be formed around the Yellow Sea eco-economic circle; In the second stage, by 2025, the Yellow Sea Wetland World Heritage Site will be scientifically protected, inherited in a living way and reasonably utilized, and the cooperation framework of the Yellow Sea ecological and economic circle will become a platform for mutual benefit and trust between regions; In the third stage, by 2035, Yancheng will have achieved significant results in urban construction, presenting a beautiful picture of harmonious coexistence between man and nature.

7.2.4 Project Information

Case Source: C+NbS Cooperation Platform.

Project Participants: Yancheng, Jiangsu Province, China.

Project Location: Yellow Sea Wetlands in Yancheng, China.

Project Duration: Since 2016.

7.3 Dongying Wetland City Construction in China

Case Highlights

Actively mobilize all kinds of social conditions, adhere to the guidance of planning, and improve the system. Social mobilization is widely carried out, and all citizens are involved in wetland protection through various activities such as “wetland school” construction and educational activities.

7.3.1 Project Background

The Yellow River Delta wetland, located on the Bohai Sea in the northeast of Shandong Province, China, is not only a rare estuarine wetland ecosystem in the world, but also the broadest, most complete and youngest wetland ecosystem in the warm temperate zone in the world. Dongying is the central city of the Yellow River Delta. Thanks to both rivers and seas, it is rich in wetland resources, with a total area of 458,100 hectares and a wetland rate of 41.58% (Figure 7-7). Under such natural conditions, Dongying was positioned as a "wetland city" at the beginning of its establishment, and devoted itself to giving full play to the advantages of wetland resources in urban construction.

Figure 7-7 Aerial View of Dongying

Due to the limitation of topography and urban infrastructure pipelines, there is prominent waterlogging problem in Dongying, with manifested wetland water shortage and degradation: On one hand, the Yellow River Delta is a typical alluvial plain, as a city built on the Yellow River Delta, the overall terrain of Dongying is flat, coupled with the relatively single drainage channel, concentrated rainfall in the flood season, serious seawater jacking and high groundwater level, leading to the fact that the city is prone to waterlogging problems, on the other hand, the wetland water in Dongying mainly depends on the Yellow River water. In recent years, due to the reduced runoff of the Yellow River and the uneven rainfall, the rain and flood resources have not been effectively utilized, resulting in the water shortage and degradation of the wetlands in Dongying.

7.3.2 Project Overview

In order to solve urban waterlogging and effectively cope with the water shortage and climate change, Dongying explores the balanced scheme for wetland protection and urban development, mainly taking the following measures.

Adhere to the guidance of planning and improve institutional systems. The *Overall Plan for Wetland Protection and Restoration in Dongying (2018-2025)* was compiled, the task target of wetland protection was established, and the wetland protection system was improved. The *Implementation Opinions on Strengthening Planning and Leading to Improve Urban Quality* was issued, which insisted on taking water as the vein, taking green as the clothing, and taking wetlands as the features, highlighting the ecological structure of the central city wetland with "two belts, three rivers, five patches, and multiple points", demonstrating the urban characteristics of "the wetlands are in the city and the city is in the wetlands", and greatly improving urban quality. The *Master Plan for Dongying Land Space (2019-2035)* was launched, promoting the "integration of multiple plans", scientifically delimiting "three districts and three lines", and rationally lays out ecological, agricultural and urban space. *Regulations on Shandong Yellow River Delta National Nature Reserve, Regulations on Wetland Protection in Dongying,* and *Regulations on Wetland City Construction in Dongying* were promulgated. To implement the responsibility of wetland protection, Dongying issued the *Implementing*

Opinions on the Overall Establishment of Forest Wetland Officer, with a total of 2,261 forest wetland officers, including five at the city level, 32 at the county level, 122 at the township level and 2,102 at the village level.

Construct national parks and building protection systems. Recently, in order to promote the construction of the Yellow River Estuary National Park, Dongying has set up a "special team for promoting the national park construction" to integrate eight existing nature reserves in the Yellow River Estuary region as a national park, covering an area of 3,554.11 square kilometers. The *Plan for the Establishment of the Yellow River Estuary National Park,* the *Investigation and Evaluation Report on the Background Resources of the Yellow River Estuary National Park,* the *Social Impact Assessment Report on the Establishment of the Yellow River Estuary National Park,* and the *Comprehensive Research and Compliance Determination Report of the Yellow River Estuary National Park*, were compiled, making every effort to promote the construction of the national park. Six water source reserves, one national urban wetland park, one national wetland park, four provincial forest parks, nine provincial wetland parks, one provincial scenic spot and 39 wetland protection areas have been established throughout the city. The city has built a wetland protection system with the national park as the main body, and initially established an ecological protection framework system taking into account both land and sea.

Enhance the storage and detention capacity and build a city without waterlogging. In order to improve the capacity of flood storage and detention, Dongying issued the *Special Planning of Sponge City in the Central City of Dongying (2016–2030)* and the *Urban Planning and Design Plan for the No-Waterlogging Central City of Dongying*, constructed the ecosystem with "large water surface, large green space, large wetland and large space"; delimited the urban blue line and green line to protect the water system and green space, and implemented a series of engineering construction, such as water accumulation point reconstruction, water system penetration, flood storage and detention, pumping station improvement, pipe network improvement, emergency drainage in communities with serious waterlogging, rainwater entering the river nearby for communities along the river, etc., to enhance the flood control and drainage capacity of the central city. During project construction, Dongying, led by the government, pays attention to attracting social capital to participate in wetland city construction, encourages market participants to participate in diversified operation of wetland city construction, and maintains operation with commercial income.

Strengthen environmental protection publicity and education and enhance ecological awareness. In recent years, Dongying City has widely carried out the construction activities of "wetland school" on the basis of ecological education bases such as Yellow River Estuary Wetland Museum and Yellow River Delta Bird Museum. Through educational activities designed by the city government, both children and parents are enabled to participate in the protection of wetlands. For this reason, Dongying compiled the wetland ecosystem moral education textbook *Beautiful Hometown: Yellow River Estuary* and set up a lecturer group for ecosystem moral education to teach courses such as *Notes on Nature* and *Letting Swallows Live in My House* deeply loved by students and parents, and "Wetland School" is booming (Figure 7-8).

Figure 7-8 Dongying Wetland School

7.3.3 Project Results

After years of efforts, Dongying has significantly increased its wetland coverage, created huge economic value by providing material products such as water supply, aquatic products and raw materials, and effectively alleviated urban waterlogging through drainage system connection and flood storage and detention projects, including:

Improve the rate of wetland protection and give full play to the ecological function of wetlands. Dongying has a total wetland area of 458,100 hectares, with a wetland rate of 41.58%, a wetland protection area of 257,000 hectares, and a wetland protection rate of 56.1%. The value of material products provided through water supply, aquatic products and raw materials reaches ¥10.224 billion, the value of tourism and leisure is ¥2.486 billion, and the cultural and scientific value is ¥141 million. The indirect ecological value of atmospheric regulation, flood regulation and storage, water quality purification, wave dissipation and revetment, biodiversity and so on reaches ¥1.325 billion (Table 7-1).

Table 7-1 Evaluation Results of Dongying Wetland Ecosystem Service Value in 2015

Unit: RMB 100 million yuan

Direct Use Value					Indirect Use Value + Option Value					Total
Material products			Tourism and Leisure	Culture and Scientific Research	Atmos-pheric Regulation	Flood Regulation and Storage	Water Quality Purifi-cation	Wave Dissipation and Revetment	Biodiversity	
Water Supply	Aquatic Products	Raw Materials								
5.41	59.56	37.27	24.86	1.41	0.29	2.70	0.84	0.27	9.15	141.76

Source of information: Dongying Municipal Government project "Value Evaluation and Research on Protection Countermeasures of Wetland Ecosystem Services in Dongying, Shandong Province".

Solve waterlogging and improve the ecological environment. Through the construction of water system connection and flood storage and detention projects, the problem of urban waterlogging has been effectively alleviated, and the fortification criterion of waterlogging in the central city has been changed from "once in 10 years" to "once in 50 years". At the same time, it also makes effective use of rain and flood resources to ensure the ecological water demand of urban wetlands. In addition, the ecological environment in the region has been improved and the biodiversity has been enhanced, and the number of bird species in the urban area has increased from less than 100 species at the beginning of the establishment of the city to 206 species.

Enhance the awareness of protection and guard the wetland city. At present, more than 40 primary and secondary schools in the Dongying city have joined the national or Dongying "Wetland School" network. There are one "International Ecological School", three "National Demonstration Schools of Ecological Moral Education for Minors", 19 "Demonstration Schools of Ecological Moral Education for Minors in Dongying", five "China Wetland Schools", and seven "Dongying Wetland Schools". It has achieved the effect of "educating a student, driving a family, affecting the whole society", and making the ecological consciousness deeply rooted in the hearts of the people.

The success of Dongying Wetland City Construction is inseparable from the systematic and scientific planning and management of the government. The government rationally laid out ecological space, agricultural space and urban space, and implemented a series of projects such as water accumulation point reconstruction, water system connection, flood storage and detention according to local conditions, which not only solved the problem of urban waterlogging, but also effectively responded to social challenges such as lack of water resources and climate change.

7.3.4 Project Information

Case Source: C+NbS Cooperation Platform.

Project Participants: Dongying Municipal Government.

Project Location: Dongying, Shandong Province, China.

Duration: Since 1983 (founding of the city).

7.4 China-ASEAN Mangrove Conservation and Restoration

Case Highlights

Select the innovative governance model of "community agreement", and through the form of protection agreement, stipulate the rights, responsibilities and interests of different stakeholders, such as government, farmers and herdsmen, NGOs and so on. Help local villages build roads, install street lights, improve the traffic environment, share the fruits of mangrove conservation with villagers, make indigenous people become "allies" in the common protection of mangroves, reduce the conflict between conservation and community development, and promote the synergic governance of the ecological environment with the participation of the government, civil society and enterprises.

7.4.1 Project Background

There are more than 4.25 million hectares of mangroves along the coast of China-ASEAN countries, accounting for about 35% of the world's mangroves, which is the region with the richest mangrove biodiversity in the world. Most ASEAN members have long coastlines, a total length of 173,000 kilometers. At the same time, with the most important marine fish spawning sites in the Asia-Pacific region—the marine Amazon and the famous "Coral Triangle", the China-ASEAN region has become the region with the richest coastal and marine biodiversity on the Earth. If the trend of mangrove disappearance and degradation fails to be reversed, almost all mangroves that are not properly protected will disappear after a hundred years. In the entire China-ASEAN region, with the exception of China, who benefited from mangrove planting and restoration projects and increased the area of mangrove, would see mangroves decreasing at a rate of 0.25% to 20% per year.

7.4.2 Project Overview

Since 2016, the Global Environmental Institute (GEI) has been exploring community independent participation in mangrove protection in China and ASEAN mangrove countries through the model of community agreement protection mechanism, so as to find a balance between protection and development. This model can not only achieve the effective protection of mangrove ecosystem, but also enable the community to continue to enjoy the well-being brought by nature in the process of participating in the protection. The Global Environmental Institute introduces the community agreement protection mechanism from abroad. Its concept refers to: in an area in need of protection, in the form of an agreement signed by two or several interested parties (governments, enterprises, local communities or individuals, etc.), the rights of protection and limited development are given to different stakeholders to solve the problem of conflicts of economic interests obtained from nature between protectors and residents and to alleviate the contradiction between resource exploitation, environmental protection and the interests of residents. Besides, due to the participation of the local community, it has increased the community residents' love and attention to their own land, changed the role of the government as a single protector in the past, and formed a new biodiversity conservation model.

Along Myanmar's Bay of Bengal, there are mangroves stretching for 2,000 kilometers of coastline. In addition to being resistant to floods and storms, mangrove ecosystems also provide important habitats for many wild animals and fish. For the local people, mangroves also provide them with everything they need, including food, wood, firewood and building materials. In recent years, however, the area of mangroves in Myanmar has been shrinking rapidly—from 659,039 hectares to 462,963 hectares, a decrease of nearly one third from 1980 to 2015, mainly due to people's production, living and economic development. For example, pond farming, rice cultivation, palm oil production and firewood deforestation all pose a threat to mangroves.

Since 2016, the GEI has been working with seven communities in Myanmar to protect mangrove ecosystems through a community agreement protection mechanism model (Figure 7-9). Through cooperation with the local government, the communities and local NGOs in Myanmar, the local communities have been promoted to sign a five year protection agreement on the responsibilities and actions of all parties to protect the mangroves around the community, and successively through publicity and training, daily patrol, cultivation of mangrove seedlings, distribution of clean energy equipment, species survey, etc. to enhance communities' protection awareness of mangroves and promote communities'

Figure 7-9 Mangrove Seedlings Cultivated in PyinBuNge and Kanti
(Source: GEI)

independent conservation actions.

Zhanjiang Mangrove National Nature Reserve occupies more than 33% of the mangroves in China, and distributes the mangroves with the largest contiguous area in China, guarding the production and living safety of eight million coastal residents (Figure 7-10). Along with population growth and community development, problems such as production and living reclamation, pond fish farming, land occupation for urbanization, sewage and garbage are the main pressures faced by mangrove protection in Leizhou Peninsula. The reserve is divided into 68 subdivisions scattered around the entire Leizhou Peninsula.

Figure 7-10 Mangrove Landscape in Zhanjiang Mangrove National Nature Reserve
(Source: GEI)

In Zhanjiang, Guangdong Province, the GEI cooperated with Zhanjiang Mangrove National Nature Reserve to explore how the community can effectively participate in the countermeasure research of mangrove protection to tackle different problems. Based on interviews with stakeholders, literature collection and site survey, the Global Environmental Institute has analyzed the main problems faced by mangrove protection in Leizhou Peninsula, and found that the following problems generally exist in the protected areas: ① in some areas, as there is no place to dispose of garbage in nearby communities, garbage or waste is directly dumped into mangroves, posing a direct threat to the health of mangrove ecosystems; ② with the demand of local development, relying on mangroves to carry out leisure tourism will also bring risks to mangrove protection. For example, Gaoqiao is the largest and most complete mangrove area in Leizhou Peninsula, and the mangrove landscape here is the most spectacular in Leizhou Peninsula, so the spontaneous tourism in this area is also the most, which brings great pressure to mangrove protection and local government management. How to balance the

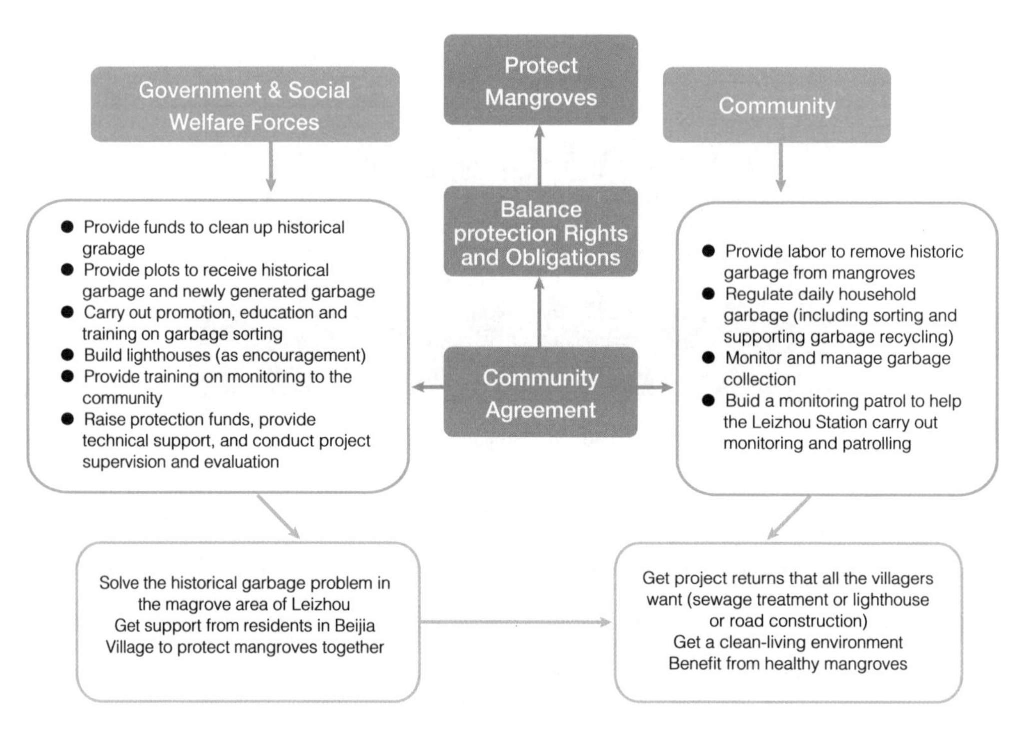

Figure 7-11 Pilot Study on Community Agreement Protection Model
(Source: GEI)

needs of the communities to benefit from tourism development, and actively mobilize the communities to participate in protection actions, so as to realize community co-management will become the key to solve the problem.

The project introduces the model of community agreement protection mechanism (Figure 7-11) to demonstrate and solve the conflict between community sustainable development and mangrove protection. A mangrove protection scheme based on community cooperation is proposed to reduce the impact of garbage on mangrove ecosystem and the impact of disorderly tourism on mangrove ecology, which provides a research basis for the implementation of typical mangrove protection in China.

7.4.3 Project Results

Since the launch of the project in 2016, mangrove conservation actions in the China-ASEAN region have achieved fruitful results, including the following three points.

Community pilot research and regional mangrove protection has achieved good results. Two pilot projects have been set up in Myanmar and China to carry out feasibility studies on community participation in mangrove conservation, which has not only achieved the goal of regional mangrove conservation, but also improved the practical benefits of residents. In Myanmar, for example, through the efforts of all parties, more than 5,000 people and 893 households have benefited from participating in the project, and a total community fund of $4,000 has been established. After the end of the second phase of the project, the average income of the community fund per household increased by 17%. In terms of conservation benefits, a 2,000-acre community protected area has been established. According to the survey, there are 28 local mangrove species, including two endangered species, and a total of 37,000 mangrove seedlings have been cultivated.

Cooperative development of mangrove ecosystem restoration guidelines. Form a set of comprehensive, detailed and professional mangrove restoration guidelines from the aspects of restoration objectives, principles, subject selection, species selection, seedling cultivation, seedling transportation and preservation, planting, returning ponds to forests, tidal flat afforestation, restoration of degraded mangroves, monitoring and evaluation, etc. At the same time, the manual will be submitted to the Restoration Department of the Ministry of Natural Resources and translated into English to support future restoration work.

Actively participate in the international cooperation of the mangrove conservation. The Global Environmental Institute has also cooperated with Global Climate Action Initiative (GCAI) to promote mangrove conservation exchanges between China and ASEAN, including the "Mangrove Conservation-Nature-based Climate Solutions Roundtable" held in Thailand in August 2019, and the official side event entitled "NbS: Mangrove Conservation and Zero-Deforestation Agricultural Production as Cases" at the UN Climate Change Conference COP25 in December 2019. All parties jointly explored how mangrove conservation can better contribute to ecosystem protection and climate change mitigation and adaptation under the NbS framework.

7.4.4 Project Information

Case Source: C+NbS Cooperation Platform.

Project Participants: Global Environmental Institute, local NGOs in Myanmar, community of the project site, etc.

Project Location: Myanmar, and Zhanjiang, Guangdong Province, China.

Project Duration: Since 2017.

Project Links: Global Environmental Institute, http://www.geichina.org/wp-content/uploads/2020/07/2019-Annual-Report_CN.pdf; China-ASEAN Environmental Information Sharing Platform, http://www.caeisp.org.cn/zh-hans.

7.5 Ocean Conservation in Small Island States in South Pacific

Case Highlights

The project provides Small Island States with funding for ocean conservation through uniting nations and local governments, indigenous villages, international organizations, multinational corporations, academics and think tanks. At the same time, the project actively explores the ecosystem payment projects including the development of ecological tourism, mangrove protection, community education and other various activities, so as to try to gradually establish a virtuous circle model to deal with climate change, improve the quality of ocean ecological environment, and promote employment and economic development.

7.5.1 Project Background

Because of their geographical location and capacity for economic development, Small Island States are fragile and most vulnerable to the immediate threat of climate change. The Independent State of Samoa is an island country consisting of two main islands Upolu and Savai'i, and eight nearby islands. It has a land mass of 2,934 square kilometers, where marine exclusive economic zone is 120,000 square kilometers. Most of its territory is covered by jungles with a tropical rainforest climate. The Republic of Fiji is located in the center of the Southwest Pacific. With land area of 18,333 square kilometers, its maritime exclusive economic zone is 1.29 million square kilometers. Fiji is made up of 332 islands, 30% of which are inhabited. It has a tropical maritime climate and suffers from hurricanes throughout the year.

7.5.2 Project Overview

"Ocean Opportunities & Small Island States (SOS-IS)" project of the World Team aims to build a network of cooperation between countries, scientific research institutes, businesses and international organizations to provide technology, finance and other resources for the sustainable development of Small Island States. The project has cooperated and built a network with a range of sectors, including United Nations Educational Scientific and Cultural Organization (UNESCO), 100% Renewable Energy (RE100), the Governments of Samoa and Fiji and relevant energy companies. This will provide an all-around funds and services for ecotourism, mangrove restoration and ocean conservation in both countries (Figure 7-12).

The SOS-IS project is dedicated to actively implementing sustainable development models for eco-friendly ocean states in different fields. It develops a series of solutions

Figure 7-12 SOS-IS Mangrove Conservation
(Source: World Team official website)

through a "point to area" model that can help oceans and islands achieve the sustainable transformation by creating jobs, enabling technology transfer, and launching new and alternative management systems. The World Team plans to use the most advanced technologies and programs to protect the ocean environment through ocean restoration projects or animal protection projects such as coral cultivation, eel grass cultivation, and turtle conservation. Among them, mangrove restoration is a key component of the World Team effort to solve the climate crisis. The SOS-IS project plans to build a series of actions mixing mangrove protection, natural education, ecological monitoring and ecotourism, at the same time also plans to unite network partners to carry out the local ocean capacity building activities. It can help people to know and care about the ocean from every drop of water to the ocean ecosystem level. So it can give people a first-hand experience of how new technologies and ideas can be used to develop and utilize the ocean resources.

In addition to mangrove conservation, this project also aims to build capacity for Small Island States in energy transition and environmental renovation, such as achieving energy independence through renewable energy micro grid systems, building clean water and electricity systems and developing sustainable transportation systems, using new renewable energy sources, such as tidal and hydrogen energies to build energy systems in conservation area, pilot hydrogen-powered homes, building vacuum toilets, carrying out the activity of using plastic limitedly, cleaning up the ocean plastics, etc.

As for the governance model, the World Team establishes regional event bases by connecting stakeholders, and works with international volunteers to maintain the ongoing operation of this project. In addition, the governance model of the project reflects the needs of different stakeholders. It is consist of the board of directors and two representatives from all partners, with an executive committee (consisting of the board of directors and major organizations) and an academic advisory committee. There is also an annual stakeholder conference attended by representatives from governments, project teams, as well as research institutes, companies and investment banks involved in new energy solutions.

7.5.3 Project Results

The SOS-IS project aims to develop locally-tailored sustainable development programs through capacity-building in Small Island States. Taking Samoa for example, in 2019, the World Team together with its partners planted more than 25 acres of nearly 1,000 trees and cleared six beaches in the Aleipata Islands with the help of Samoa's Department of Natural Resources, which provided habitats for local species like snakes and lizards. These effectively protect coastal communities and ecosystems from flooding, and enable them to play a role in purifying water and providing carbon sinks. In addition, the community can obtain a stable income through the project, which will boost the local economy. Compared to other ocean restoration efforts, mangrove restoration is more cost-effective because of the greater carbon sequestration in mangroves.

An ecotourism project based on a mangrove park can effectively boost the local economy. Combined with the local circular economy and renewable energy solutions, the project provides a quality infrastructure guarantee for Small Island States. It can promote the tourism industry and show the world indigenous cultures and build sustainable island marine ecosystems at the same time. In Fiji, for example, tourism directly contributed about $1 billion to GDP in 2019. Its GDP has grown by 7.43% annually since the start of the SOS-IS project in 2017, and its annual unemployment rate has dropped from 4.32% in 2016 to 4.10%. Therefore, tourism is an important source of income for Small Island States as well as the economic lifeline of the country. The development of ecotourism can provide a large number of job opportunities, thus the coordinated development of ecological protection, social progress and economic growth can be possible.

The SOS-IS project is also actively exploring its broader impact at the global level in legal system, technology and other areas, such as affecting the laws of the sea at the United Nations level through expert groups. During the implementation of the project, the network of ecological observation station provides a large amount of local data on biological conservation. With the promotion of the project, wastewater treatment technology, which converts local waste into energy, has revolutionized local water treatment mode. The project has also established a number of small-scale integrated systems to provide sustainable solutions, demonstrating the interaction and synergy of culture, education, transportation, agriculture, energy, policy and legal equality.

7.5.4 Project Information

Case Source: UNEP - NbS Contributions Platform.

Project Participants: World Team, and local governments and communities in Samoa and Fiji.

Project Location: Samoa and Fiji.

Project Duration: Since 2017.

Project Links: World Team's official website, https://worldteamnow.org/blogs/sos-is/; United Nations Ocean Conference, https://oceanconference.un.org/commitments/?id=21714.

7.6 Case Summary

Peatlands are one of the most important carbon sinks. Although peatlands account for only 3% of the Earth's land area globally, they store one third of the terrestrial carbon, and twice as much as global forests. The 2015 fires in Indonesia brought the unprecedented international attention to the protection of peatlands. To reduce the impact of fires and protect this fragile ecosystem, Indonesia's President announced a moratorium on the drainage and conversion of virgin peatland, and set up a government department to enforce regulations and restore 2.6 million hectares of degraded peatland in Indonesia. Indonesia, working with a number of international agencies, developed a dedicated information disclosure and monitoring system to track the current status of peatlands and monitor the progress and impact of restoration activities. At the same time, the Indonesian Government has raised funds through multiple channels to carry out the project "Preventive Measures for Improving Tropical Peatland Use" and the Peatland Restoration Project in Sebangau National Park in Kalimantan Tengah, turned passiveness into activeness and improved peatland safety management by strengthening monitoring and evaluation systems.

Coastal mudflat refers to the littoral zone between the high tide and low tide of the coastal spring tide, which is called "intertidal zone" in geomorphology. The Yellow Sea Wetlands in Yancheng, China, is the largest coastal mudflat in China. It was successfully listed on the World Heritage List at the 43rd session of the World Heritage Committee in 2019, became the first intertidal wetland world heritage site in China and the second in the world. In addition to the ecological functions of water conservation, water purification, carbon absorption and storage, the Yellow Sea Wetlands in Yancheng, China, are rich in biodiversity resources, and boast 14 species under first class national protection such as red-crowned cranes, white-headed cranes, elks, Chinese sturgeons and white sturgeons. The Yellow Sea Wetlands are also the world's largest wintering ground for red-crowned cranes and the world's largest gene pool for elks, which is of great tourism value. Over the years, Yancheng has provided a local excellent case example of wetland ecological restoration in China echoing with the international community by strengthening policies, supporting scientific research and trying a top-down model.

The Yellow River Delta wetland is a rare estuary wetland ecosystem in the world. It is the most widely preserved, the most perfect and the youngest wetland ecosystem in the warm temperate zone in the world, which is located on the shore of Bohai Sea in the northeast of Shandong Province. Dongying City in Shandong Province is the central city of the Yellow River Delta. It is rich in wetland resources, and the wetland rate reaches 41.58%. However, restricted by topography and urban basic pipelines, the urban waterlogging problem in Dongying City is prominent, and the phenomenon of wetland water shortage and degradation has appeared. After years of effort, Dongying has significantly increased the rate of wetlands in the city. It creates huge economic value by providing material products such as water resources, aquatic products and raw materials, and effectively alleviates the problem of urban waterlogging through the connection of water system and the construction of flood storage and detention projects. The success of Dongying Wetland City Construction Project can not be separated from the government's systematic and scientific plans and management, as well as the reasonable layout of ecological space, agricultural space and urban space. In addition, the government has also implemented a series of construction projects in accordance with local conditions, such as the reconstruction of water accumulation points, connection of water systems, and flood storage and detention. These measures can not only solve the problem of urban waterlogging, but also effectively deal with social challenges such as water shortage and climate change.

Mangroves are known as "coastal guardians" because their dense root systems help improve surface roughness, increase frictional resistance, slow down the flow of water and encourage the deposition of sediment, and thus reduce coastal erosion. Complex mangrove root systems also filter out nitrates, phosphates and other pollutants from water, improving water quality into estuaries and ocean environments. Mangroves are also an important carbon sink, which captures large amounts of carbon dioxide and other greenhouse gases from the atmosphere and store them for thousands of years in carbon-rich waterlogged soil. The carbon, which is stored underwater in coastal ecosystems such as mangroves, seagrass beds and salt marshes, is known as “blue carbon”. Ocean is the largest and most active carbon pool on the Earth, with a capacity about 50 times that of the atmospheric carbon pool and 20 times that of the terrestrial carbon pool. Yet the global blue carbon sink is in sharp decline. Since the 1940s, coastal eutrophication, land reclamation, coastal engineering and coastal urbanization have eliminated most of the blue carbon from the Earth. About one third of the world’s seagrass area has disappeared, 25% of the salt marsh area has ceased to exist, and 35% of the mangrove area has been destroyed with the disappearance rate about 1%-3% per year (Nellemann C et al., 2009). As the global attention to biodiversity and climate change is increasing, the importance of blue carbon resources including mangroves is gradually increasing in different regions. According to the American Association for the Advancement of Science (AAAS), from late 20th century to early 21st century, mangrove loss around the world decreased by nearly an order of magnitude from 3% to 0.3%-0.6% per year, and this was largely due to mangrove conservation efforts worldwide. The China-ASEAN Mangrove Conservation and Restoration Project of the Global Environmental Institute and the Ocean Opportunities & Small Island States of World Team share experience of NGOs that explore the win-win results of mangrove conservation and community development through innovative conservation models at the regional level.

As an important part of the socio-ecological system, cities are the focal point of population growth and economic development. At the same time, cities often face conflicts between economic development and ecological environment, While dealing with multiple social challenges such as climate change, food and water security and potential disasters. Cities only account for 2% of the land mass, while gathering 80% of the population, consuming over two-thirds of the world's energy, and accounting for more than 70% of global CO_2 emissions.[①] Cities are major contributors to greenhouse gas emissions, but they can also be participants and contributors in providing solutions. At present, on one hand, the global NbS in cities focus on emission mitigation, such as urban greening and other projects to improve the living environment of residents, promote the sustainable use of urban resources and energy, and promote the process of urban carbon neutrality, on the other hand, focusing on adaptation, the *Horizon 2020 Report: Nature-Based Solutions and Re-Naturing Cities* published by European Union in 2015 emphasized that NbS can be used in the transformation of wetlands or aquatic ecosystems to enhance their urban ecology and ornamental value, while explores how NbS can be utilized to reduce disaster risks related to climate change from the disaster containment point of view.

Cities vary in their geographical locations, scales, population density, economic development stages, and functionality. Thus, it is necessary to design and implement

① Data source: C40 Cities, https://www.c40.org/why_cities.

NbS given based on the city's characteristics. Therefore, in this chapter, six cities representing different development stages are selected to illustrate various scenarios of NbS implementation. From the perspective of strengthening urban green infrastructure and mitigating climate change, projects in Milan (Italy), London (UK) and Chengdu (China) are selected to showcase how to use NbS to achieve multiple goals such as greenhouse gas emissions reduction, energy efficiency improvement and urban ecological environment; from the perspective of improving urban adaptive infrastructure and climate resilience, projects in Rotterdam (Netherlands), New York Manhattan (USA) and the San Francisco Bay Area (USA) are selected to demonstrate the use of NbS to improve urban disaster prevention and mitigation capabilities.

8.1 Milan-NbS for Urban Regeneration

Case Highlights

Through top-level design, regional development planning and private capital utilization are combined to encourage people to participate in the transformation and implementation of urban gardens. The Vertical Forest integrates buildings and greening projects and becomes the landmark building of the city with NbS.

8.1.1 Project Background

As the second largest city in Italy, Milan is the main economic and industrial region for Italy, and one of the most developed and vibrant cities in Europe. However, like other typical commercialized cities, it also faces similar challenges such as traffic and pollution caused by the accelerated urbanization process.

8.1.2 Project Overview

Since 2006, Milan has begun to redevelop its old town, whose population is over 4.1 million. First, Milan releases the *Carta of Milan*, a strategic plan to tackle urban environment issues, and establishes a transformation strategy in line with the goal of environmental protection, sustainable social development and improvement of social welfare, with green infrastructure as the main approach. At the same time, the Lombardia, where Milan is located, also rolled out green infrastructure initiatives for ecological connections and creation of ecosystems to ensure the connectivity between the Alps and the Po Valley within that area. This comprehensive plan provides customized guidelines for different municipalities in consideration of their management, creation of ecosystems, and funding mechanisms.

Based on this, Milan has implemented NbS as part of its architectural and urban renewal strategies, such as Bosco Verticale (Vertical Forest), Parco Agricolo Sud (Green Rays and Green Belt), Gorla Water Park, and Urban Gardening.

1. Bosco Verticale (Vertical Forest)

Bosco Verticale (Vertical Forest) was launched in 2004, aimed at increasing the use of NbS in urban block planning and renovation, with a total investment of more than €2 billion. It consists of two residential towers with heights of 110 meters and 76 meters respectively. 900 trees (each no taller than 10 meters), and more than 20,000 plants (various shrubs and flowers) have been planted in each tower, designed and distributed according to the sunlight exposure of the external walls. It is estimated that the ecosystem benefits (carbon sink, air quality, biodiversity improvement, etc.) provided by the plants in the two towers are equivalent to the carbon sink provided by two hectares of forest (Figure 8-1).

Figure 8-1 Vertical Forest in Milan

2. Parco Agricolo Sud (Green Rays and Green Belt)

In the *Carta of Milan*, there is a proposal to establish a green corridor that connects the green spaces and parks of the entire city. Under this proposal, the Parco Agricolo Sud (Green Rays and Green Belts) is a park network providing local residents with agricultural, forestry, cultural and recreational activities. It can restore the ecological environment while protecting the landscape, and connect the external area with the urban green corridor to improve the ecological balance of the metropolitan area. To ensure the protection of biodiversity, some areas have been dedicated to rebuilding ecosystems and reintroducing local animal species that are becoming rarer (such as the Lombard Spadefoot toad, the Lataste's frog, the swamp turtle, and the river prawn). At the same time, peri-urban farming is of primary importance, not only for soil conservation, but also for food production. More and more people prefer locally-grown food, and peri-urban farming can meet this need.

3. Gorla Water Park

Gorla Water Park is a multi-functional infrastructure which can be used for sports education, cycling, running, picnics, animal observation, and other recreational activities. There are a number of educational services and information boards on local animals (waterfowl and small amphibians) and flora (especially plants with water purification function) in the park.

4. Urban Gardening

Urban gardening in cities has grown in popularity over the last decades. In 2012 the City Council of Milan joined hands with non-profit organizations to create new urban gardens in the municipality, aiming for quick in construction with low maintenance costs. To achieve this goal, the municipal government has introduced an incentive mechanism to encourage residents to participate in urban greening and fix up abandoned green areas. This was a way to promote gardening as a hobby and encourage social contacts. At the operational level, the government provides public and municipal land specifically for growing fruits and vegetables for private consumption. Users of these plots must abide by municipal regulations. Municipal authorities will provide water and other resources for most plots, while charging only a small fees every year.

The Lombardia, where Milan is located, provides more than €4 million of financial support for the transformation of the upgrade and enhancement of the integrated ecosystem in the old town. Individuals and entities from the public and private sector can apply for an initial subsidy of up to €30,000 per hectare, and a maintenance cost subsidy of €4,000 per hectare annually for a maximum of three years. Subsidies are mainly used to forest ecosystems and soil restoration. From 2007 to 2013, the Lombardia had invested a total of €61.2 million, of which about 20% are funds from the EU (Environment and Climate Action LIFE Plan, European Agricultural Fund for Rural Development, European Regional Development Fund), 60% are local, and rest includes private funds as well. Milan encourages individuals, private and semi-private companies to cooperate in the maintenance of the green areas. The municipal government launched the "adopt a green area" initiative to encourage local residents to participate in green space management, and seek sponsorship to help the city's finances. As of April 2016, there are 396 cooperation agreements, managing more than 60% of Milan's territory.

8.1.3 Project Results

The vegetation in the Vertical Forest can reduce the surrounding particulate concentration. Inhalable particles and total suspended particles have been reduced by an average of 20% to 30%. These results have confirmed the effectiveness of trees and green barriers (shrubs and hedges) in removing particles in air. In addition, Milan has also benefited a lot from the attractiveness and popularity of the Vertical Forest, resulting in numerous awards, including the High-rise Award 2014 and the Council on Tall Buildings and Urban Habitat (CTBUH) Award 2015. The Vertical Forest is proved to be extendable and replicable. Starting in 2017, Boeri, the company responsible for the development of the project, began the construction of the vertical forest project in Nanjing. This will be the first of its kind to be built in Asia. In addition to carbon dioxide absorption, it is expected to produce approximately 60 kg of oxygen per day.

Currently 80% of the total area of 47,000 hectares of Green Rays and Green Belts has been developed into agricultural land. The park has attracted more than 1,000 villagers. At the same time, linking the park history with the architectural heritage in the park makes an important way of connecting leisure with cultural. Maintaining these architectural heritages (small villages, castles, villas, and monasteries in the park) will be another way to attract tourists.

Gorla Water Park is a comprehensive park covering an area of about three hectares, including a flood control area (one hectare), a pollutant cleaning area (0.4 hectares of a phragmites reed bed and 0.3 hectares of natural-like multispecies wetland) and a leisure and recreational area (1.3 hectares of park). Four sand filter vertical beds have been built to deal with the first flush caused by the combined sewer overflow mechanism. At the same time, an extended reservoir was built for the second flush and slow release in the river. Green infrastructure basically solves the problems of pollution prevention and flood control. After a third-party evaluation, the construction of Milan's Gorla Water Park reflects how the EU's research and innovation funding strategy serves the city management, highlighting the positive impact on environmental protection and social participation.

In Milan, most people who were interviewed shared their preference to consume healthier food and pursue a healthy lifestyle, thus their motivation to participate in Milan's Urban Gardening project and experience urban gardening. Urban Gardening provided many policy recommendations for the municipal government to carry out localized urban gardening projects, such as encouraging gardeners to carry out more interventions to improve soil quality, providing soil testing services for gardeners, developing real-time city soil quality map, and researching on potential pollution source near the garden.

8.1.4 Project Information

Case Source: Oppla Platform - EU City NbS Case Studies.

Project Participants: The Milano government and related parties.

Project Location: Milan, Italy.

Project Duration: 2006-2016.

Project Link: Oppla Platform - EU City NbS Case Studies, https://oppla.eu/casestudy/19446.

8.2 London - NbS for a Leading Sustainable City

Case Highlights

The London Sustainable Development Commission (LSDC) has been established to provide advice and strategies to policy makers, giving frontier theories such as NbS space for implementation. In the early stages of the sustainable development projects, special emphasize were placed on the use of financial tools and market models to promote commercial participation and ensure the sustainability of the project.

8.2.1 Project Background

London is one of the pioneer cities in the world for sustainable development. London had established the London Sustainable Development Commission (LSDC) before 2002, responsible for suggestions and strategies to the mayor regarding sustainable development in London. LSDC plays a critical role in urban planning and construction in London, covering topics including urban art and culture, business and economy, youth and education, environment and health, community, funding, employment, transportation, etc. It has conducted research on energy conservation, emission reduction, green and low-carbon development, and has trained many leaders or activists in sustainable business and communities, laying a good foundation for the sustainable development of London. In 2018, London was listed first in the Arcadis "Sustainable Cities Mobility Index".

8.2.2 Project Overview

The Mayor of London, Sadiq Khan, is committed to building the London into one of the greenest and healthiest cities in the world. He launched the *London Sustainable Development* project to improve the overall environment of London, aimed at benefiting all London residents. The project covers seven key areas including air quality, green infrastructure, climate change, waste management, urban resilience, environmental noise and circular economy. At the same time, London has also formulated a long-term sustainable development strategy and pledge to achieve carbon neutrality by 2050. In addition, London is determined to achieve 80% of the city's regional travel done by walking, cycling, or public transportation by around 2040.

It is worth noting that there are a large portion of NbS initiatives and actions in the "London Sustainable Development" project, which can effectively mitigate surface water flooding, improve air quality, reduce urban heat island effect, provide walking and cycling opportunities, and improve urban landscapes to enhance biodiversity and ecological resilience. NbS plays an important role in helping London become a green city and achieve sustainable development goals. In its newly released

London Infrastructure Plan 2050, London emphasizes the importance of urban green infrastructure, and the government needs to actively seek commercial solutions to assist the planned facilities. London plans to use two financial and market models: ① The developer's fee collection model ensures long-term maintenance of good green infrastructure; ② The co-funded model is used for the construction of green infrastructure in large public areas such as the Thames River, Olympic Park, and other green regions. At the same time, the London city government established the Green Infrastructure Working Committee in 2019, responsible for adopting a more strategic and long-term approach to invest and promote the construction of green infrastructure. This will bring new economic opportunities and create job opportunities for London, and help improve the development of social green industries and the EU's green recovery plan in the post-pandemic era. Funded by the EU's *Transitioning towards Urban Resilience and Sustainability* project, London NbS projects include the following points.

1. Natural Water Retention Measures (NWRM) - River Quaggy ①

Measures such as restoration of natural river channels and flood control have been implemented in the River Quaggy in southeast London to counter the reduction of natural river valleys and floodplains, while reducing the risk of flooding. This project is a renovation of the River Quaggy channel implemented from 1990 to 2005, repositioned and upgraded its flood protection facilities. The above measures help improve the water quality of London, reduce the cost of urban sewage treatment, increase the coverage of green infrastructure, and the quantity of populations. In addition to carbon sequestration and emission reduction, it helps reduce the risk of droughts and floods, and enhance urban water resources management and ecological cycles.

① The River Quaggy is located in the eastern part of the Thames.

2. Brownfield restoration, Barking Riverside

Brownfields refer to the land and above-ground structures that have a certain degree of pollution, have been abandoned, or have not been fully utilized due to pollution. Barking Riverside is a 443 acre brownfield site located at the borough of Barking and Dagenham, to provide new residences for 20,000 residents in 20 years (Figure 8-2). The planning process recognized the ecosystem service potential of the Barking Riverside from the start, to ensure the sustainable development of the site. These included: the conservation of its valuable biodiversity; the retention of 40% of the site as green space; the development of a comprehensive Sustainable Urban Drainage (SUD) system and a community wildlife garden. Before the development of the Barking Riverside, the government has determined the precious invertebrate species and their habitat features of the brownfield, and incorporated them into the design of the community garden. The garden was then used to introduce residents to the wildlife that was associated with the site prior to development, and residents were shown how they could recreate these features in their own gardens.

Figure 8-2 Rendering of Barking Riverside

3. Green roof project

London requires the construction of green roofs on 40% of the houses in the city to integrate with marshes, rain gardens and ponds, and build a comprehensive sustainable urban drainage system. The EU's *Transitioning towards Urban Resilience and Sustainability* project (hereafter the Project) has carried out a series of green roof work on the Barking Riverside. By comparing the performance of different green roof systems in terms of ecological services (including habitat provision, water saving and thermal insulation, etc.), the traditional industrial-led roof construction is transformed into a green roof system constructed of flowers and plants with highly diverse biodiversity (Figure 8-3). In the construction of the Barking Riverside, *Transitioning towards Urban Resilience and Sustainability* project created a number of wetland roofs for ecological simulation experiments: investigate the potential for recreating key habitats associated with the site's pre-development brownfield state, and create habitat test sites on the green roofs by manipulating the drainage, using different aggregates and varying the substrate depths. Finally, the project team researchers will monitor the roof to assess the impact of the habitat for its overall biodiversity. In addition, "Transitioning towards Urban Resilience and Sustainability" project also experimented in evaluating the biodiversity value of an existing 0.25 hectares green roof of the Queen Elizabeth Olympic Park, to which solar panels had been added, and the niche[①] method is used to quantify the implementation effect. The study found that compared with other green roofs in London, the Queen Elizabeth Olympic Park green roof presents a higher diversity of invertebrates and endemic species.

---

① Niche refers to the position of a population in time and space in the ecosystem and its functional relationship and role with related populations.

Figure 8-3 Green Roofs in London

4. The Beetle Bump project

The Beetle Bump project is a rescue attempt for what was probably Britain's rarest insect, the streaked bombardier beetle (Brachinus sclopeta). The University of East London (UEL), in cooperation with Buglife staff, used ecomimicry design principles to make the Beetle Bump suitable for beetle life and reproduction from 65 tons of recycled materials (hard core, chalk, brick and topsoil), and planted flowers in the highest quality open habitat in the region. The Beetle Bump was designed as a multifunctional space combining art, landscape design, and habitat conservation. Beetles rescued from the construction site were released at UEL where the site was used as an open-air laboratory to study the behavior and habitat requirements of the beetles.

The urban renewal project in Thamesmead, southeast of London, is another successful sustainable development project. It uses effective policy planning and management to integrate landscapes, vegetation and natural ecosystems to one place for piloting. Thamesmead is a modern community, home to more than 45,000 residents, and its apartments and houses are surrounded by an extensive network of green spaces, lakes and canals, built in the 1960s on the floodplain of the Thames. To prevent large-scale tidal floods, the lakes and canals of Thamesmead manor were designed with the function of flood storage. However, for many years, due to problems such as the route design and direction of entry and exit, the lakes and green spaces in this area have not been fully utilized. The huge green space has a single function, which is difficult to attract tourists and residents. It uses nature-based urban transformation to integrate more natural elements into the city.

8.2.3 Project Results

The sustainable development in London has achieved a multiple positive effect on the city, including:

(1) The solid theoretical research foundation provides exemplar for the world. Studies in London have proved that green roofs have multiple benefits, including reducing runoff, alleviating urban heat island effects through transpiration, improving air quality and providing new leisure spaces, and so on. At present, London has developed tools to assess the potential of urban green roofs to help policy makers identify high-potential areas that can be developed into green roofs during the planning phase. It is important to note that the green roof strategy poses potential risks. Generally, areas with high development potential belong to areas with higher income, while low-income areas are more susceptible to the impact of climate change because of the limited green roof area capable for development, this may intensify the poverty gap. The above research achievements provide valuable references for green planning in other regions of the world.

(2) NbS in London will achieve multiple goals. For example, the Mayor's London Plan announced the planting of two million trees by 2030, an increase of 5% of green space, and another 5% by 2050. The Street Tree Planting plan is for known overheating

areas. The London Plan also has some measurable objectives, including a no net loss of designated Sites of Importance for Nature Conservation over the plan period, a reduction in CO_2 emissions to 23 % below 1990 levels by 2016, no net loss of functional flood plain, and the production of 945 GW·h of energy from renewable sources by 2010.

8.2.4 Project Information

Case Source: Oppla Platform - EU City NbS Case Studies.

Project Participants: The London municipal government and related parties.

Project Location: London, UK.

Project Duration: 2018–2030 (first stage); 2030–2050 (second stage).

Project Link: The London Municipal Government, https://www.london.gov.uk/about-us/organisations-we-work/london-sustainable-development-commission.

8.3 Chengdu Park City Construction

Case Highlights

Combining President Xi Jinping's ecological civilization theory, at the initial stage of the Park City, Chengdu adopted the strategy of highlighting the characteristics of park cities, and considering ecological values. This ensures that Chengdu's Park City can not only achieve multiple benefits, but also produce long-term sustainable benefits at the economic and employment levels.

8.3.1 Project Background

Located in the southwestern region of China and the western part of the Sichuan Basin, City of Chengdu is the capital of Sichuan Province. In a humid subtropical monsoon climate, it has a flat terrain, complicated vertical and horizontal river networks, rich in produce, and developed agriculture. It has been entitled the “Land of Abundance” in ancient times. As of 2019, Chengdu has a built-up area of 950 square kilometers, with a permanent population of around 16 million, a population density of 1,180 people per square kilometer, and an urbanization rate of about 75%. Chengdu is an international sustainable development pilot city of UN-Habitat, and part of China’s third batch of Low-carbon Development Pilot Cities. It has clarified its development positioning as “core economic growth pole with global comparative advantages, national speed advantages, and western China high-end advantages”. However, the rapid development of urbanization and industrialization in Chengdu in recent years has continuously intensified the siphonic effect on surrounding cities. The total population of Chengdu continues to increase, the number of high-rises buildings keeps growing, and the number of car ownership has ranked second in China. However, problems such as sharp decrease of forest land, decrease of arable land, water pollution, fragmentation of the ecological sector, and reduction of biodiversity have gradually become prominent.

8.3.2 Project Overview

In February 2018, Chinese President Xi Jinping visited Sichuan Province, and gave a clear development direction to “highlight the characteristics of park cities and incorporate ecological benefit into development”. As the first city to put forward the concept of Park City, Chengdu is accelerating the pace of building a “Beautiful and

Livable Park City", and has released the *Planning: Building a Beautiful and Livable Park City (2018-2035)*, determined to shape itself into a beautiful park city and rural scenery in which residents can "open the window to see fields, push the door to see greens". Through protection, restoration and sustainable management of ecosystems, effectively respond to various social challenges, and provide a better life for residents. It strives to create a high-quality all-for-one park system that is accessible, participatory, perceivable, readable, appreciable and consumable, while provides a "Chengdu case" for others in urban NbS application.

Park City is an advanced form of active and coordinated protection of nature and urban development in the ecological civilization, and a new model for sustainable urban construction in the new era. It has the following four characteristics: First, highlight the development concept led by ecological civilization; second, highlight the people-centered values; third, highlight the ecological concept of building a living community of mountains, rivers, forests, fields, lakes and grasses; fourth, highlight the harmony form between modern city with people, the city, ecosystems and industry. Under this concept, the urban construction model should go through three changes: The first is the shift from "industry, city, and people" to "people, city, and industry". Return from industry-based logic to human centric logic, relying on a good ecological environment and public welfare to attract talents and enterprises, drive industrial prosperity, and achieve the harmonious development of people, city and industry; Second, shift from "building a park in a city" to "building a city in a park". Urban construction should adhere to requirements of the park construction on ecological, aesthetic, cultural, economic and morphological, and integrate the park morphology with the urban space; Third, shifting from "space construction" to "scene creation". Focusing on people's needs, through facility embedding, functional integration, scene introduction to constructing comprehensive scenes for urban lifestyle, consumption and innovation to enhance the sense of spatial belonging. As an important practice model responding to the needs of the human settlement environment in the new era, also shaping the competitive advantage of the city, the park city has a series of important values that reflect the characteristics of present, including the ecological value of "clean water and green mountains", the aesthetic value of poetic dwelling, the humanistic value of cultural people, the economic value of green and low-carbon development, the value of simple and healthy life and the social value of a better life. The Park City planning and construction consists of four parts: forest parks, country parks, urban parks and green spaces, and Tianfu greenways (Figure 8-4).

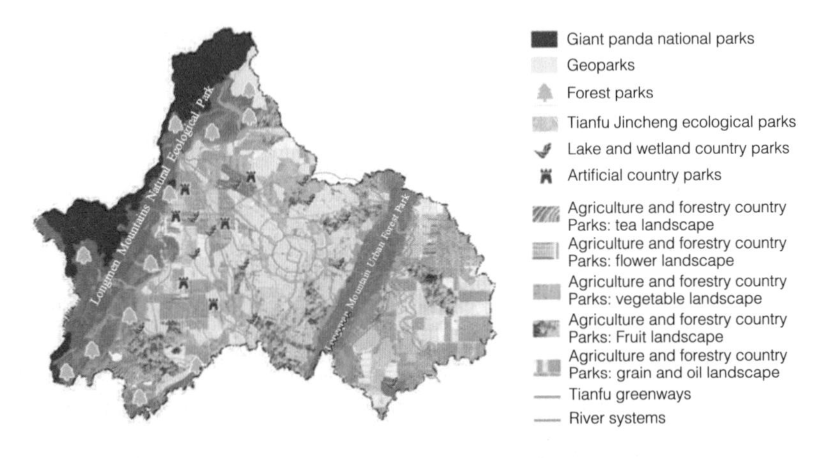

Figure 8-4　Planning of Chengdu Park City
(Source: Chengdu Tianfu Park City Research Institute)

In addition, Chengdu is implementing a carbon inclusive mechanism through the *Low Carbon Tianfu* initiative, aiming to quantify the energy-saving and carbon-reduction behaviors of small and micro enterprises, communities, households and individuals, by establishing a mechanism combining policy encouragement, business incentives and carbon emission reduction trading to influence resident's behavior, thus mobilizing the enthusiasm of the whole society to practice green and low-carbon development behavior from the bottom-up, and then actively connecting with the national strategy to tackle climate change. According to the *Chengdu Low Carbon Tianfu Management Measures (Trial)*, it is not a copy of other cities, but the first in China to propose the construction of a "double paths" incentive mechanism for the public, enterprises and institutions. The two paths: public carbon emission reduction credit and project carbon emission reduction development and operation; for the prior, Chengdu will focus on the public's basic need of clothing, food, housing, transportation and travel, and guide enterprises to create low-carbon scenarios and implement low-carbon management by formulating low-carbon evaluation specifications for restaurants, supermarkets, scenic spots and hotels. Operating entities can generate carbon credits based on the low-carbon behavior data of the public in the scene through the operating platform conversion and issue them to personal accounts. The carbon credits held by the public can be exchanged for inclusive goods or services on the operating platform. For project carbon emission reduction development and operation, Chengdu will focus on guiding the development of carbon sinks in major ecological projects such as Longquan Mountain Urban Forest Park, Tianfu Greenway, Western Sichuan Linpan, and lake wetlands, and systematically formulate project methodologies with great carbon emission reduction potential and environmental benefits. The carbon emission reductions of various projects developed according to the corresponding methodology can be traded with those incentivized to be carbon neutral in the Sichuan United Environment Exchange, thereby integrating nature-based ecological construction projects and energy conservation by enterprises and institutions. The environmental benefits of carbon reduction show its economic value in carbon, for example, according to the Low Carbon Tianfu methodology, Jincheng Greenway can reduce 4,200 tons of carbon dioxide equivalent emission each year for carbon neutral trading and realize ecological value.

8.3.3 Project Results

Since 2018, Chengdu has been implementing the planning concept of "building the city in the park, a city full of parks, the harmony between park and city, and the harmony between the city and its people". It connects forests, country sides and urban park green spaces with greenways and water networks to form seamless green spaces transition and a connected park system. By 2035, 1,275 square kilometers of Longquan Mountain Urban Forest Park and 16,900 kilometers of Tianfu Greenway will be built, 3,689 kilometers of greenway at all levels will be built, 38.85 million square meters of green space will be added, the city's forest coverage rate will reach 39.93%, and the green coverage rate for built-up area will reach 43.5%.

Take Tianfu Greenway as an example, as the longest greenway in the world, 3,689 kilometers has been built as of 2020 with a total investment of ¥34.1 billion. Chengdu is connecting the ecological value transformation path of "building a city through roads, promoting business with roads, and making people happy with roads" with *Greenway* +, and will realize the eight functions of ecological landscape, slow traffic, leisure tour, urban and rural integration, cultural creativity, sports, landscape agriculture and emergency respite (Figure 8-5).

The revolutionary layout of Tianfu Greenway is the test of Chengdu's NbS long-term operation, maintenance and sustainable development capabilities. Tianfu Greenway adheres to the strategy of government-led, market-based and commercialized, and invites global investments by methods such as facility leasing, joint operations and

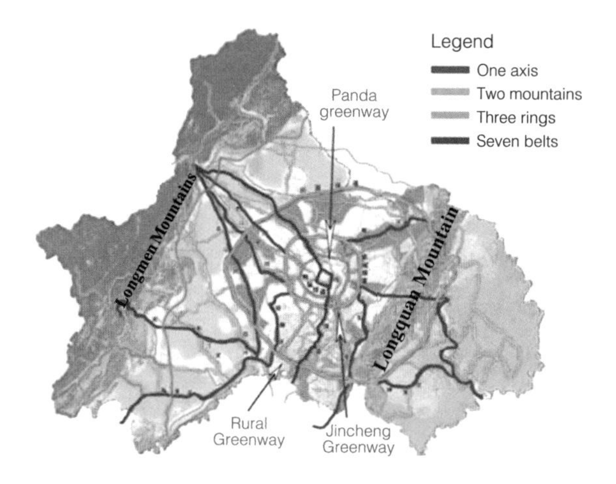

Figure 8-5 Regional Greenway Planning of Chengdu
(Source: Chengdu Tianfu Park City Research Institute)

resource equity participation. The social investment in Tianfu Greenway accounts for more than 70% of its total value. Take Jincheng Greenway as an example, it drove the value of surrounding land by hundreds of billions, radiate the surrounding modern service industry clusters of hundreds of billions, attract hundreds of millions of tourists every year, create more than 100,000 jobs, attract high-quality talents, and promote innovation and entrepreneurship. In addition, according to the Institute of Mountain Hazards and Environment, Chinese Academy of Sciences, Chengdu, and the preliminary estimates of 18 ecological service value indicators, including climate regulation, carbon fixation and oxygen release, soil conservation, and water conservation, the annual ecological service value of Jincheng Greenway is about ¥26.9 billion. It is expected to produce sustainable benefits for more than 40 years, with a total value of more than ¥1 trillion.

8.3.4 Project Information

Case Source: C+NbS Cooperation Platform.

Project Participants: Chengdu Municipal Government, and all sectors of society.

Project Location: Chengdu, Sichuan Province, China.

Project Duration: 2018-2035 .

8.4 Rotterdam-Climate-resilient Infrastructure in Netherlands

Case Highlights

Adopt effective and innovative governance model, on one hand, it is aware of the vulnerability of its geographical location to future climate change, on the other hand, it actively responds to climate change. It integrates community life with disaster prevention and mitigation and improves resilience to achieve efficient and intensive use of resources.

8.4.1 Project Background

Low sea level is the most prominent feature of the Dutch terrain. A quarter of the land in the Netherlands is less than one meter above sea level. The area near the satellite city of Alexanderstad in the northeast is the lowest in the whole country, around six to seven meters below sea level with a population of 175,000 residents. Similar coastal cities are affected by ocean tides as well. For example, Amsterdam, the largest city in the Netherlands, is an actual water city dominated by rivers and waterways in particular, featured with very low altitude, making the city vulnerable to rain and flood damage.

Therefore, living with water has become an important concept for the development of cities in the Netherlands.

Rotterdam is the largest port in Europe, located in the delta of the Rhine and Maas. In 2008, the Rotterdam City Council approved the *Rotterdam Climate Proof* and in 2013, it passed the *Rotterdam Climate Change Adaptation Strategy*. The strategy aims to strengthen the flood defense system in the Netherlands, improve urban space utilization through comprehensive planning, enhance urban resilience, and explore opportunities brought about by climate change, such as economic development, improved quality of life, and increased biodiversity. Rotterdam has adopted a tailor-made "inner/outer dikes" approach. Most areas of Rotterdam, including the main port, are located in the outer dike area. There are about 40,000 inhabitants living in the outer dike area (3 to 5.5 meters above sea level), which is vulnerable to rising sea level and temporary flooding. In outer dike area, it will be transformed by the combination of innovative technologies and traditional approaches. Most of the inner dike area is below sea level and consists of a system of dikes drained by water outlets and pumps, and protected by smaller secondary dikes. Flood will catastrophically destroy these dikes.

8.4.2 Project Overview

Rotterdam's goal stated in the *Rotterdam Climate Change Adaptation Strategy* aims to achieve 100% resistance to climate risks by 2025. It emphasizes that by 2025, measures will be taken to ensure that each area is least affected by climate change in the current and future decades. In addition, all city plans in Rotterdam will consider long-term foreseeable risks of climate change, and will take actions in the following three aspects.

1. Water storage

Climate change will cause frequent extreme rainfall and increase the risk of disruption and damage caused by flooding, especially in areas with insufficient water storage capacity, dense buildings and insufficient pavement permeability. It is necessary to adjust and improve the drainage and storage capacity of urban water supply systems to cope with extreme rainfalls in the future. Suggested measures include rainwater collection and storage, delayed drainage, and restoration of the city's "sponge function". These measures include increasing green area on the roof and facade, reducing pavement hardening, increasing vegetation in public streets and communities, building green infrastructure in water squares and infiltration zones. These measures will be especially effective in highly populated, built-up areas with little open space.

2. Delta plan

Establishment of multi-layered flood protection in outer-dike Rotterdam based on adaptive construction and design: flood-proof buildings, flood-proof public spaces, floating communities and building with nature. Special attention is being paid to the port and essential infrastructure, which are well protected from flooding. The protection of inner-dike Rotterdam focuses on prevention. Storm surge barriers will be optimized and dikes reinforced as needed and made multifunctional, to merge into the city as recreational routes, natural embankments or as part of new development.

3. Tidal Park

Climate change adaptation needs to be combined with ecology improvements of the city. In the tidal park program (Figure 8-6), several outer-dike areas are being developed with alternatives from solid construction materials to prevent flooding at high water levels. For example, in the flood plains of Esch and Mallegat, green and blue corridors have been created to ensure high-quality natural environments, which can improve biodiversity and landscape connectivity, as well as increase carbon storage and regulate the local microclimate. The construction of green and blue corridors can be combined with the new entertainment area, where residents can use as spot for sports and leisure activities. At the same time, bicycle lanes have been built to encourage green lifestyle. The establishment of the tidal park enables the restoration and improvement of the ecosystem services and functions related to the wetland, which can be used for water storage, water flow regulation and water quality filtering. The city's blue-green resilience facilities not only help the city to withstand climate risks, but also make it more attractive and livable.

Figure 8-6 Tidal Park in Rotterdam

8.4.3 Project Results

Rotterdam's climate change adaptation program is not focused solely on the use of nature-based solutions but combines "grey" solutions with "green" and "blue" solutions. For example, through the construction of water storage space, including a museum park with a capacity of 10,000 cubic meters, underground storage reservoirs for parking lots, blue-green corridors and other projects to reduce the threat of heavy rainfall. These blue-green corridors, including waterways and stagnant areas, can promote groundwater replenishment, reduce urban waterlogging, increase biodiversity, and improve the urban life quality. In 2014 alone, Rotterdam installed more than 185,000 square meters of green roofs, and launched a 100% resilient neighborhood demonstration project, a landmark that Rotterdam is becoming an adaptive delta city.

In addition, Rotterdam has also developed innovative urban flood control models, such as the Benthemplein Water Square built in 2013. The Benthemplein Water Square can play a water storage function during peak rainfall and relieve the pressure on the urban sewage treatment system to a certain extent. The Benthemplein Water Square is the world's first full-scale water square, which can accommodate up to 1.7 million liters of rainwater on the square. After rainfalls, the rainwater will seep back into the soil or be pumped to other canals of the city. In addition, it helps to reduce the huge cost incurred for the transformation of the underground water network, as well as for sports and entertainment purposes, and provides space for basketball, skateboarding and performing arts.

8.4.4 Project Information

Case Source: Oppla Platform - EU City NbS Case Studies.

Project Participants: Government of Rotterdam.

Project Location: Rotterdam, Netherlands.

Project Duration: 2008–2025 .

Project Links: Oppla Platform - EU City NbS Case Studies, https://oppla.eu/casestudy/19457; C40 Cities: https://www.c40.org/case_studies/benthemplein-water-square-an-innovative-way-to-prevent-urban-flooding-in-rotterdam.

8.5 Lower Manhattan Coastal Resiliency (LMCR) in the US

Case Highlights

First, to identify high-risk areas, then formulate reasonable actions based on scientific adaptation and community conditions; then through community research, risk analysis and future urban planning, comprehensive actions are taken to improve urban resilience.

8.5.1 Project Background

On October 29, 2012, Hurricane Sandy hit New York, flooding 17% of the city's land mass, claiming 44 lives, causing about 125 deaths across the United States, and economic losses exceeding $19 billion. In Lower Manhattan alone, the impact of Hurricane Sandy was devastating, damaging thousands of homes and interrupting several critical transportation hubs. At present, the scientific community has reached consensus that climate change will increase the frequency and intensity of hurricanes and heavy rainfalls. Rising sea levels may submerge parts of Lower Manhattan and put critical infrastructure and jobs serving all of New York and the region at risk. This includes the subway and ferry network, the sewer system, 10% of the city's jobs, and many historical, cultural and community assets.

8.5.2 Project Overview

Based on the above factors, New York is taking actions, investing $500 million in climate adaptation projects to protect high-risk areas of Manhattan, especially Lower

Manhattan. The principle of adaptation is to take into account the specific circumstances of the community, through research, risk analysis, and future urban planning to establish comprehensive actions to improve urban resilience. It aims to reduce the risk of coastal storms and sea level rise to Lower Manhattan. At the same time, Manhattan is utilizing this opportunity to come up with comprehensive and innovative planning and design to improve the community.

Lower Manhattan Coastal Resiliency (LMCR) can be divided into the following parts: Battery Park City Resilience Projects, Battery Coastal Resilience, Financial District and Seaport Climate Resilience Master Plan, and BK Bridge-Montgomery Coastal Resiliency (BMCR).

Battery Park City Resilience Projects and the Battery Coastal Resilience: The Battery Park City Authority is developing four coastal landscape restoration projects to reduce the risk of storm surge and sea level rise to Battery Park and surrounding areas. These projects will be implemented in phases. Measures like increasing parks and green space, building fixed flood walls, and arranging temporary flood control facilities, which can reduce the risk of storms and floods in coastal areas. Resilience actions in the southern part of Battery Park include the construction of landscape dikes and flood walls. The Battery Coastal Resilience aims to rebuild the wharf and promenade that have been in disrepair, and heighten their level to prevent the daily tidal siltation in the park due to rising sea levels by 2100. As a part of independent project, the Battery Park City Authority will also build a landscape protection road and green walls behind the park to reduce the coastal floods' invasion of the area.

Financial District and Seaport Climate Resilience Master Plan: As a part of the whole LMCR, New York launched the Financial District and Seaport Climate Resilience Master Plan in 2019. The main purpose is to determine what actions to take to reduce the risk of climate change in the region. Those main climate risks include the rising sea levels can cause frequent floods of four to six feet depth in parts of Lower Manhattan; changes in rainfall caused by climate change may lead to flooding and worsening river pollution, and floods caused by coastal storm surges will threaten Lower Manhattan. To counter this, New York focuses on development and design of flood control infrastructure to improve drainage capacity, carry out rainwater collection, and manage wastewater. In addition, it will also provide some funding and financing channels for the infrastructure projects. The first phase of the plan will be completed in the fall of 2021.

BK Bridge-Montgomery Coastal Resiliency (BMCR): The integrated flood control system consists of fixed and deployable mobile barriers located above the Marina Square. These barriers can be used to protect the area near the two bridges before a typhoon or heavy rainfall. These movable barriers will be installed on a raised platform to solve the problem of tide water leakage caused by sea level rise. The BMCR also includes the reconstruction of drainage facilities to solve the problem of inland water seepage caused by rainfall. On one hand, the movable barrier will remain as a coastal landscape and open to the community to visit, on the other hand, it can protect the community from coastal storms. The project will kick off in 2021.

In addition, temporary flood protection barriers have been set up at the two bridges in the financial district and the seaport to protect these areas from more frequent and severe storms. The New York municipal government is studying the potential expansion of this plan, while part of the construction has been completed in 2019.

8.5.3 Project Results

In March 2019, New York completed and released *Lower Manhattan Climate Resilience Study*, a comprehensive overview of the current and future climate risks and impacts on Lower Manhattan. The study assessed a broad range of climate impacts, including chronic conditions, sea level rise, groundwater table rise, tidal inundation; climate events, storm surge, extreme precipitation and heat waves. The study found that by 2050, 37% of the buildings in Lower Manhattan will be at risk from storm surge; by 2100, almost 50% of the buildings will be at risk from storm surge, while 20% of Lower Manhattan's streets will be at the risk of daily flooding due to over six feet of sea-level rise; groundwater table rise is projected to put 7% of buildings at risk of destabilization and 39% of streets with underground utilities will be exposed to corrosion and water infiltration.

The findings from the study helped the New York to identify permanent adaptation projects and develop an overall strategy for the climate resilience of Lower Manhattan. These projects will protect 70% of Lower Manhattan's coastline and will be in construction by 2021. Through strengthening the coastline protection and improving the utilization of public space, the project transforms the five feet waterfront from Manhattan's Two Bridges neighborhood to Battery Park City into a comprehensive flood control area to protect vulnerable coastal communities. The use of innovative landscape dikes and native plants not only protects the Manhattan coastline from flooding, but also provides cultural, recreational and ecological benefits, and enhances community cohesion. In addition, the green infrastructure of the project, such as bio-swales, rain gardens and street trees, will help increase local biodiversity and reduce the urban heat island effect.

8.5.4 Project Information

Case Source: C40 Cities.

Project Participants: The New York municipal government and related parties.

Project Location: New York, USA.

Project Duration: Since 2014.

Project Link: NYCEDC in New York, https: //edc.nyc/project/lower-manhattan-coastal-resiliency.

8.6 San Francisco Bay Area Greenprint in the US

Case Highlights

The Bay Area Greenprint emphasizes innovative governance, supports decision-making through the development and construction of visual models, and establishes a set of guidelines based on the protection of natural resources to achieve the cycle of "problem identification - planning (adjustment) - implementation - scientific evaluation".

8.6.1 Project Background

The San Francisco Bay Area (Bay Area in short) is located in Northern California, with a total area of 18,000 square kilometers and a population of approximately 7.7 million. As one of the three major bay areas in the world, it is one of the world's major centers for high-tech R&D, science, education and culture, and the most important financial center on the west coast of the United States. But like other bay areas, San Francisco Bay Area faces a series of challenges due to population growth, urban infrastructure development, increased demand for natural resources, and climate change. Against the backdrop of rapid urbanization, the land and transportation planning of most cities in the United States usually does not give full consideration and attention to natural resources.

8.6.2 Project Overview

In light of the above-mentioned challenges, in 2017, the San Francisco Bay Area of California launched the Bay Area Greenprint (Greenprint in short), aimed to help the area establish a set of guidelines based on the protection of natural resources, so as to promote its sustainable development and balance the needs of nature, agricultural resources and urban development. The main organizers of the project include The Nature Conservancy, Greenbelt Alliance, GreenInfo Network, American Farmland Trust, and TOGETHER Bay Area. It focuses on nine values and benefits of natural and agricultural resources: food production, water supply, water quality, water disaster risk reduction, carbon sequestration, outdoor recreation, high-priority habitats, habitat connectivity, and buffering species and habitats. In addition, the Greenprint also considers climate change, showing the potential threats and opportunities that climate change poses to the above-mentioned nine natural values and benefits. The construction and process of Greenprint are shown in Figure 8-7 (Feng Yijia et al., 2015).

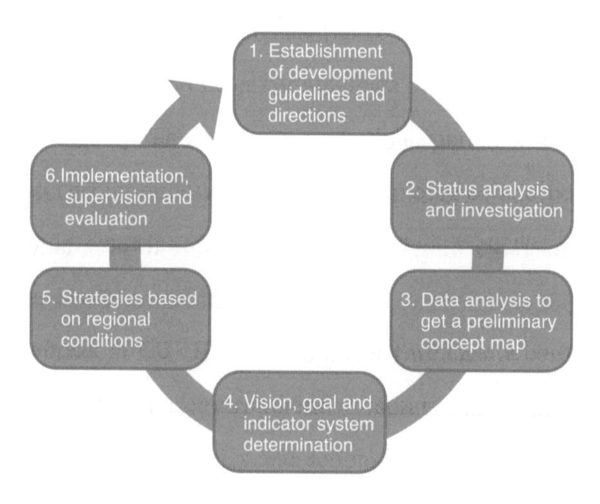

Figure 8-7 Greenprint Flow Chart

Establishment of development guidelines and directions: The Greenprint is developed according to regional characteristics and plan goals, its long-term concerns, and through the public opinion solicitation and summarization of key areas that need to be improved and changed. Common directions include increase in park green space, building new greenways, enhancement of the accessibility of green spaces, and improvement of the environment and recreational facilities. Based on the above characteristics, corresponding guidelines will be drawn up in response to the existing special problems in the region.

Status analysis and investigation: A series of analyses are used to understand the status of the region, including stakeholder surveys, geographic information system data analysis, existing parks and green space analysis, and funding source analysis.

Data analysis for preliminary concept map: Geographic information system is used to graph the data into a series of current drawings. The natural features of the area, such as large rivers and undulating mountains, are used as the skeleton to identify the green infrastructure and build it into a green network. The green network is used as the base, and relevant factors are superimposed separately according to the problems being studied (such as population, employment, housing, transportation and cultural resources) to obtain the initial concept of the Greenprint.

Vision, goal and indicator system determination: The establishment of the target system is based on the Greenprint Planning Committee, community meetings and field surveys. Once the goal system is established, questionnaires will be issued to the community to develop understanding of the needs, priorities and recognition of the residents. Based on the feedback, analysis needs to be conducted for the city resources, indicators are superimposed, and values are assigned through the geographic information system spatial analysis software to illustrate the corresponding targets.

Strategies based on regional conditions: In general, there are multiple strategies, covering all aspects of the plan. When the area is large or the target is complex, the implementation strategy will also be formulated under the sub-target, including policy changes, the formulation of regulations, and the actions that need to be taken.

Plan implementation and benefit evaluation: After the Greenprint is completed, and there is no objection from all parties, the local government will assist the Greenprint Planning Committee in its implementation and management. The main tasks include budget estimation and fund raising, setting phased goals and developing action plans, and regular benefit evaluations.

8.6.3 Project Results

The Bay Area Greenprint has been developed as a web-based application tool (Figure 8-8). The toolkit can be used by various stakeholders including policy makers, city managers, environmental protection organizations, and the general public. The tools use visual atlases to show the distribution of various natural resources and land use. Through personalized reading and analysis of the land, it can help solve local challenges and plan according to local conditions with great convenience. For example, the natural and agricultural resources in the Bay Area can be introduced in detail through the Natural Resources Panel, and the distribution of natural and agricultural resources in the Bay Area, land protection status, and development risks can also be displayed. Furthermore, it can also make an overall assessment of the various natural values and benefits which users pay attention to, and display the overlapping areas and degrees of overlapping of these values and benefits on the map, supporting management in making wise coordination or weighing decisions in planning.

The Bay Area Greenprint is being adopted by departments in government

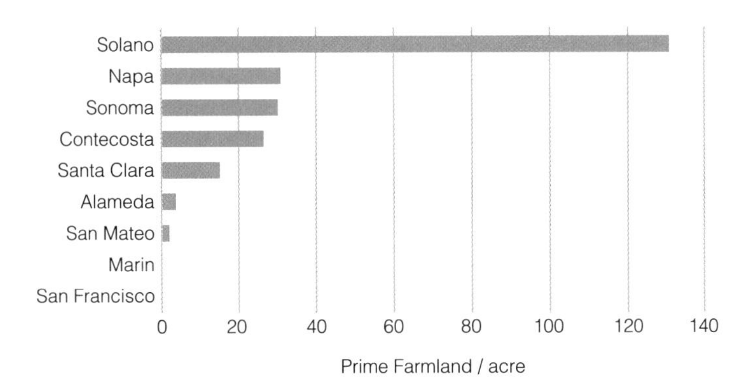

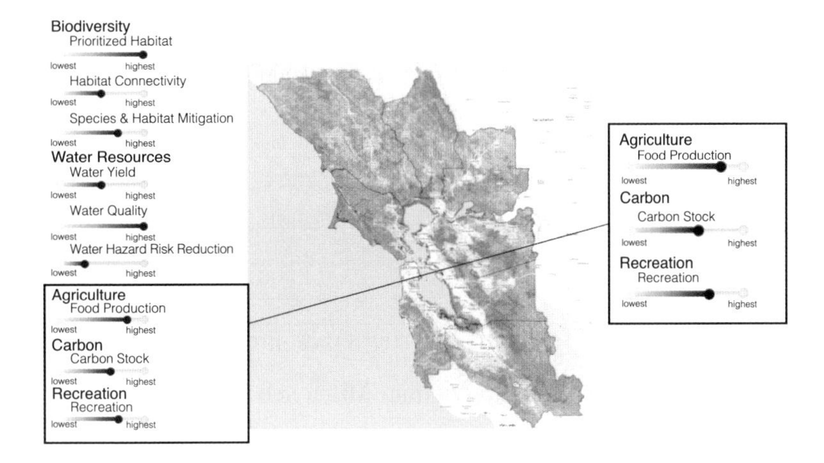

Figure 8-8 Schematic Diagram of the Natural Resources Dashboard (top) and Conservation Assessment (bottom) of the Bay Area Greenprint

and nature conservation organizations in the Bay Area. For example, the bay area transportation planning department is trying to use this tool to quickly analyze the ecological footprint of the regional transportation plans. The California State Coastal Conservancy is trying to use this tool to guide the conservation efforts for natural, recreational and agricultural resources in the Bay Area. City of Antioch is incorporating the natural resource value information from the tool into its key area planning, and promoting local protection for important natural and agricultural resources to make it more suitable to live in.

8.6.4 Project Information

Case Source: TNC Global NbS Case Database.

Project Participants: San Francisco Bay Area Government and relevant parties.

Project Location: San Francisco Bay Area, California, US.

Project Duration: Since 2017.

Project Link: Official Website of Bay Area Greenprint, https://www.bayareagreenprint.org/.

8.7 Case Summary

Among the three cases emphasizing urban green infrastructure construction and greenhouse gas emission reduction, Milan and London face similar situation. Urbanization has made the old town of the two cities on the decline and it is urgent to find new opportunities for their transformation and upgrade.

Milan puts NbS into practice through urban regeneration. Through regional ecological construction planning, multi-funding and citizens' participation, projects such as Vertical Forest, Parco Agricolo Sud, Gorla Water Park and Urban Gardening have resulted with outstanding achievements. These projects have fulfilled the various needs, such as enhancing urban landscape, improving air quality and local microclimate, maintaining biodiversity, and disaster prevention and mitigation.

London integrates NbS into its existing urban construction projects. The old town optimization and brownfield restoration projects have preserved the local species and biodiversity. In project implementation, a large number of ecological bionic designs have been adopted, such as the research on green roof, the protection of the local unique striped beetles and the construction of solar panels based on wetland parks, etc., which fully integrate natural laws with modern technology. These are new attempts that others can refer to for old city transformation and green innovation.

Chengdu of Sichuan Province is an international sustainable development pilot city released by UN-Habitat and one of the third batch of low-carbon pilot cities in China. The city government regards "Park City" as a new model for sustainable urban construction in the new era. Through systematic planning and layout arrangement in the governance process, it aims to create a beautiful, big City Park and a rural scenery in which residents can "open the window to see fields, push the door to see greens". In addition, the three cities have tried different forms to further integrate the lives of community residents. Milan attaches great importance to residents' needs. London incorporates local design inspiration and actions into its renovation projects. Chengdu has used the "Low Carbon Tianfu" as the brand carbon inclusive mechanism and echoes with the national climate change strategy by mobilizing the whole society to implement green and low-carbon development actions.

Among the three cases emphasizing urban adaptive infrastructure and climate resilience improvement, Rotterdam, a representative of a low-lying delta city, has long been the showcase for adaptation technology and the innovative testing ground. In accordance with the *Rotterdam Climate Change Adaptation Strategy* issued in 2013, it adopts NbS to upgrade urban flood control facilities to reduce the risks caused by sea tides, storms and rain flood disasters. It strengthens the flood defense system in the Netherlands through the combination of "grey, green and blue" infrastructure, and explore economic and employment opportunities brought about by climate change.

Affected by Hurricane Sandy, New York has selected an area with higher disaster risk, namely the coastal area of Lower Manhattan, to carry out the pilot program, and rolled out the *Lower Manhattan Coastal Resiliency* (LMCR) for high-risk areas to enhance its adaptability.

Scientific planning is essential to the successful implementation of NbS. The Bay Area Greenprint in San Francisco shows the strategic planning tools using mathematical models and geographic information system guides community development and promote a balance between nature, farmland and the needs of regional expansion, becoming a practical medium for resolving the contradiction between urban development and the natural environment.

9 Country

Addressing climate change requires the participation of all parties, among which, the state is one of the most important leaders and promoters, and the public sector plays a vital role in the regional and global promotion of NbS. The implementation of NbS is of great significance to developing countries. On one hand, NbS can break through traditional thinking and models, and promote mitigation and adaptation actions through diversified and innovative approaches, on the other hand, it is also conducive to the integration of the 2°C target set by the *Paris Agreement* and the carbon neutrality target proposed by the countries to businesses in various industries. The important measures for countries to promote the development of NbS include strengthening the awareness and understanding of NbS by various government bodies, issuing policies and regulations conducive to the development of NbS, launching national-level pilot projects, incorporating NbS into the Nationally Determined Contributions (NDCs) commitments and relevant national strategic plans. This chapter selects national NbS projects and strategies in Costa Rica, China and Brazil to show how the country can play an important role in the design and promotion of NbS.

9.1 Costa Rica National Ecological Protection Plan

Case Highlights

The attention given by the presidents at the national level and the support of multiple ministries are essential to the success of Forever Costa Rica; a new government-enterprise cooperation model was established, the new financial mechanism of "permanent project financing" has been tested, and a number of public welfare and corporate foundations have set up trust funds to carry out protection work.

9.1.1 Project Background

Costa Rica is a well-known biodiversity hotspot country in the world (with a large number of highly threatened endemic species). Although with the land area similar to Switzerland, the number of species is equivalent to that of North America combined. A total of 26% of Costa Rica's land area and 1% of the national waters (including 17% of the territorial sea) are classified as protected areas. However, these terrestrial and marine protected areas and their surrounding ecosystems are facing enormous challenges brought about by development. Overfishing, illegal fishing, underregulated tourism development, urbanization, logging, water pollution, coral reef degradation and depletion of fishery resources all pose threats to the protected areas and adjacent land and waters. Despite its great efforts, there is still a significant gap between the actual protected area in Costa Rica and the protected area and target set out in the *Convention on Biological Diversity*.

9.1.2 Project Overview

To bridge the above-mentioned gap and realize the goals of the *Programme of Work for Protected Areas of the United Nations Convention on Biological Diversity*, the Costa Rican government has established the "Forever Costa Rica" national strategy through its National System of Protected Areas (SINAC). The launch and implementation of this strategy are inseparable from the attention and strong support rendered by two presidents. President Oscar Arias launched his visionary "Peace with Nature" philosophy, advocating the provision of continuous funds to carry out nature protection work. President Laura Chinchilla Miranda also provided endorsement of the effort, enabling the project to be successfully launched in 2010. In addition, ministers from the Ministry of Environment, Energy and Telecommunications, the Ministry of Natural Resources Conservation, and the Ministry of Foreign Affairs, have also provided a lot of support for the project. In addition to the strong government support, it has also pioneered the government-enterprise cooperation model, which ensured the responsibilities and obligations of the participants by establishing trust funds and contract management models. Generally speaking, long-term protection for the protected areas is the most direct way to achieve biodiversity. However, developing countries usually face challenges such as the lack of protection resources and insufficient protection. At present, few developing countries have completed the goals mentioned in *Convention on Biological Diversity*. The successful implementation of the Forever Costa Rica strategy has the chance to make Costa Rica the first developing country in the world to achieve the protection goals. It provides valuable experience for other developing countries.

Forever Costa Rica was officially launched at the 10th Conference of the Parties of the Convention on Biological Diversity (CBD COP10) in 2010. At the call of the then President Laura Chinchilla, Linden Trust for Conservation worked with the Gordon and Betty Moore Foundation, the Walton Family Foundation and The Nature Conservancy to establish a long-term funding mechanism to carry out protection work. At the same time, a steering committee was established at the cabinet level to monitor and evaluate the implementation of the project. It is worth mentioning that Forever Costa Rica has set specific phase goals and made clear provisions on the rights and responsibilities of the participants, which includes the following: Achieving the goals of the *Convention on Biological Diversity* and estimating the costs through the National System of Protected Areas; ensuring that funding is available at the initial stage, with the government specifying the proportion and the amount of funding it will provide; realizing $50 million financing and establishing a trust fund for government-enterprise cooperation, for which the trust fund and the government will sign a cooperation agreement to clarify the fund budget appropriations.

The strategic goal is to "establish and maintain, by 2010 for the terrestrial areas and by 2012 for the maritime areas, comprehensive national and regional systems of effectively managed and ecologically representative protected areas". Costa Rica is one of the pioneer countries that have successfully implemented a payment system for environmental services across the country. The National Financing Fund launched the ecosystem service payment system in 1997 to benefit small and medium landowners who are suitable for forestry activities and promote the protection and restoration of forests. Costa Rica has achieved the goal of reforestation of 6,500 hectares, sustainable management of 10,000 hectares of natural forest, and protection of 79,000 hectares of private natural forest by investing $14 million for ecosystem services. In terms of funding, Forever Costa Rica has cooperated with partners such as the Linden Trust for

Conservation, the Gordon and Betty Moore Foundation, and The Nature Conservancy to establish a trust fund and create a "permanent project financing" approach and learning from the financing model of private companies. This financing model respects the interdependence of participants while restricting them, so that multiple investors can make large investments at the same time and get paid based on the realization of goals. This is a common practice in the private sector to finance large projects such as power plants, but it is rarely used in the non-profit sector.

9.1.3 Project Results

The success of "Forever Costa Rica" is inseparable from the international popularity of biodiversity conservation, a stable domestic political situation, more than 40 years of protection experience, and strong government implementation. It sets goals in three dimensions.

Firstly, increasing ecological representativeness in the protected area: With the support of The Nature Conservancy, the Costa Rican government has determined the conservation goals of its important terrestrial, fresh water and marine ecosystems and flagship species. These goals focus on filling the missing areas in the terrestrial protected area system and increasing the protection of marine ecosystems. Specifically, the expansion or creation of 12 areas has doubled the total marine protected area. The newly added areas highlight priority areas to conserve based on threats from climate change (temperature increase, rising sea levels and acidification). In addition, the terrestrial protected areas will also expand from 26% to 26.5% of Costa Rica's land mass.

Secondly, strengthening protected area management: Assessing and improving the management effectiveness of protected areas is the top priority for the government and partners. This includes updating and managing the objectives of each protected area, as well as combining multiple tasks to evaluate these objectives. The National System of Protected Areas has formulated a unified *Towards Efficient Administration of Protected Areas: Policies and Indicators for Monitoring*, which contains 37 indicators covering five areas, i.e., social, administrative, cultural and natural resource management, policies and regulations, and financial management. These management tools and methods are of great significance to existing and newly established protected areas in the future. The SINAC has also made a commitment to improve institutional efficiency, and will establish a special marine department and a unit to evaluate the effectiveness of protected area management.

Thirdly, responding to climate change: At present, all parties have realized that climate change will seriously threaten the biodiversity of Costa Rica. The Costa Rica government took into account climate change impact on the protected area for the first time in the implementation of the strategy. At the beginning of the project, it clarified the threat of climate change to the ecosystem and species in the protected area, as well as possible mitigation and adaptation measures; during project implementation, the impact of climate change was continuously monitored and evaluated, and dynamic adaptation mechanisms and measures were established; in the later stage of the project, the strategic plan was summarized as to the development of biodiversity and ecosystem adaptation to climate change and extreme weather events in the protected area, and pilot implementations were carried out.

9.1.4 Project Information

Case Source: TNC Global NbS Case Database.

Project Participants: Government of Costa Rica, Linden Trust for Conservation, Gordon and Betty Moore Foundation, Walton Family Foundation, The Nature

Conservancy.

Project Location: Costa Rica.

Project Duration: 2010–2015.

Project Link: More information about "Forever Costa Rica", https://www.geofunders.org/documents/534.

9.2 China Ecological Conservation Red Line

Case Highlights

"The Ecological Conservation Red Line" is an important means of coordinating economic development and environmental protection at the national level. It emphasizes protection priority and adapts measures to local conditions, starting from the actual situation of each region, and dividing the ecosystem with scientific methods. In addition to regulations and policies, big data and visual platforms are used to enhance the scientific nature of decision-making. China is exploring ecological protection and climate change solutions suitable for the developing countries, which can benefit those underdeveloped regions, their inhabitants and the endangered species.

9.2.1 Project Background

With the rapid development of industrialization and urbanization, China's resources and ecological environment are facing severe threats. China's environmental protection and construction efforts have been on the rise each year. But in general, as resource constraints continue to increase, environmental pollution continues to worsen, ecosystem degradation is still serious, ecological problems become more complex, and the trend of environmental and ecological deterioration has not been reversed. Due to overlapping protected areas, low efficiency of ecological protection, lack of integrated protection, lax enforcement, and the spatial pattern to ensure national and regional ecological, economic and social coordinated development has not yet been formed (Li Ganjie, 2014).

9.2.2 Project Overview

Ecological Conservation Red Line (ECRL) refers to the areas that have special and critical ecological functions within the ecological space, thus must be strictly protected. It is the bottom line and lifeline to ensure national ecological security, which includes key ecological areas for water conservation, biodiversity, soil and water conservation, windbreak and sand fixation and coastal ecological stability. It also covers other ecologically sensitive and fragile areas such as those with water and soil erosion, land desertification, rocky desertification and salinization.[①] It is an institutional innovation for land and space planning and management and another "lifeline" mentioned at the national level after the "red line of 180 million hectares of arable land". The concept was first proposed in 2011, and was subsequently incorporated into the *Environmental Protection Law of the People's Republic of China (2015)*. In 2017, the Chinese government issued the *Opinions on Delineating and Strictly Observing the Ecological Conservation Red Line* to start its nationwide implementation. As of 2020, the area within the ECRL accounts for about 25% of China's total land area. It aims to include all endangered species and their habitats in the protection, while preventing floods and

① The definition comes from the *Opinions on Delineating and Strictly Observing the Ecological Conservation Red Line* issued by the General Office of the Central Committee of the Communist Party of China and the General Office of the State Council.

sandstorms, and providing clean water and other ecosystem services to achieve the synergy of climate change response and biodiversity protection.

ECRL is an important step to coordinate economic development and environmental protection. At the beginning of the 21st century, the results of the "National Ecological Environment Survey and Assessment System" showed that China's forest, wetland, grassland and other ecological spaces were severely occupied. Regarding the question of how to change the situation and achieve the goal of protecting the ecological environment while promoting economic development, Chief Expert Gao Jixi, Director of the Nanjing Institute of Environmental Sciences of the Ministry of Ecology and Environment and Chairman of the National Ecological Conservation Red Line Delineation Expert Committee, put forward the concept of "Ecological Conservation Red Line" to protect those important ecological areas, and the areas outside the protected areas can be used for industrial expansion and urbanization.

The Opinions on Nature Reserves issued by China in June 2019 requires that "all types of nature reserves with important ecological functions, sensitive and fragile ecological environments, and other types of nature reserves that need to be strictly protected should be included in the Ecological Conservation Red Line". In November of the same year, the Central People's Government issued the *Guiding Opinions on the Overall Planning and Implementation of Three Control Lines in the Land and Space Planning*, in which the "three areas and three lines" are considered as the core of China's land and space planning. The "three areas" are based on three types of space: urban space, agricultural space, and ecological space. The "three lines" correspond to the three control lines of urban development boundary, permanent basic farmland protection red line, and ecological conservation red line (Wang Yinglin et al., 2020).

In addition, China issued the *Master Plan for Major Projects of National Important*

Ecosystem Protection and Restoration (2021-2035) in 2020, which clarified the protection goals by 2035, including a forest coverage rate of 26%, the forest volume of 21 billion cubic meters, a natural forest area stable at about 200 million hectares, and 60% comprehensive vegetation coverage of grasslands; maintaining the current wetland area with protection rate increased to 60%; 56.4 million hectares of newly added area under comprehensive soil erosion control, and more than 75% of sanded land to be brought under control; the deterioration of marine ecology should be fully reversed, and the retention rate of the natural coastline shall not be less than 35%; the nature reserve with national parks as the main body shall occupy more than 18% of the land area, and endangered wild animals and plants and their habitats should be fully protected. Evolution of China Ecological Conservation Red Line (ECRL) shown in Figure 9-1.

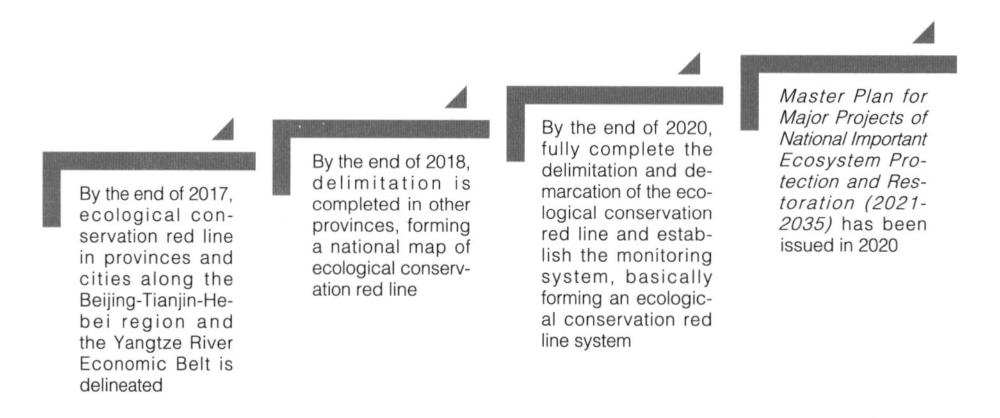

Figure 9-1 Evolution of China Ecological Conservation Red Line (ECRL)

In addition, the Chinese government has provided tremendous support for the implementation of the ecological red line. In October 2017, the National Development and Reform Commission officially approved the launch of the supervision platform for ECRL, with a total investment of ¥286 million. The red line supervision database has been designed, with 67 types of data in four categories. The supervision platform will rely on satellite remote sensing and terrestrial ecosystem monitoring sites to form a space-air-ground integrated monitoring network to assess and warn ecological risks timely. It provides real time monitoring of human interference activities, and punishment will be meted out immediately if the behavior violates the red line to ensure that the ecological function is not reduced, the area is not reduced, and the nature is not changed.

9.2.3 Project Results

Through the joint efforts of the Ministry of Ecology and Environment, the National Development and Reform Commission, the Ministry of Natural Resources and other relevant provinces, the project adopts a combination of top-down and bottom-up approaches, whereby the central government is in charge of planning and the local authorities are in charge of enforcement and feedback. Through scientific evaluation, departmental coordination, planning coordination, regional coordination, land-sea coordination, expert review and inter-ministerial coordination group review, the delineation of ecological conservation red lines in 15 regions, including the three regions of Beijing, Tianjin and Hebei, 11 provinces (cities) in the Yangtze River Economic Belt, and Ningxia Hui Autonomous Region, has been carried out successfully. As of 2018, the total area of ecological conservation red lines delineated by the 15 regions (including Beijing) is about 610,000 square kilometers, accounting for about a quarter of the total land area of the 15 regions. They are mainly ecologically important and ecologically sensitive and fragile areas, covering national and provincial-level nature reserves, scenic spots, forest parks, geological parks, world cultural and natural heritage, wetland parks and other protected areas. In general, it has realized its full delineation. China implemented ecological restoration and other measures in the designated ecological red line areas. The carbon sequestration in the region accounts for 45% of the total carbon sequestration, which plays a key role in climate mitigation.

9.2.4 Project Information

Case Source: C+NbS Cooperation Platform.

Project Participants: Chinese government departments at all levels.

Project Location: China.

Project Duration: Since 2017 .

Project Links: News report | Gao Jixi: Forming "a National Map" of Ecological Conservation Red Line. November 19, 2018, http://www.chinacses.org/hjkp_23038/hbkpxw/201811/t20181127_675124.shtml; News report | 1/4 Land Area of the 15 Regions Will Be Guarded by the Ecological Conservation Red Line. February 13, 2018, http://www.gov.cn/xinwen/2018-02/13/content_5266384.htm.

9.3 Brazil's Oasis Project for the Protection of Water Sources

Case Highlights

To achieve the goal of environmental protection through economic means, Payment for Environmental Services (PES) can be used in areas such as biodiversity protection, climate change, soil protection, nutrient recovery and water source protection; trying to develop an economic pricing model to enable farmers to obtain tangible economic incentives to increase the forest area, improve forest quality and adopt the best agricultural practices that can protect the ecological environment on their land.

9.3.1 Project Background

Climate change and water security are the common social challenges faced by the world. In many regions, changes in precipitation and melting of snow and ice caused by climate change are significantly affecting the water resources and their quality. At the same time, the destruction of ecosystems caused by climate change also poses severe threats to water security. Currently, about 2.2 billion people in the world do not have safe drinking water, 4.2 billion people do not have access to safe environmental health services, and 3 billion people lack basic hand-washing facilities. The *World Water Development Report 2018* of the United Nations pointed out that the global water demand is increasing at a rate of 1% per year, and it will increase by one third in the next 30 years. In addition, since the beginning of the 20th century, about two thirds of forests and wetlands have disappeared or degraded (UNESCO, 2018). Insufficient supply is the main reason for the increasing challenges of water management. Water security can be improved by protecting and restoring water-related ecosystems, including mountains, forests, wetlands, rivers, aquifers and lakes. Take forests as an example, according to statistics from the United States Department of Agriculture, forests provide drinking water for 180 million Americans in 68,000 communities and are the largest source of drinking water in the United States. Forests managed by the US Forest Service can provide water for 66 million people in 3,400 communities in 33 states. The cost of drinking water treatment decreases with the increase in forest coverage in the watershed.

9.3.2 Project Overview

Located in the Amazon Basin, Brazil is abundant in water resource and biodiversity. However, since the 1970s, in the Brazilian economic miracle, its natural environment has paid a heavy price for economic growth. Urban expansion, population growth, and production based on energy and material have caused serious threats to its environment. The government has issued a series of environmental protection laws and policies, including economic attempts to achieve environmental protection goals. Payment for Environmental Services (PES) is one of them. PES is related to NbS, which aims to encourage local people to protect the ecosystem through economic compensation. The logic of PES is very simple: Increase income from economic activities that are compatible with environmental protection to encourage the sustainable use of natural resources and punish predatory activities. PES can be used for biodiversity protection, climate change response, soil protection, nutrient recovery, water source protection, and so on. In recent years, academia has continuously advocated the use of economic means to complement the existing regulatory approach in terms of environmental protection, and PES is the product of this promotion. Brazilian PES is one successful practice of payment for ecosystem services.

Since 2006, Brazil has been implementing the Oasis Project. It is led by the Boticário Group Foundation, a non-profit organization dedicated to nature conservation, and under the guidance of the local government, also joining hands with enterprises (such as the local water company and the Mitsubishi Foundation). The Oasis Project has been implemented in Apucarana, Sao Paulo, and São Bento do Sul, Brazil. Its goal is to protect the watershed by protecting the native forest on private land.

The Oasis Project in Sao Paulo, Brazil, for example, the water company donated 1% of the income from the city to the local government's environmental protection fund, which the government would use to reward private farmers for their excellent work in native forest protection. The Boticário Group provided technical support in this project. Private lands were graded with their unique scoring mechanism and evaluation criteria, and the amount of rewards is calculated based on the financial conversion formula. Specifically, the reward given to the farmers can be expressed as

$$\text{Reward} = Z \times X \times (1 + G1 + G2 + G3)$$

where X represents subsidy per unit area, Brazilian real per hectare. Z represents the land area (hectare), and $G1$, $G2$ and $G3$ are three bonus items, representing freshwater protection, natural ecosystem protection and sustainable agricultural behavior respectively. Each bonus item has a different weight: G1 (0~1), G2 (0~2.5), G3(0~1.5).

In addition, the Boticário Group also measures the project through rating and questionnaires. Through the economic pricing model, farmers can obtain tangible economic incentives to increase the forest area, improve forest quality and adopt the best agricultural practices that can protect the ecological environment on their land.

9.3.3 Project Results

Such incentives can improve the health of the river basin and the stability of water supply in the entire region, which can directly benefit the water company, thereby a multi-win closed loop has been formed. According to estimates by the Boticário Group, the initial implementation of the Oasis Project in São Bento do Sul has achieved an additional conservation of 1,620 hectares of forest and restoration of 3,239 hectares of degraded pastures. It means a reduction in the soil loss of 54%, a reduction of soil turbidity of 44%, saving from 13% to 26% in water treatment costs and an increase of 2.8% in water production. The reduction in water treatment costs, regional benefits (reduce flood losses), and global benefits (carbon sinks) enable the plan to recover its investment within six years. The Boticário Group also predicts that the value of the project in terms of carbon sinks will reach $ six million. Its adaptability enhancement and social and environmental contribution include improving adaptability through economic benefits, social influence (increasing income, reducing poverty, climate equity, etc.), food security, reducing species extinction and ecological loss, protecting biodiversity.

9.3.4 Project Information

Case Source: UNEP - NbS Contributions Platform.

Project Location: Apucarana, Sao Paulo and São Bento do Sul, Brazil.

Project Duration: Since 2006.

Project Link: United Nations Environment Programme, https://wedocs.unep.org/bitstream/handle/20.500.11822/28898/NBS_water_security.pdf?sequence=1&isAllowed=y.

9.4 Case Summary

Costa Rica is a pioneer country in global biodiversity and ecological protection. Through the government's forward-looking planning and deployment, it successfully implemented the "Forever Costa Rica" strategic cooperation project in 2010, making it the only tropical country that reversed forest degradation and possibly the first developing country to achieve the vision of protected areas under the *Convention on Biological Diversity*. The successful launch and implementation of the "Forever Costa Rica" project cannot be possible without the attention and support given by the presidents at the national level. The project not only received support from the Ministry of Environment, Energy and Telecommunications, the Ministry of Natural Resources Conservation, and the Ministry of Foreign Affairs and other ministries, but also established a new model of government-enterprise cooperation, pioneering the "permanent project financing" mechanism, which brings together public welfare and corporate foundations to establish trust funds for the protection work. It is worth mentioning that the "Forever Costa Rica" protection strategy has made clear provisions on the responsibilities of each participant, and has set clear milestone targets as the basis for fund allocation. Its success is inseparable from the international attention to biodiversity conservation, a stable domestic political situation, more than 40 years of protection experience, and strong government implementation. It has set goals in three dimensions, including increasing ecological representativeness in the protected area, strengthening protected area management, and responding to climate change.

China is committed to the construction of ecological civilization, aiming to coordinate development under the premise of protecting the environment and responding to climate change. At the United Nations Climate Action Summit 2019, China was invited to jointly promote global NbS work with New Zealand. During this period, China's "Ecological Conservation Red Line" (ECRL) policy has caught the attention and has been understood by more and more countries. Meanwhile, China is also constantly exploring and innovating. The achievements of ECRL that originally aimed at protecting areas with important ecological functions and fragile areas and restoring wild animal and plant populations may be developed into potential resources for emission reduction and carbon sequestration, so as to build a bridge of communication and cooperation between scientific protection of nature, collaborative governance of ecological environment and response to climate change, setting a precedent for the international community. The ECRL emphasizes the priority of protection and adapts measures to local conditions, starting from the actual situation in each region, and dividing the ecosystems with scientific methods. At the same time, in terms of management measures, the ECRL advocates a top-down design, emphasizing the differentiated management of the ecological conservation red line in multiple provinces. With the support of policies and funds from the state and ministries, China is already actively using big data and visual platforms to enhance the scientific nature of decision-making. In the future, China can continue to innovate and explore ecological protection and climate change solutions suitable for developing countries, so as to benefit more underdeveloped regions, their inhabitants and the endangered species.

In order to protect the precious freshwater resources and the Amazon forest ecosystem, Brazil has launched the Oasis Project and conducted pilot projects in the cities of Apucarana, Sao Paulo, and São Bento do Sul. The Oasis Project is guided by the concept of "Payment for Environmental Services" (PES) and encourages local people to carry out work to protect the ecosystem through economic compensation. The project is led by non-governmental organizations, bringing together local governments, foundations, and water companies to encourage local communities to participate through the development of simple and effective incentive mechanisms to achieve ecological protection of forests, degraded pastures, and river basins, and promote healthy basins and stable water flows throughout the region.

10 Platform & Initiative

As an emerging, global and interdisciplinary issue, NbS requires cross-border communication and collaboration among all countries and stakeholders in the world. Gathering the power of global governments, international organizations, enterprises, think tanks, and individuals to spread the concept of NbS, promote the practice of NbS, and promote the development of NbS is inseparable from the publicity, coordination and integration role of related NbS platforms and initiatives. This chapter selects NbS communication and promotion platforms in different periods and in different regions, including Nature4Climate, which played a fundamental role in the initial stage of NbS development, the "Countdown" global initiative that emerged after NbS gained global attention in 2019, and the C+NbS cooperation platform initiated by China, to analyze the active role played by the cooperation platform and global climate action initiatives in promoting the development of NbS.

10.1 Nature4Climate

Case Highlights

With the help of major global conferences and platforms such as WEF, UNFCC and Climate Action Summit, N4C works to promote the huge potential of nature in the field of mitigation and adaptation, specifically, N4C amplifies the voice of NbS in the field of climate change through effective communication strategies; N4C encourages more NbS practices by creating a global NbS action community. N4C visualives the mitigation potential of NbS globally and by region with the help of visual maps to further increase public's awareness and understanding of NbS.

10.1.1 Project Background

Nature4Climate (Figure 10-1) is a multi-stakeholder alliance, which was jointly initiated and established in 2017 by the United Nations Development Programme, the United Nations Environment Programme, the *Convention on Biological Diversity*, the International Union for Conservation of Nature, The Nature Conservancy and other organizations. It hopes to establish a cooperation platform to convene more institutions and organizations to promote the potential and possible opportunities of NbS, thereby increasing the contribution of NbS to global climate actions. The main targets of Nature4Climate (N4C) include decision makers, policy makers, the private sector and opinion leaders. Currently, Nature4Climate has more than 20 partners worldwide (Figure 10-2), aiming to gain support from foundations, UN agencies, civil organizations, and the private sector to increase actions and investments in NbS and help realize the goals of the *Paris Agreement*.

Figure 10-1 Logo of Nature4Climate

Figure 10-2 Global Partners of Nature4Climate
(Source: Nature4Climate)

10.1.2 Project Overview

Nature4Climate pointed out that through the protection, restoration and sustainable management of ecosystems, by 2030, NbS may contribute about 30% of emission reductions to achieve the 2°C or 1.5°C temperature rise target specified in the *Paris Agreement*. Nature4Climate hopes to carry out activities in the following three areas to achieve the goal of "cutting 10 billion tons of carbon dioxide equivalent through NbS by 2030":

①Raise public awareness and understand the important potential and role of NbS;

②Enhance and strengthen the actions of countries to achieve national voluntary emission reduction commitments through NbS;

③Increase climate financing to enable more funds to flow to NbS.

The work of Nature4Climate includes demonstrating the potential of climate mitigation in the field of land use, advocating excellent scientific methods of land management, sharing excellent global cases, promoting multi-party dialogue, and disseminating scientific knowledge. Nature4Climate brings the story of NbS to as many audiences as possible through high-level dialogue and cooperation, strategic mobilization, corporate cooperation, youth mobilization, and innovative communication.

10.1.3 Project Results

Since its launch in 2017, Nature4Climate has spoken out at major international events or conferences every year, which has increased the number of topics related to NbS by three times, and the share of NbS in the field of climate change has increased from 1% to the current 12%. The formation of a NbS community further expands the scope of NbS's influence. Climate financing has been directed more towards NbS related fields, and the proportion of NbS in net investment has increased from 3% to 8%. The field of theoretical development of NbS has also advanced by leaps and bounds. In 2015, there were only three academic articles mentioning NbS, and by 2019, there were more than 100 NbS-related papers. The communication and cooperation strategy of Nature4Climate has also achieved great results. Through the establishment of an integrated communication and exchange platform, more companies and young talents are absorbed into the NbS cooperation.

10.1.4 Project Information

Case Source: C+NbS Cooperation Platform.

Project Participants: The United Nations Development Programme, the United Nations Environment Programme, International Union for Conservation of Nature, World Wide Fund for Nature, The Nature Conservancy and other institutions.

Project Location: Worldwide (headquartered in London, UK).

Project Duration: Established in 2017.

Project Link: The official website of Nature4Climate, https://nature4climate.org.

10.2 "Countdown" Global Initiative

Case Highlights

Relying on the TED platform and its subordinate TEDx Program, this initiative gathers NbS cases presented from around the world, and shares them with the world. It amplifies the voice and contribution of NbS and successfully advocates for climate change. "Countdown" focus on different levels of climate governance actors, ensuring broad multi-stakeholder participation. It leverages TED'S reach to bring climate change issues to the attention of more people around the world.

10.2.1 Project Background

To reverse the trend of climate change, all sectors of society need to make the most positive response, including strong and decisive political leadership, a transformative business vision, and the active participation of all citizens. To this end, TED and the Future Stewards (a coalition of partners) jointly launched the "Countdown" global initiative, aimed at advocating and accelerating the proposal of solutions to the climate crisis, and turning green ideas into practical actions (Figure 10-3). The ultimate goal of the initiative is to reduce greenhouse gas emissions by at least half by 2030 through the efforts of all parties, creating a safer, cleaner and fairer world for mankind.

Figure 10-3 "Countdown" Global Initiative Logo

10.2.2 Project Overview

The "Countdown" initiative was launched on October 10, 2020. It raised five interrelated issues for the human world and tried to draw a blueprint for a clean future by advocating the discussion of these five issues. From these five issues, five sub-themes in the "Countdown" initiative were derived: Energy, transportation, materials, catering and nature.

The initiative invites multiple partners to participate in the "Countdown" initiative. Local organizations, companies, cities and countries, as well as citizens from all over the world can sign up to participate. "Race to Zero" is a substantive global movement proposed in the "Countdown" initiative, trying to mobilize companies, cities, regions and investors to participate in actions to reduce carbon emissions ahead of the 26th Conference of the Parties of the United Nations Framework Convention on Climate Change (UNFCCC COP26) in November 2021. All the organizers are invited to share their own cases, experience and ideas on climate governance. Among them, different organizers have different ways of participation: Business entities are invited to make practical climate commitments (Figure 10-4) to minimize the impact on climate change; cities and regions can cooperate with international organizations related to the platform to carry out carbon emission reduction practices; the activities organized by non-profit organizations are mainly sharing of scientific knowledge.

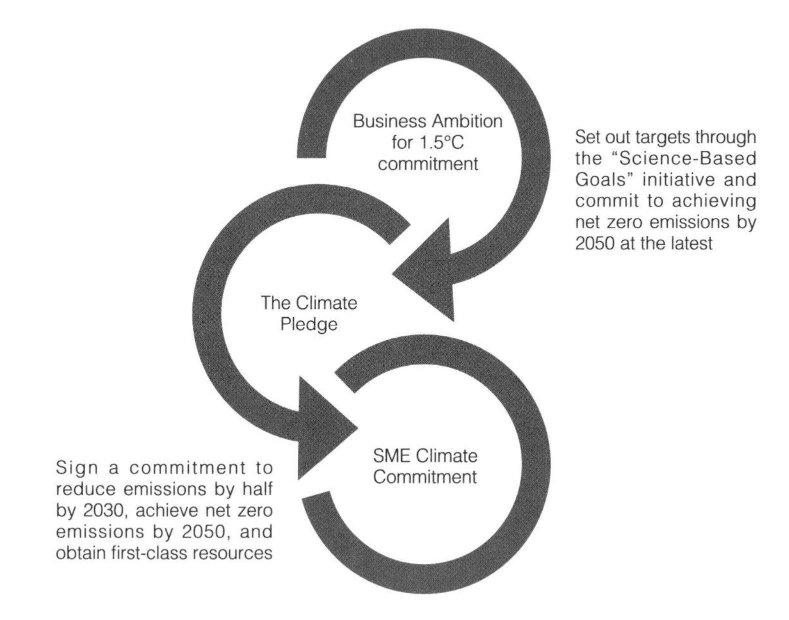

Figure 10-4 Path of Companies Participating in the "Countdown" Global Initiative
(Source: Countdown official website)

The initiative is expected to last for nearly one year. On October 10, 2020, the "Countdown" initiative was officially launched globally through an online press conference. The Global Kick-off Conference was held simultaneously on the video platform YouTube. More than 50 thinkers and actors in different fields shared their views on climate change under five themes, calling on leaders and citizens everywhere to take action, and also developed multiple practical ideas with research values from different perspectives. From the start of the conference to October 2021, the

"Countdown" initiative has entered the "accelerated development" stage, and the baton was handed over to the local organizers. Organizers focus on solving the five interrelated issues of the climate crisis, combined with the actual local situation to carry out specific local activities. The form of activities includes but not limited to watching and sharing with the audience the videos of speeches and dialogues of the global launch of the "Countdown" initiative, hosting local climate dialogues and lectures, organizing and participating in "Countdown" creative projects, etc. From October 12 to 15, 2021, a four-day "Countdown Summit" was held in Edinburgh, Scotland, combining lectures, case studies and seminars to comprehensively summarize and consolidate the "Countdown" initiative activities from 2020 to 2021.

10.2.3 Project Results

The biggest goal of the TEDx Program is to bring the spirit of "ideas worth spreading" to communities around the world in line with TED's mission, advocating and helping them to host TED-style sharing activities. The "Countdown" global initiative, as the flagship platform newly created by the TEDx Program, also aims to share and disseminate the solutions and local NbS experience of actors at all levels in response to climate governance, and to gather ideas from all over the world to solve the climate crisis on the platform. At present, more than 600 entities around the world have participated in the "Countdown" initiative. In China, as of December 2020, nearly ten "Countdown" landing events have been held. Hosts include first-tier cities, think tanks, non-profit charity organizations, enterprises, etc. Thousands of people participated in the Countdown global initiative as sharers or audiences. This has aroused extensive reports and attention from Chinese social media and academic circles.

The "Countdown" initiative is expected to lead the sharing and dissemination of widespread concerns about the climate change crisis through the platform. At the Global Kick-off Conference, five interrelated issues were raised to the world. Based on these five themes, it is urgent to participate in climate change governance. In the coming year, climate experts, policy makers, visionary business leaders, and thousands of local organizers will take these five key issues into consideration when discussing, debating, and formulating strategies locally. In addition, the initiative also provides a platform for brainstorming ideas for solutions by bringing together the governance forces of all parties. The "Countdown" initiative aims to promote and improve what climate leaders have been doing, rather than completely overthrowing and recreating it. So far, hundreds of climate governance bodies have put forward proposals on climate issues and will share their ideas and solutions in the coming weeks. Scientists, activists, entrepreneurs, city planners, farmers, CEOs, investors, artists, government officials and others gather to find the most effective and well-founded ideas.

The ultimate goal of the "Countdown" initiative is to gather ideas and propose solutions through continuous communication and sharing, and to further activate the solutions through summits and follow-up actions in the future. Based on the ideas shared at the global press conference on October 10, 2020, the initiative cooperates with partners to summarize the local experience, cases and ideas put forward by various governance bodies during the year, and supports a series of multi-divisional projects aimed at addressing climate-related challenges. Based on the five interrelated issues raised at the kick-off conference, the initiative will set specific and bold goals to provide answers from all over the world at the 2021 "Countdown" summit, and try to encourage governance bodies to make new commitments and concrete actions before the UNFCCC COP26, and strive to achieve the ultimate goal of reducing greenhouse gas emissions by at least half by 2030.

10.2.4 Project Information

Case Source: C+NbS Cooperation Platform.

Project Participants: Initiated by TED and Future Stewards, with response from global partners.

Project Location: Worldwide.

Project Duration: 2020-2021.

Project Link: "Countdown" initiative official website, https://countdown.ted.com/.

10.3 C+NbS Cooperation Platform

Case Highlights

As China's first platform to focus on the new international topic of NbS, it faces the needs of the world from a China perspective, and strives to provide Chinese solutions and wisdom to the international arena. It is committed to effectively solving sustainable development challenges in China and the world with more robust NbS. Local practice and international experience complement each other by integrating the academic research of world-class experts and field studies of enterprises. The results and consensus obtained from each workshop have increased the theoretical and practical understanding of NbS among all stakeholders, enabling Chinese stakeholders to quickly integrate with international standards.

10.3.1 Project Background

In September 2019, the United Nations Climate Action Summit in New York identified NbS as one of the nine major action areas, and invited China and New

Zealand to jointly take the lead. A delegation from the Institute of Climate Change and Sustainable Development of Tsinghua University learned about Nature-based Solutions at the summit. They learned that advancing Nature-based Solutions requires interdisciplinary and cross-field exchanges and cooperation. The systematic research of the Nature-based Solutions is still relatively lacking, and there is no relevant platform organization in China to gather the experience and knowledge of different parties.

10.3.2 Project Overview

In order to promote interdisciplinary exchanges on climate change and biodiversity at home and abroad, as well as systematic research on NbS, the Institute for Climate Change and Sustainable Development of Tsinghua University began to organize and build a NbS cooperation platform to address climate change (abbreviated as C+NbS Cooperation Platform, C represents both Climate and China) in April 2020.

The C+NbS cooperation platform (Figure 10-5) aims to "see clearly the 'identity', understand the connotation, understand the 'past and present' of NbS, and form preliminary judgments and suggestions". It mainly makes efforts in the following two aspects (Figure 10-6) .

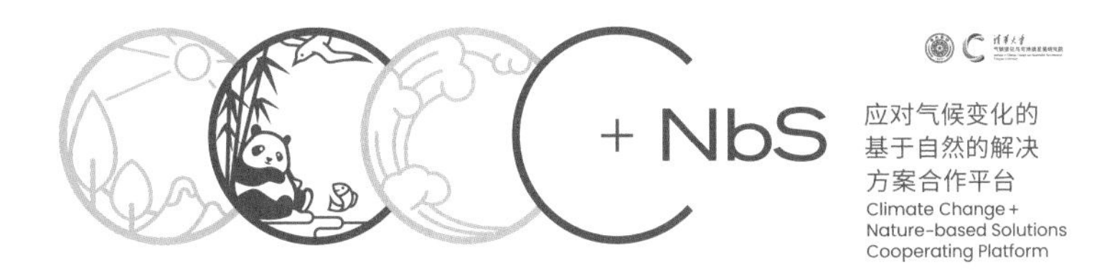

Figure 10-5 C+NbS Cooperation Platform Logo

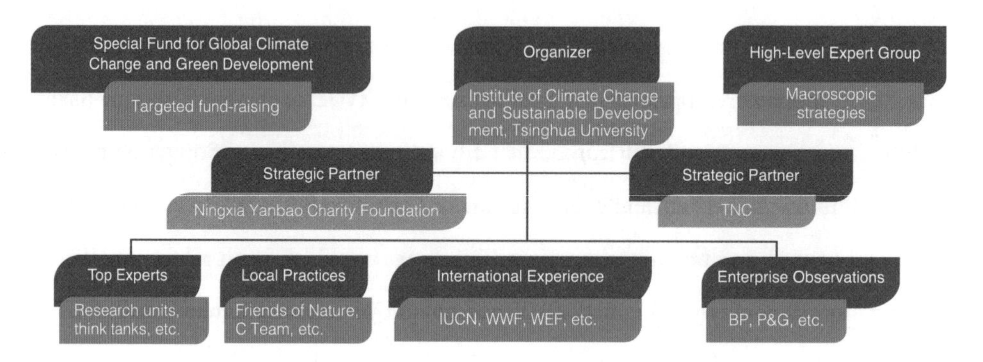

Figure 10-6 C+NbS Cooperation Platform Architecture

On one hand, the platform systematically reviews the existing research foundation of NbS, sorts out the "past" stories, forms a comprehensive report, produces a comprehensive literature review of NbS from the perspective of climate change, and simultaneously advances comprehensive research and global case studies. The results were officially released at COP15.

On the other hand, by organizing monthly workshops, the platform quickly gathers domestic and foreign counterparts and partners in NbS-related fields, and obtains the latest developments in the NbS field through exchanges and interactions, so as to know its "present" stories, integrate resources, and seek cooperation. The platform has held a total of nine monthly workshops in 2020. Focusing on the core topic of NbS, the platform has set different discussion topics for each workshop, including the concept, methodology, international standards and experience of NbS, practices from China and the world, etc., aiming to promote discussion and cooperation in theory and practice, and stimulate innovative research (Table 10-1). According to statistics, the international online participants taking part in the NbS workshop cover four continents, come from 21 countries and regions around the world, and involve multiple fields and industries.

Table 10-1 Monthly Workshop Content of C+NbS Platform

S.N.	Time	Theme	Introduction
1	April 29	A Preliminary Study of NbS Theory	Online closed-door meeting, inviting more than 30 representatives from government departments, scientific research institutes, financial institutions, international organizations, non-governmental organizations and the private sector to discuss the value, positioning, content and domestic and international development potential of NbS topics and cooperation platforms
2	May 20	NbS Biodiversity	In order to welcome the International Day for Biological Diversity, more than 50 representatives were invited to participate in an online closed-door meeting to deeply interpret the connotation of "Our solutions are in nature" proposed on this day
3	June 24	Focus on NbS Methodology	Online closed-door meeting, inviting more than 50 representatives to participate, and conduct interdisciplinary and multi-angle in-depth discussions on the current status of carbon emissions in the ecosystem and related fields, the potential for reducing sources and increasing sinks, and evaluation methodology
4	July 20	NbS International Experience Ⅰ	First time using open registration, inviting Chinese and foreign experts to focus on sharing international experience, attracting more than 200 representatives from government departments, scientific research institutes, international organizations, private institutions, global universities, private sectors, and the media to participate in online learning and have lively discussions
5	August 20	NbS International Experience Ⅱ	Online public event, inviting Chinese and foreign experts to share wonderful views on the core value of NbS and pioneering practices around the world, attracting more than 2,000 representatives from all over the world to participate in the online live broadcast and learn together
6	September 24	NbS in China	Online public event, focusing on China case sharing, inviting domestic representatives to share practical actions and exchange local NbS experience, and attracting more than 1,600 representatives from around the world to participate in online live learning and discussion
7	October 27	NbS & Carbon Neutrality	Offline and online integration, Professor He Jiankun, Deputy Director of the National Expert Committee on Climate Change and Director of the Academic Committee of the Institute of Climate Change and Sustainable Development of Tsinghua University, lectured on China's long-term low-carbon development strategy and transformation path, with more than 120,000 hits worldwide

(continued Table)

S.N.	Time	Theme	Introduction
8	November 16	TED Youth Event	Offline and online integration, the TEDx ICCSD C+NbS Youth Event & Closing Ceremony of the 2nd Graduate Forum of Global Alliance of Universities on Climate were held to help young students broaden their horizons and understand the international frontier topics of NbS, with more than 320,000 hits worldwide
9	December 18	NbS Annual Conference	With an online and offline combination, the platform co-organized NbS annual conference "2020 Review and Outlook" with the World Economic Forum, aiming to strengthen the cooperation between all parties and contribute to COP15

10.3.3 Project Results

In the past year, there was a growing trend of internationalization in the emerging topic of NbS for C+NbS Cooperation Platform, which has aroused widespread attention and attracted active participation both at home and abroad (Figure 10-7). Over the past year, the C+NbS cooperation platform has formed a regular domestic and international cross-border dialogue mechanism represented by monthly workshops. The timely adjustment of ways of participation has deepen the international exchanges. The first three workshops adopted a closed-door invitation approach, and the guests were scholars, officials, and experts from domestic and foreign research institutions and think tanks related to the NbS field. Online registration was opened for the fourth workshop to cater to the growing needs, which attracted the attention of national ministries and commissions, local governments, relevant research institutions, international organizations, news media and universities and colleges in China. Dozens of representatives signed up for the discussion. To further expand its coverage, the follow-up salon activities were live-streamed in both Chinese and English, which facilitated the participation of more teachers and students from overseas colleges and universities.

More than 2,000 people watched live broadcasts, replays, and participated in discussions, which was an increase of nearly 66 times. The seventh workshop started the global live broadcast, and the number of clicks and viewers reached 120,000. The eighth workshop in November and the closing ceremony of the Graduate Forum of Global Alliance of Universities on Climate received 320,000 hits worldwide. In December, the C+NbS cooperation platform was invited to co-organize the annual conference of "NbS: 2020 Review and Outlook" with the World Economic Forum. Through dialogues and exchanges, the C+NbS made its contribution to the 15th Conference of the Parties of the Convention on Biological Diversity in 2021. The workshop in December 2020 was broadcast live on the official Facebook account and Weibo account of the Global Alliance of Universities on Climate, which attracted widespread attention both in China and internationally.

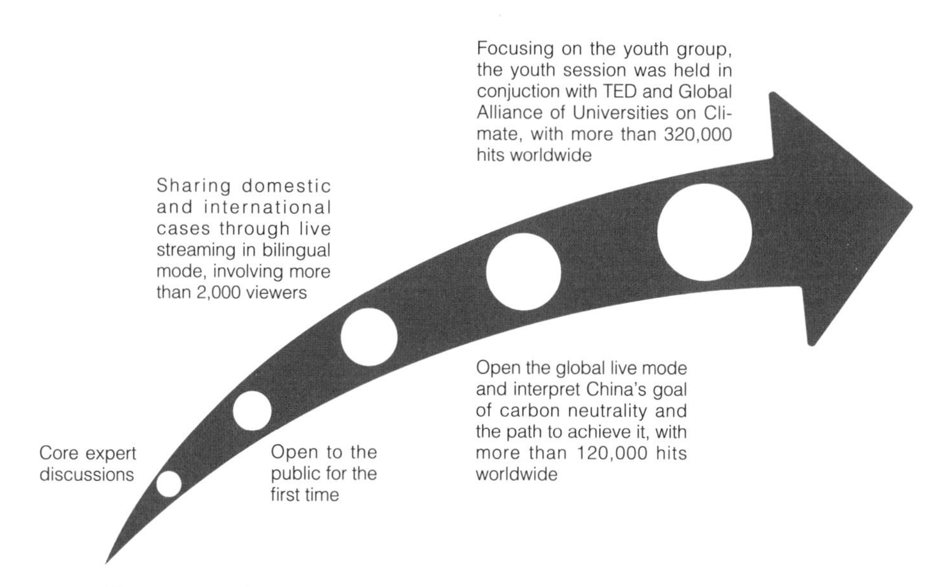

Figure 10-7 C+NbS Cooperation Influence Development Schematic Diagram

As of December 2020, the C+NbS cooperation platform has organized nine monthly workshops and made its effort for global live broadcasting. More than 400 domestic and foreign institutions have signed up to participate in the conference. The online viewers for a single session have reached a maximum of 500,000, and the cumulative number has hit 620,000. With its rapid international influence, the platform was invited to cooperate with major conference platforms to promote the dissemination and exchange of opinions in the field of climate change and biodiversity protection. In October 2020, the platform became the core partner of the "Countdown" global initiative by TEDx in China, and linked with the TED platform to launch the "TEDx ICCSD" activities within two months, spreading China's climate voice to the world and build a bridge between them with the help of the TED platform.

10.3.4 Project Information

Case Source: C+NbS Cooperation Platform.

Project Participants: Led by the Institute of Climate Change and Sustainable Development, Tsinghua University.

Project Location: Worldwide (headquartered in Beijing, China).

Project Duration: Since 2020.

10.4 Case Summary

Established in 2017, Nature4Climate plays an important role in both theoretical and practical exploration of NbS. As one of the first international platforms dedicated to enhance the potential of NbS to address climate change, Nature4Climate has long been committed to increasing NbS awareness among policymakers, the private sector, and the public, and advocates more climate funds to support NbS.

Launched on October 10, 2020, the "Countdown" global initiative provides a global platform for exchanging and sharing of climate crisis governance experience. It gathers diverse entities in climate governance and provides an important channel to enhance the attention of NbS projects. As the most influential international initiative at present, its initial attention was derived from the global influence accumulated by its initiator, TED: TED's YouTube account has 17.9 million subscribers, and its TEDx Talks has 26.6 million subscribers. Its Twitter account has 11.438 million followers and TEDx Talks has 553,000 followers. Generally speaking, the viewership of its videos can easily reach 10 million. Leveraging the huge international, diversified platform of TEDx, the "Countdown" global initiative speaks for climate solutions such as NbS. It provides a direct, professional, and certified path for the international community to understand NbS. At the same time, many international organizations, non-profit institutions, university think tanks and advocacy coalitions participate in the activities. Governance entities at different levels can reach out to more companies, governments, and non-profit organizations through the TED platform and Future Stewards to increase the interaction between participants and international NbS pioneers.

The C+NbS cooperation platform was established in April 2020. Its regular exchange mechanism represented by monthly workshops has greatly promoted the NbS communication and exchanges for China. It adheres to its foothold in China's actual condition, and cooperates with international organizations and platforms (such as the United Nations Environment Programme, the International Union for Conservation of Nature, the Nature4Climate, and The Nature Conservancy) to speak out the voice of China and enhance China's NbS competence in the global arena. On one hand, it gathers consensus, contributes Chinese wisdom to the world, and proposes Chinese solutions to climate change and global governance, on the other hand, the platform actively learns from the international advanced NbS scientific theory and practical experience, internalizes them to promote the progress and development of China's NbS, and actively seeks coordination with relevant policies in different fields. Through the exchange results and strategic consensus of each workshop, it lays the ideological foundation for the future joint academic research of global think tanks, universities and research institutions as well as creating conditions for NbS global implementation.

11 Corporation

Enterprises are important stakeholders of Nature-based Solutions and the backbone to putting NbS into practice. Their participation in the design, investment and implementation of NbS projects demonstrates their willingness to assume social responsibility and their response to social challenges. It is also an important approach to improve the sustainability of their production and operations, expand their business and increase their profits. In this section, Procter & Gamble, Nestlé, and Baofeng Energy (Ningxia, China) are chosen as examples for multinational companies and Chinese local companies that are inclined to participate in NbS to address climate change and promote green and low-carbon development.

11.1 P&G Embraces Nature-based Solutions to Make Operations Carbon Neutral

Case Highlights

As a global consumer goods giant, P&G has made many attempts in sustainable development and corporate social responsibility. Through innovative design, it has integrated sustainable development into its product design and packaging. It is also one of the pioneers and leaders in advocating enterprises' participation in NbS in China and has carried out field projects with multiple international organizations to enhance the global confidence in fighting climate change. P&G's climate strategy and its NbS projects have provided valuable experience for other enterprises in NbS project implementation.

11.1.1 Project Background

Founded in 1837, P&G is one of the largest consumer goods companies in the world, with nearly 110,000 employees worldwide. It is well-known in the daily chemicals market. Its products ranges from shampoo, hair care products, skin care products, cosmetics, baby care products, feminine hygiene products, medicine, textiles, household care, personal cleaning products. To accelerate the global response to climate change, P&G announced its 2030 climate commitment on July 16, 2020, aiming to promote its internal reforms and measures to protect, improve and restore the ecological environment and realize carbon neutrality in global operations by 2030.

11.1.2 Project Overview

P&G is committed to reducing absolute greenhouse gas emissions through a comprehensive energy transition (Figure 11-1). At present, 70% of P&G's global operations are using clean renewable electricity. By 2030, it will achieve 100% renewable electricity coverage in its production, while it will reduce greenhouse gas emissions by 50%. P&G will adopt NbS and other measures to offset the emissions (about 30 million tons) that cannot be completely eliminated by 2030 due to technological restrictions. As one of the global consumer goods giants, P&G spreads NbS knowledge through cross-border exchanges and international dialogues to promote global climate actions. In July 2020, P&G cooperated with National Geographic Society to host "It's Our Home" roundtable, inviting business leaders, NGOs and youth representatives to discuss the contribution of global climate action and NbS. It also cooperates with the C+NbS to speak out for companies participating in NbS in the platform's monthly workshops and TED youth sessions.

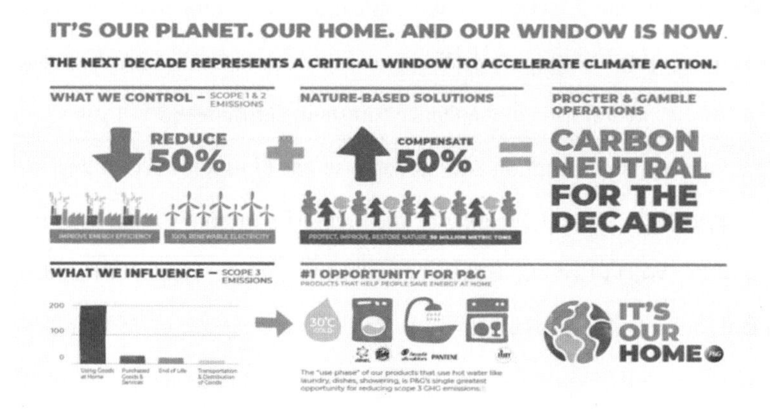

Figure 11-1　Procter & Gamble 2030 Climate Action Roadmap

P&G announces that it will work with CI and WWF to identify and find a series of projects aimed at protecting, improving and restoring ecosystems such as forests, wetlands, grasslands and peatlands. In addition, to offsetting carbon emissions, it aims to bring synergy benefits to the environment and social economy, thereby protecting the natural environment and improving the lives of local residents. P&G is developing a detailed project portfolio and investing in projects across the globe. Projects already identified include:

Philippines Palawan Protection Project with CI—To protect, improve and restore Palawan's mangroves and critical ecosystems. Palawan is the world's fourth most "irreplaceable" area for unique and threatened wildlife.

Atlantic Forest Restoration Planning with WWF—In the Atlantic Forest on Brazil's east coast, laying the groundwork for forest landscape restoration with meaningful impacts on biodiversity, water, food security and other co-benefits for local communities.

Evergreen Alliance with Arbor Day Foundation—Bringing corporations, communities and citizens together to take critical action to preserve the necessities of life affected by climate change - including planting trees to restore areas devastated by wildfires in Northern California and enhance forests in Germany.

To tackle energy consumption and greenhouse gas emissions arised from domestic water use, at the 2020 Annual Meeting of the World Economic Forum in Davos, with the support of the World Economic Forum, the World Business Council for Sustainable Development, and the World Bank 2030 Water Resources Group, P&G took the lead in launching the “50L Home Coalition”, committed to the use of innovation and technology to achieve the goal of limiting household water consumption to 50 liters per person per day (in some countries, per capita household water consumption can reach up to 500 liters per day), and promote innovations in household water systems to help solve urban water crises. The Coalition plans to carry out pilot projects in one to two cities in 2020, and implement innovative pilot projects in three to four cities between 2020 and 2025, to determine the full-scale implementation route and promote it in the globe. If the 50L Home Coalition can be promoted in urban areas of China, it could potentially save 14 billion cubic meters of water every year and reduce 15 million tons of carbon dioxide equivalent greenhouse gas emissions. In addition, P&G is also reducing resource consumption and greenhouse gas emissions through green information disclosure, sustainable supply chain reform, and promotion of recyclable packaging within the company’s production and operations.

11.1.3 Project Information

Case Source: C+NbS Cooperation Platform.

Project Participants: Procter & Gamble, Conservation International (CI), World Wildlife Fund (WWF), Arbor Day Foundation and other partners.

Project Location: Procter & Gamble's global factories as well as nature reserves in the Philippines, Brazil, the United States and other nature regions.

Project Duration: Since 2020.

Project Link: Procter & Gamble official website, https://www.pg.com.cn/environmental-sustainability/.

11.2 Nestlé Net-Zero Roadmap by Promoting Regenerative Agriculture and Afforestation

Case Highlights

As the world's largest food and beverage manufacturer, there is a natural link between its corporate efforts in sustainability and climate change and NbS. Among them, the vast majority (95%) of Nestlé's greenhouse gas emissions come from its supply chain activities. This allows NbS to directly contribute to its long-term carbon neutral strategy. Nestlé's Net-Zero commitment to the globe has set a good example for all companies as to how to achieve carbon neutrality.

11.2.1 Project Background

Founded in 1867, Nestlé is the world's largest food and beverage manufacturer and one of the largest multinational companies, with more than 500 factories worldwide. Originated as a baby food manufacturer in Switzerland, Nestlé is famous for producing chocolate bars and instant coffee. It boasts dozens of products including instant coffee, condensed milk, milk powder, baby food, cheese, chocolate products, candy and instant tea. Foreign markets contribute to 98% of Nestlé's sales, and thus it is called "the most internationalized multinational corporation". Over the past ten years, Nestlé has been committed to reducing greenhouse gas emissions related to food and beverage production and distribution by improving energy efficiency, using cleaner energy sources, and investing in renewable resources. In 2013 and 2012, Nestlé received the highest scores in the Climate Disclosure Leadership Index and Climate Performance Leadership Index of Carbon Disclosure Project (CDP), and has been acclaimed as "world leader" by the agency.[①] In September 2019, Nestlé announced its goal on "net-zero by 2050" and promised to strive for Business Ambition for 1.5°C at UN Climate Action Summit in New York.

11.2.2 Project Overview

In December 2020, Nestlé released the Net-Zero Roadmap. With 2018 as the base year, it commits to the goal of 20%, 50% greenhouse gas reduction and net zero emission by 2025, 2030, and 2050, respectively. In terms of strategies, Nestlé focuses on supporting farmers and suppliers to promote the development of regenerative agriculture and will plant hundreds of millions of trees in the next 10 years to achieve 100% use of renewable electricity by 2025 to meet the goal of net-zero.

Nestlé's 2050 Net-zero commitment follows the Science-based Targets initiative and has worked with external consultants to calculate the carbon footprint in the base year (Figure 11-2)[②]. According to statistics, its total greenhouse gas emissions in the

---

① CDP is an international non-profit organization that provides companies and cities with a system for measuring, disclosing, managing and sharing important environmental information. The CDP index measures efforts of the top 500 companies in the FTSE Global Equity Index in reducing carbon emissions and their performance in transparency of information disclosures.

② The Science-based Targets initiative was started by the CDP Worldwide, the World Resources Institute, the World Wide Fund for Nature, and the United Nations Global Compact. It aims to provide companies with a clear guidance framework for setting emissions reduction targets based on climate science to ensure their efforts in greenhouse gas emission reduction are consistent with the target of controlling global temperature rise to less than 2°C in the *Paris Agreement*.

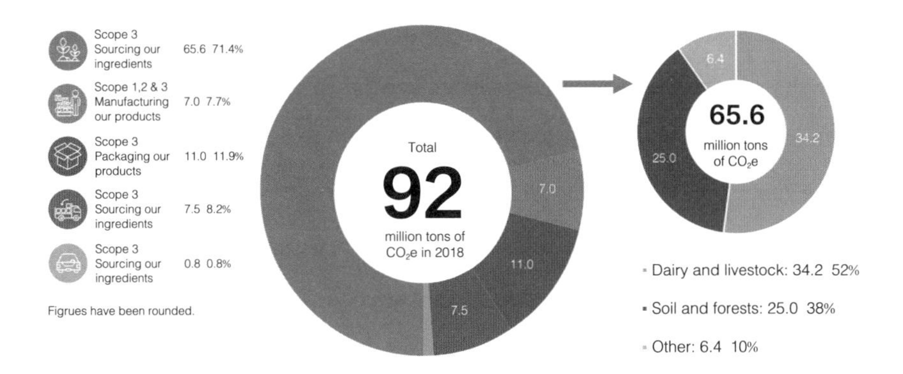

Figure 11-2 Nestlé's In-Scope GHG Emissions by Operation in 2018 (million tons of CO_2e)

base year of 2018 are 113 million tons in carbon dioxide equivalent, including Scope 1 (direct emissions such as emissions from burning coal, natural gas, and fuel used by the company's fleet), Scope 2 (indirect emissions such as electricity production) and Scope 3 (all other indirect emissions including carbon emissions generated during the consumption process of purchased and sold products). Among them, carbon emissions from direct operations (Scope 1 and Scope 2) only account for 5% of its total greenhouse gas emissions, while majority (95%) come from its supply chain activities (Scope 3). Nestlé's 2050 Net-zero commitment includes carbon emissions generated by raw material procurement, product production, product packaging, logistics management, travel and employee commuting, etc. These activities emitted a total of 92 million tons of carbon dioxide equivalent of greenhouse gases in the base year of 2018.[①] Among which, the purchase of raw materials from the upstream of the supply chain emitted 65.6 million tons of carbon dioxide equivalent of greenhouse gas, accounting for 71.4% of

① The current Nestlé's Net Zero commitment does not include the following carbon emissions: 1) Consumers use sold products (12.7 million tons of carbon dioxide equivalent); 2) Purchased services, leased assets, capital goods, investment (8.6 million tons of carbon dioxide equivalent).

the emissions of its commitment. 90% of the raw material procurement emissions come from the dairy industry, livestock and poultry industry, soil and forests, which provides an opportunity for Nestlé to develop NbS to mitigate climate change.

As an important food and beverage producer in the world, nearly two thirds of Nestlé's greenhouse gas emissions come from agriculture. Therefore, the implementation of NbS such as regenerative agriculture and afforestation is an important strategy for Nestlé to achieve net-zero. It plans to increase carbon sinks to remove or offset 13 million tons of carbon dioxide equivalent greenhouse gas emissions from the atmosphere by 2030. The measures include:

①Plant vegetation around water sources and wildlife ecological corridors to capture carbon while improving water quality;

②Plant trees in the pasture to make the forage grass grow better, increase the output of feed, and achieve synergy;

③Use organic fertilizers as much as possible and use local compost (raw materials such as coffee pulp) for organic matter accumulation and improve soil structure and its carbon storage potential;

④Adopt more sustainable agricultural practices and implement agricultural technologies such as no-tillage, crop rotation and cover crops to avoid nitrogen depletion, reduce soil erosion, and control plant diseases, insects and weeds;

⑤Plant trees and shrubs to form a natural protective barrier to protect crops from severe weather and soil erosion;

⑥Carry out agroforestry measures, using shade trees to protect coffee and other crops from heat, while increasing the organic matter in the soil and improving the soil's ability to retain water and store carbon;

⑦Restore forests and peatlands, increase carbon sinks while maintaining groundwater levels and reducing fire risks (Nestlé invested 2.5 million Swiss Francs to protect and restore forest resources in Côte d'Ivoire).

In regenerative agriculture, Nestlé has been cooperating with more than 500,000 farmers and more than 150,000 suppliers to support them in adopting renewable agricultural practices, improving soil health, and maintaining and restoring ecosystem diversity. To achieve this, Nestlé has adopted incentives such as premium prices, increased purchase quantities, and joint investment. It is expected that, by 2030, it will purchase more than 14 million tons of raw materials from renewable agriculture, thereby driving market demand and encouraging farmers to develop renewable agriculture. At the same time, Nestlé is also expanding its "reforestation program" and will plant 20 million trees every year in its raw material sources in the next decade. Through agroforestry, trees will provide more shade for crops to prevent high temperature damage and obtain higher crop yields. Reforestation can also significantly increase carbon sinks and improve biodiversity and soil health. Nestlé is committed to "zero deforestation" by 2022 in its supply chain of major commodities such as palm oil and soybeans.

In terms of its operations, Nestlé expects to achieve 100% renewable electricity in its 800 workplaces in 187 countries by 2025, and strive to achieve low-emission transportation transformation. Nestlé has also taken measures to protect and regenerate water resources, and strive to eliminate food waste in operations. In addition, it proactively carries out technological innovation, promotes new plant-based foods and beverages, and adjusts formulas to make products more environmentally friendly.

11.2.3 Project Information

Case Source: C+NbS Cooperation Platform.

Project Participants: Nestlé and its global factories and suppliers.

Project Location: Nestlé's global factories and sources of raw materials.

Project Duration: Since 2020.

Project Related Link: Nestlé official website, https://www.nestle.com/sites/default/files/2020-12/nestle-net-zero-roadmap-cn.pdf.

11.3 Baofeng Energy's Agricultural-Photovoltaic Integration and Green Hydrogen Innovative Practice

Case Highlights

Baofeng Energy has contributed innovative practices for enterprises to participate in NbS, and provided a development direction that can be promoted on a large scale to achieve carbon neutrality. The Agricultural-Photovoltaic Integration Project fully integrate local resources and local environment. It accelerates the development of new energy to replace fossil energy and has achieved multiple benefits such as employment promotion. The Hydrogen Production and Energy Storage by Solar-powered Electrolysis of Water & Application Demonstration Project have brought enormous green raw materials to the chemical industry. It breaks down the traditional raw material restrictions, creates a larger environmental capacity, and unlocks more potential of nature.

11.3.1 Project Background

As one of the underdeveloped provinces, Ningxia Hui Autonomous Region is located in the inland area of northwest China. It has a temperate continental arid and semi-arid climate and is short in water resources. However, it is rich in coal and other minerals. At the same time, it is one of the regions rich in solar energy. With its terrain high above sea level and the long sunlight duration, it is abundant in photovoltaic resources.

With chemical industry, new energy and modern agriculture as its main business, Baofeng Energy carries out innovative practices to replace fossil energy with new energy and promote carbon neutrality. By 2020, it has built China's largest "methanol, olefin, polyethylene, polypropylene, fine chemicals, new energy" integrated circular industrial cluster and the world's largest centralized photovoltaic power station.

11.3.2 Project Overview

At present, Baofeng Energy can reduce 552,500 tons of carbon dioxide emissions per year. At the same time, it has pledged to achieve 50% carbon emission reduction in a decade, and to achieve corporate carbon neutrality within 20 years. Echoing China's pledge to "become carbon neutral by 2060", it attaches great importance to NbS to solve environmental problems. Based on source governance, it has pioneered in the new path to carbon neutrality of "sustainable ecological management of desertification-new energy industry innovation - electrolysis of water by new energy to produce hydrogen production and modern chemical industry integrated and coordinated development". The new path to carbon neutrality establishes a technical route transforming end treatment into source treatment, and obtains economically feasible solutions through scientific exploration and technological innovation. Among them, the Agricultural-Photovoltaic Integration Project and the Demonstration Project for Hydrogen Production and Energy Storage by Solar-powered Electrolysis of Water are examplar representation.

It utilizes photovoltaic resources of Ningxia Hui Autonomous Region and combines photovoltaic power generation with agricultural production to provide large-scale promotion of the "Agricultural-Photovoltaic Integration Project" (Figure 11-3). Through measures such as vegetation planting and soil improvement, ecological governance has been carried out for 160,000 mu of desertification on the east bank of the Yellow River in Yinchuan, with vegetation coverage rate increased from less than 30% to 85%. In addition, Ningxia's characteristic wolfberry industry has been fully developed into a 10,000 mu of high-quality wolfberry base. At the same time, economic forests, grasses and other crops have been planted according to local conditions. To make full use of land and light resources, in 2016, it built an one GWp[①] solar power project above the wolfberry base, creating a new model for the development of a characteristic industry of "on-board power generation, under-board planting" and "multiple use in one place, agriculture and light integration", and has achieved the "three wins" goal of economic, ecological and social benefits.

Figure 11-3 Picture of Baofeng Photovoltaic + Agricultural Industry

① GWp (Giga Watt peak) is the unit of measurement for solar photovoltaic cells. 1 GWp refers to the peak power generation of a photovoltaic project is one gigawatt.

Technical advantages: Frontier technologies such as Huawei's smart photovoltaic solutions, the most efficient monocrystalline silicon modules, and the world's leading inclination flat single-axis tracking technology are used in all solar power generation projects. The solar power generation capacity of the world's largest centralized photovoltaic power station is 20% higher than that of the traditional solar power generation facilities, with a high conversion rate. The project employs a new model by integrating intelligent monitoring and management platform, with drone inspection to timely detect failures, troubleshooting, and ensure the safe, efficient and stable operation of those photovoltaic power plants.

Environmental benefits: The project can directly supply "green power" instead of thermal power to enterprises and meet some social consumption needs. It saves 557,000 tons of standard coal, reduces carbon dioxide emissions by 1.693 million tons, sulfur dioxide emissions by 51,000 tons, nitrogen oxide emissions by 26,000 tons, and dust emissions by 462,000 tons, equivalent to planting approximately 90 million trees. It increases the environmental capacity by about 2.23 million tons per year for local traditional energy consumption, which is conducive to energy consumption and emission reduction, and accelerates the development of new energy to replace fossil energy. At the same time, the shading of photovoltaic panels reduces evaporation by about 70%, increasing the carbon sink capacity of the local agricultural ecosystem.

Social significance: The Agricultural-Photovoltaic Integration model translates ecological benefits into economic benefits and industrial advantages. The cleaning and maintenance of photovoltaic power generation components and wolfberry harvesting require labor service of 80,000 person annually, which can contribute to each household by more than ¥40,000 Yuan. Hence these projects consolidate the gains of poverty allieviation, and promote the overall revitalization of the surrounding villages. The combination of the wolfberry industry and new energy projects (Figure 11-4) has accelerated the development of new technologies and new industries, spawned new forms of employment and wealth creation, bringing additional value to the society.

Figure 11-4 Ningxia Baofeng Energy's Wolfberry Industry Poverty Allieviation Project

While engaging in traditional energy transformation and ecological restoration, Baofeng Energy also spearheads in science and technology innovation. It has developed the "Hydrogen Production and Energy Storage by Solar-powered Electrolysis of Water & Application Demonstration Project" with "green electricity for green hydrogen" as its core (Figure 11-5). In the global shift to clean and low-carbon energy, as a renewable energy source that is flexible and efficient with wide sources and zero-emissions, hydrogen is regarded as the most ideal and most promising clean energy, which can play an important role in reducing greenhouse gas emissions, dealing with climate change and realizing low-carbon industry. Baofeng Energy adopts the world's leading technology, with the optimal combination of "solar power generation + hydrogen production by electrolysis of water". It uses solar energy to produce green power, and then uses it to produce "green hydrogen" and "green oxygen" through water electrolysis.

It replaces the traditional fossil fuels and directly supply green energy to the chemical industry to produce hundreds of high-end chemical products such as polyethylene and polypropylene, which forms a complete "carbon neutral" industrial chain. The "National Comprehensive Demonstration Project for Hydrogen Production by Solar-powered Electrolysis of Water" is currently known as the water electrolysis hydrogen production project with the world's largest single plant and maximum capacity by the single unit. Upon its completion, it can produce 200 million standard cubic meters of "green hydrogen" and 100 million standard cubic meters of "green oxygen" each year.

Figure 11-5 Bird's Eye View of Facilities for Hydrogen Production by Electrolysis of Water

Technological innovation: Hydrogen production by solar-powered electrolysis of water is a process in which solar energy is used as the primary energy and water is used as the medium to produce hydrogen as the secondary energy. The hydrogen produced under the optimal combination of "solar power generation + hydrogen production by electrolysis of water" is truly "green hydrogen". The project introduced a single set of electrolytic cells, gasification separators, and hydrogen purification systems with a production capacity of 1,000 standard cubic meters per hour, which has reached the domestic advanced level with its low energy consumption and high conversion rate. It ensures the quality of hydrogen in terms of technology, process and use, and the purity of hydrogen are in line with the international standard for high purity of hydrogen (99.999%), meeting the needs of fields with challenging requirements. The project is a showcase for the promotion and development of clean energy.

Cost advantage: The project improves the conversion rate through scientific and technological innovation. It realizes the long-term high-load hydrogen production and improves the utilization rate of the equipment, while effectively reducing the overall cost of hydrogen production. The power consumption of hydrogen production system is 4.5-5kW·h/standard cubic meter of hydrogen, and the cost of green hydrogen can be reduced to ¥0.7/standard square, which realizes the effective conversion of renewable energy to high-end chemical new materials.

Environmental benefits: Upon its completion, the project can reduce coal resource consumption by 317,500 tons and carbon dioxide emissions by approximately 552,500 tons per year. In addition, wind and solar power curtailment[①] are currently important factors restricting the development of renewable energy in China, especially in Northwest China. The implementation of hydrogen production and energy storage by solar-powered electrolysis can effectively absorb the curtailed solar energy in this region with significant environmental benefits.

11.3.3 Project Information

Case Source: C+NbS Cooperation Platform.

---

① Wind and solar power curtailment refers to the forced abandonment of wind, water and solar energy due to some reasons by stopping the corresponding generator set or reducing its power generation. It can also be that the power generation of the photovoltaic power station is greater than the maximum transmission power of the power system + the load consumption. The main reasons of power curtailment lie in three system elements, i.e. power supply, power grid and the load.

Project Participants: Ningxia Baofeng Energy Group Co., Ltd.

Project Location: Baofeng industrial base in Ningxia Hui Autonomous Region, China.

Project Duration: Baofeng Agricultural-Photovoltaic Integration Project (since 2013); Hydrogen Production and Energy Storage by Solar-powered Electrolysis of Water & Application Demonstration Project (since 2020).

11.4 Case Summary

As a global consumer goods giant with factories and branches in more than 80 countries around the world, Procter & Gamble operates more than 300 brands in home care, hairdressing and beauty, baby and home care, food and beverages, and its products are marketed in more than 160 countries and regions. Its 2030 climate commitment aims to promote P&G to become carbon neutral in global operations by 2030 through internal reforms and measures to protect, improve and restore the ecological environment. It disseminates carbon neutrality and NbS through platforms such as National Geographic and C+NbS, and provides solutions for enterprises to participate in NbS. Through cooperation with the Conservation International and the World Wildlife Fund, it has invested in various ecological protection projects in the Philippines, Brazil, the United States, Germany and other places. It participates in advocating global nature-based climate action through corporate commitments, platform collaboration, and cooperative investment.

Nestlé is the world's largest food manufacturer and one of the largest multinational companies. It is famous for chocolate products, instant coffee, baby food, milk powder and other products. Given its product types and raw materials, Nestlé's greenhouse gas emissions mainly come from agricultural production. To strengthen climate action to mitigate climate change, Nestlé announced the 2050 Net Zero commitment in September 2019, and published its Net Zero Roadmap in December 2020. Scientific tools are used to evaluate Nestlé's greenhouse gas emissions, with the focus on animal husbandry, agriculture, forestry and other sectors. Furthermore, it carries out other nature-based emission reduction actions such as regenerative agriculture and reforestation. Nestlé provides a valuable reference for companies to scientifically assess their carbon footprint, improve information disclosure and transparency, and formulate a clear roadmap to be carbon neutral.

Ningxia Baofeng Energy's innovative practice based on fossil energy substitution and ecological restoration is a showcase of the ecological civilization of "adhering to the harmonious coexistence of man and nature": Relying on the existing natural resources, its "Agricultural-Photovoltaic Integration Project" promotes the replacement of fossil energy by new energy, which not only adds impetus to resource conservation and environment protection, but also provides employment opportunities for locals, which carries great social significance. Its Demonstration Project for Hydrogen Production and Energy Storage by Solar-powered Electrolysis of Water uses natural resources, sunlight and water, which can be easily obtained to produce hydrogen to achieve energy storage. It conforms to the future development trend of the international and domestic energy markets. Furthermore, it will achieve cost reduction, efficiency improvement, energy saving and emission reduction, and lead the high-quality development of the green hydrogen industry. It carries great practical significance in green hydrogen operation, demonstration and promotion.

Bibliography

- Acharya P. "New Pathways for NbS to Realise and Achieve SDGs and Post 2015 Targets: Transformative Approaches in Resilience Building," in Dhyani, Shalini, eds, "Nature-based Solutions for Resilient Ecosystems and Societies," Springer, Singapore, 2020: 435-455.
- Aryn Baker, Mbar Toubab, Senegal. Can a 4,815-Mile Wall of Trees Help Curb Climate Change in Africa [EB]. Time, 2019.
- Bhattarai B. Community forest and forest management in Nepal [J]. American Journal of Environmental Protection, 2016, 4(3): 79-91.
- BSDC. Valuing the SDG Prize in Food and Agriculture [R]. 2016.
- Cohen-Shacham E, Walters G, Janzen C, et al. Nature-based solutions to address global societal challenges [R]. IUCN: Gland, Switzerland, 2016.
- Dohong A. Strategy and Approach in Restoring Degraded Peatlands in Indonesia [A]// The Institute of Foresters of Australia (IFA) Biennial Conference [C]. Tropical Forestry: Innovation and Change in the Asia Pacific Region, 2017.
- European Union. Towards an EU research and innovation policy agenda for nature-based solutions & renaturing cities [R]. 2015.
- FOLU. Growing Better Global Report [R]. 2019.
- Galab S, Prudhvikar Reddy P, Sree Rama Raju D, et al. Impact Assessment of Zero Budget Natural Farming in Andhra Pradesh – Kharif 2018-19: A comprehensive approach using crop cutting experiments [R]. Telangana, India: Centre for Economic and Social Studies, 2019.
- Goffner D, Sinare H, Gordon L J. The Great Green Wall for the Sahara and the Sahel Initiative as an opportunity to enhance resilience in Sahelian landscapes and livelihoods[J]. Regional Environmental Change, 2019, 19(5):1417-1428.
- Griscom B W, Adams J, Ellis, P W, et al. Natural climate solutions[J]. Proceedings of the National Academy of Sciences, 2017, 114(44): 11645-11650.

- IUCN. No time to lose: make full use of nature-based solutions in the post-2012 climate change regime [R]. Gland, Switzerland: IUCN, 2019.
- IUCN. Global Standard for Nature-based Solutions: A user-friendly framework for the verification, design and scaling up of NbS [R]. Gland, Switzerland: IUCN, 2020.
- Lieth H, Whittaker R H, et al. Primary productivity of the biosphere [J]. Springer Science & Business Media, 2012(14): 203-215.
- Ministry of Forest and Soil Conservation. Persistence and change: review of 30 years of community forestry in Nepal [R]. Kathmandu: Ministry of Forest and Soil Conservation, 2013.
- Nellemann C, Corcoran E, et al. Blue carbon: the role of healthy oceans in binding carbon: a rapid response assessment [R]. UNEP, 2009.
- Pardo R Back to the future: Nepal's new forestry legislation [J]. Journal of Forestry, 1993, 91 (6): 22-26.
- Pokharel R K, Rayamajhi S, Tiwari K R. Nepal's community forestry: need of better governance [R]. Global Perspectives on Sustainable Forest Management, 2012.
- Rythu Sadhikara Samstha. Andhra Pradesh Zero Budget Natural Farming (APZBNF): A systemwide transformational programme [R]. Andhra Pradesh: Department of Agriculture, Government of Andhra Pradesh, 2019.
- Sunderland T, Abanda F, de Camino R V, et al. Sustainable forestry and food security and nutrition [R]. CFS-HLPE/FAO, Technical Report 11, 2013.
- World Bank. Nepal Overview [DB]. Washington, DC: World Bank, 2014.
- WWF. Concept Note: Heritage Colombia (HECO): Maximizing the contributions of sustainably managed landscapes in Colombia for achievement of climate goals [R]. Green Climate Fund, 2019.
- Ye Q. Ways of training individual ecological civilization under nature social conditions[R]. Scientific Communism, 1984.
- Feng Yijia, Zhao Jing, Wang Xiangrong. Exploration of the construction and driving force of the Greenprint as a medium for urban development in the United States [J]. Landscape Architecture, 2015(9): 62-69.

- He Qingtang. The Construction of Ecological Civilization and Nature-based Solutions [J]. China Forestry Industry, 2019(3): 77-80.
- Compiled by UNESCO. The United Nations World Water Development Report: Nature-based Solutions for Water [M]. Translated and Edited by China Water Resources Strategy Research Association (Global Water Partnership China). Beijing: China Water & Power Press, 2018.
- Liao Maolin. Theoretical Cognition and Practical Path of Constructing Global Ecological Civilization [J]. Enterprise Economy, 2020, 39(7): 131-137.
- Li Pengyu. Planting trillions of trees: addressing climate change and contributing to China's carbon neutrality [J]. China Sustainability Tribune, 2020, 19(10): 45-46.
- Liu Jielong. On Targeted Poverty Alleviation in the Era of Green Development [J]. Journal of the Party School of Guizhou Province, 2016(5): 92-98.
- Liu Jing. Research on the Construction of Socialist Ecological Civilization with Chinese Characteristics [D]. Beijing: Party School of the Central Committee of the Communist Party of China, 2011.
- Lu Feng. On Nature-based Solutions (NbS) and Ecological Civilization [J]. Journal of Fujian Normal University (Philosophy and Social Sciences Edition), 2020(5): 44-53, 169.
- Osieyo, Marion A. Nature In All Goals 2020 [R]. WWF, 2020.
- Raymond, et al., A Framework for Assessing and Implementing The Co-Benefits Of Nature-Based Solutions In Urban Areas[J]. Environmental Science & Policy, 2017, 77: 15-24.
- Wang Yinglin, Zhao Zhicong. Research on the relationship between natural reserve and ecological conservation red line [J]. Chinese Landscape Architecture, 2020, 36(8): 20-24.
- Zou Haigui. Intergenerational Justice and Concern for the Interests of Social Vulnerable Groups: Analysis Based on the Moral Legitimacy of the Modern Social Assistance (Guarantee) System [J]. Journal of Central South University of Forestry & Technology: Social Sciences, 2012(1): 43-46.

Appendix 1: IUCN Global Standard for Nature-based Solutions

The *IUCN Global Standard for Nature-based Solutions* (hereinafter referred to as the *Global Standard*) issued by the International Union for Conservation of Nature (IUCN) in July 2020 provided great reference for the selection of the cases. The *Global Standard* aims to provide a user-friendly framework for the verification, design and scaling up of NbS. IUCN encourages national governments, city, local governments, planners, companies, donors and financial institutions including development bank and non-profit organizations to apply the *Global Standard*.

The *Global Standard* consists of eight criterions and 28 indicators, displayed in the following table. The eight criteria are interrelated, and jointly provide clear evaluation criteria for the comprehensive ability of NbS to play a role in the environment, economy and society, transparency in the implementation process, and supervision and safeguard measures, the figure below shows the details. The Standard is intended to be a simple yet robust hands-on tool that enables the translation of the NbS concept into targeted actions for implementation, reinforcing best practice, and addressing and correcting shortfalls.

Table 6-1 Eight Criteria and 28 Indicators of the *Global Standard*

Criteria		Indicators	
1	NbS effectively address societal challenges	1.1	The most pressing societal challenge(s) for rights-holders and beneficiaries are prioritized
		1.2	The societal challenge(s) addressed are clearly understood and documented
		1.3	Human well-being outcomes arising from the NbS are identified, benchmarked and periodically assessed
2	Design of NbS is informed by scale	2.1	The design of the NbS recognizes and responds to interactions between the economy, society and ecosystems
		2.2	The design of the NbS is integrated with other complementary interventions and seeks synergies across sectors
		2.3	The design of the NbS incorporates risk identification and risk management beyond the intervention site
3	NbS result in a net gain to biodiversity and ecosystem integrity	3.1	The NbS actions directly respond to evidence-based assessment of the current state of the ecosystem and prevailing drivers of degradation and loss
		3.2	Clear and measurable biodiversity conservation outcomes are identified, benchmarked and periodically assessed
		3.3	Monitoring and periodic assessments of unintended adverse consequences on nature arising from NbS
		3.4	Opportunities to enhance ecosystem integrity and connectivity are identified and incorporated into the NbS strategy
4	NbS are economically viable	4.1	The direct and indirect benefits and costs associated with the NbS, who pays and who benefits, are identified and documented
		4.2	A cost-effectiveness study is provided to support the choice of NbS including the likely impact of any relevant regulations and subsidies
		4.3	The effectiveness of the NbS design is justified against available alternative solutions, taking into account any associated externalities
		4.4	The design of the NbS should consider resourcing options such as market-based, public sector and voluntary commitments, and take actions to support regulatory compliance

(continued Table)

Criteria		Indicators	
5	NbS are based on inclusive, transparent and empowering governance processes	5.1	A defined and fully agreed upon feedback and grievance resolution mechanism is available to all stakeholders before an NbS intervention is initiated
		5.2	Participation is based on mutual respect and equality, regardless of gender, age or social status, and upholds "the right of Indigenous Peoples to Free, Prior and Informed Consent" (FPIC)
		5.3	Stakeholders who are directly and indirectly affected by the NbS have been identified and involved in all processes of the NbS intervention
		5.4	Decision-making processes document and respond to the rights and interests of all participating and affected stakeholders
		5.5	Where the scale of the NbS extends beyond jurisdictional boundaries, mechanisms are established to enable joint decision-making of the stakeholders in the affected jurisdictions
6	NbS equitably balance achievement of their primary goal(s) and the continued provision of multiple benefits	6.1	The potential costs and benefits of associated trade-offs of the NbS intervention are explicitly acknowledged and inform safeguards and any appropriate corrective actions
		6.2	The rights and responsibilities to land and resources of different stakeholders are acknowledged and respected
		6.3	The established safeguards are periodically reviewed to ensure that mutually-agreed trade-off limits are respected and do not destabilize the entire NbS
7	NbS are managed adaptively, based on evidence	7.1	Strategy is established and used as a basis for regular monitoring and evaluation of the intervention
		7.2	A monitoring and evaluation plan is developed and implemented throughout the intervention lifecycle
		7.3	A framework for iterative learning that enables adaptive management is applied throughout the intervention lifecycle
8	NbS are sustainable and mainstreamed within an appropriate jurisdictional context	8.1	The design, implementation and lessons of the NbS are shared to trigger transformative change
		8.2	The NbS enhances facilitating policy and regulation frameworks to support its uptake and mainstreaming
		8.3	The NbS contributes to national and global targets for human well-being, climate change, biodiversity and human rights, including the *United Nations Declaration on the Rights of Indigenous Peoples* (UNDRIP)

Conceptual Schematic Diagram of NbS
(Source: IUCN)

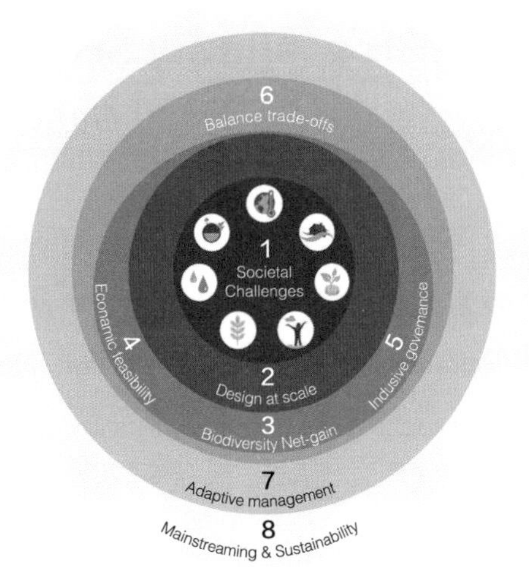

Interrelationship of Eight Criteria of *Global Standard*
(Source: IUCN)

Appendix 2: Primary Screening Indicator Scores of Selected Cases

The table below shows the performance of each indicator of the 28 final selected cases in eight categories in the preliminary screening stage. For the detailed description of each indicator and scoring standard, please refer to Section 3.2. It is worth noting that due to the different functions and characteristics of different types of NbS cases, the comparison and analysis of the scores of various indicators is limited to the horizontal comparison within the unified case category.

Primary Screening Indicator Scores of Selected Cases

Case Category	Case Name	Score for Mitigating and Adapting Climate Change	Score for Preserving Biodiversity	Score for Addressing Multiple Societal Challenges	Score for Synergy of Multiple Goals	Score for Promoting Local Economic Development	Score for Caring Disadvantaged Populations	Total
Forestry	National Forest Conservation Program of Colombia	3	3	3	3	2	2	16
	Community Forestry Campaign in Nepal	3	3	3	3	3	3	18
	"Trillion Trees" Initiative	3	3	1	3	1	1	12
Grassland	Great Green Wall in Africa	3	3	3	3	3	3	18
	Desertification Control of the Mu Us Desert in China	2	3	3	3	3	3	17

(continued Table)

Case Category	Case Name	Score for Mitigating and Adapting Climate Change	Score for Preserving Biodiversity	Score for Addressing Multiple Societal Challenges	Score for Synergy of Multiple Goals	Score for Promoting Local Economic Development	Score for Caring Disadvantaged Populations	Total
Agricul-ture	Zero Budget Natural Farming (ZBNF) in India	3	1	3	3	3	3	16
	Climate-Smart Staple Crop Production in China	3	1	3	2	3	1	13
	Practice of "Three Goods Agriculture" in Hangzhou, China	1	1	3	3	3	1	12
Wetland	Peatland Conservation in Indonesia	3	1	2	2	2	1	11
	Ecological Restoration of Yellow Sea Wetlands in Yancheng, China	1	3	1	2	2	0	9
	Dongying Wetland City Construction in China	1	2	2	2	2	0	9
	China-ASEAN Mangrove Conservation and Restoration	2	2	3	3	2	2	14
	Ocean Conservation in Small Island States in South Pacific	1	1	2	2	3	1	10
City	Milan - NbS for Urban Regeneration	2	1	3	3	1	0	10
	London - NbS for a Leading Sustainable City	3	3	3	3	1	2	15
	Chengdu-Park City Construction	3	0	3	3	3	0	12
	Rotterdam-Climate-resilient Infrastructure	3	1	3	3	0	0	10
	Lower Manhattan Coastal Resiliency (LMCR) in USA	3	1	3	3	0	0	10
	San Francisco Bay Area Greenprint	3	1	3	3	0	0	10

(continued Table)

Case Category	Case Name	Score for Mitigating and Adapting Climate Change	Score for Preserving Biodiversity	Score for Addressing Multiple Societal Challenges	Score for Synergy of Multiple Goals	Score for Promoting Local Economic Development	Score for Caring Disadvantaged Populations	Total
Country	Costa Rica National Ecological Protection Plan	3	3	2	3	2	1	14
	China Ecological Conservation Red Line (ECRL)	3	2	2	3	1	1	12
	Brazil's Oasis Project for the Protection of Water Sources	3	1	3	3	3	2	15
Platform & Initiative	Nature4Climate	3	2	1	3	1	1	11
	"Countdown" Global Initiative	3	2	1	3	0	1	10
	C+NbS Cooperation Platform	3	3	3	2	1	1	13
Corporation	P&G Embraces by Promoting Regenerative Agriculture and Afforestation Solutions to Make Operations Carbon Neutral	3	2	2	3	3	2	15
	Nestlé Net-Zero Roadmap	3	1	2	3	3	3	15
	Baofeng Energy's Agricultural-Photovoltaic Integration and Green Hydrogen Innovative Practice	3	1	2	3	3	3	15

Appendix 3: Advanced Evaluation Dimension Performance of Selected Cases

On the basis of preliminary screening, this book summarizes and scores the highlights of each case based on the three advanced evaluation dimensions of governance model, financial and market mechanism, and participation of diverse actors and women (see Section 3.3). The table below shows the results of this analysis.

Highlight Analysis of Selected Cases based on Advanced Evaluation Dimensions

Case Category	Case Name	Governance Models	Finance & Market Mechanisms	Diversity and Inclusion
Forestry	National Forest Conservation Program of Colombia	Top-down approach; government-enterprise cooperation	Permanent project financing; revocable transition fund; ecosystem service payment; ecological compensation mechanism; carbon tax	Ensure the participation of diverse actors and women in national and local governments, non-governmental organizations, enterprises, universities and communities; pay attention to issues of climate equity and gender equality
	Community Forestry Campaign in Nepal	Bottom-up Community Forestry User Groups	Government departments, local governments and non-governmental organizations provide funding; forest user groups generate income	Election mechanisms to ensure women's participation; encourage the participation of poverish communities, aboriginals and Dalits
	"Trillion Trees" Initiative	A platform that integrates all projects and provides funding and political support	—	Participation of global enterprises, international organizations, and governments, organizations and individuals from 56 countries

(continued Table)

Case Category	Case Name	Governance Models	Finance & Market Mechanisms	Diversity and Inclusion
Grassland	Great Green Wall in Africa	Organic combination of top-down and bottom-up mode	—	Participation of African countries, funders (World Bank, European Union, United Nations, etc.), and communities
	Desertification Control of the Mu Us Desert in China	Top-down approach; accurately implement policies and innovate sand prevention and control models; prevention-use combination	—	The Central Government of China, the governments of Inner Mongolia Autonomous Region, Shaanxi Province and Ningxia Hui Autonomous Region and the local residents
Agricul-ture	Zero Budget Natural Farming (ZBNF) in India	From top to bottom, promote renewable agricultural measures, large household training and capacity building	—	Governments, NGOs, research institutes, local farmers, female farmers
	Climate-Smart Staple Crop Production in China	From top to bottom, transform and redefine the development direction of the agricultural system	—	Cooperation between the Ministry of Agriculture and Rural Affairs of China, the World Bank, and the Global Environment Facility
	Practice of "Three Goods Agriculture" in Hangzhou, China	Promote from green consumption, develop innovative business models, and explore long-term mechanisms	Charitable trusts have added more diversified and influential investment channels	The government, non-gov ernmental organizations, enterprises and farmers
Wetland	Peatland Conservation in Indonesia	Community partnership, unified early warning system	—	Community partnership mobilizing communities
	Ecological Restoration of Yellow Sea Wetlands in Yancheng, China	Overall coordination from top to bottom	—	—

(continued Table)

Case Category	Case Name	Governance Models	Finance & Market Mechanisms	Diversity and Inclusion
	Dongying Wetland City Construction in China	Establish and improve the indicator system for development	—	—
	China-ASEAN Mangrove Conservation and Restoration	Community agreement protection mechanism, unified restoration guide	—	The community agreement protection system guarantees the joint participation of the government, civil society and enterprises
	Ocean Conservation in Small Island States in South Pacific	Set up a committee to coordinate operations	—	NGOs, local government and community cooperation
City	Milan - NbS for Urban Regeneration	—	—	"Allotment Gardens" and other community participation projects
	London - NbS for a Leading Sustainable City	Indepth integration of top-down and bottom-up modes	Developed financing + co-financing	Community skills training
	Chengdu- Park City Construction	—	Facility leasing, and social investment	Residents' participation in "Low Carbon Tianfu"
	Rotterdam- Climate-resilient Infrastructure	Government conducts adaptive management for the city	—	—
	Lower Manhattan Coastal Resiliency (LMCR) in USA	—	Government financing	—
	San Francisco Bay Area Greenprint	NGO-led; establish a special committee	NGO fundraising + farmland trust	—
Country	Costa Rica National Ecological Protection Plan	Government-enterprise cooperation	Trust funds, payment for ecosystem services	Clarify appropriation for performance target
	China Ecological Conservation Red Line (ECRL)	Adjust measures to local conditions; big data usage	Overall coordination from top to bottom	—
	Brazil's Oasis Project for the Protection of Water Sources	—	Payment for environmental services	—

(continued Table)

Case Category	Case Name	Governance Models	Finance & Market Mechanisms	Diversity and Inclusion
Platform & Initiative	Nature4Climate	High-level communication mechanism	Nature4Climate strategic platform	Emphasize youth mobilization
	"Countdown" Global Initiative	Local specialized lectures and seminars	Rely on the TED platform	Lectures and seminars are available online
	C+NbS Cooperation Platform	Cross-border workshop	Relying on the resources of Tsinghua Universing	Simultaneous webcast of forums
Corporation	P&G Embraces Natural Climate Solutions to Make Operations Carbon Neutral	Roundtable discussion, and joining hands with NGOs	Full coverage of the industry chain	Take the lead in encouraging multi-stakeholder participation
	Nestlé Net-Zero Roadmap	Evaluation by external institutions, and collaboration across the entire production chain	Premium prices, increased purchase quantities and joint investment for renewable agriculture	Encourages farmers to change their planting methods by surrendering part of the profits
	Baofeng Energy's Agricultural-Photovoltaic Integrationand Green Hydrogen Innovative Practice	Technology supports the development of hydrogen energy	—	Employment and poverty alleviation

Appendix 4: Abbreviations

Abbreviations

Abbreviation	Full Name in English
AU	African Union
BSDC	Business and Sustainable Development Commission
CBD	*Convention on Biological Diversity*
CFUGs	Community Forestry User Graups
CI	Conservation International
COP	Conference of Parties
EU	The European Union
FAO	Food and Agriculture Organization of the United Nations
FOLU	Food and Land Use Coalition
GCAI	Global Climate Action Initiative
GEF	Global Environment Fund
GEI	Global Environmental Institute
GIS	Geographic Information System
IADB	Inter-American Developments Bank
ICCSD	Institute of Climate Change and Sustainable Development, Tsinghua University
IUCN	International Union for Conservation of Nature
LULUCF	Land Use, Land Use Change and Forestry
N4C	Nature4Climate
NbS	Nature-based Solutions
NDCs	Nationally Determined Contributions
NGO	Non-Governmental Organization
PES	Payment for Ecosystem Services
PFP	Project Finance for Permanence

(continued Table)

Abbreviation	Full Name in English
RAMSAR	Convention on Wetlands of International Importance Especially as Waterfowl Habitat
SBTi	Science Based Targets Initiative
SDGs	Sustainable Development Goals
SIFF	Sustainable India Fimance Facility
TNC	The Nature Conservancy
TSP	Total Suspended Particulate
UN	The United Nation
UNCCD	*United Nations Convention to Combat Desertification*
UNDP	United Nations Development Programme
UNEP	United Nations Environment Programme
UNESCO	United Nations Educational, Scientific and Cultural Organization
UNFCCC	*United Nations Framework Convention on Climate Change*
UNOPS	United Nations Office for Project Services
WB	World Bank
WCC	World Conservation Congress
WCS	Wildlife Conservation Society
WEF	World Economic Forum
WRI	World Resources Institute
WWF	World Wildlife Fund
ZBNF	Zero Budget Natural Farming

清华大学气候变化与可持续发展研究院出品

李　政　王彬彬 / 主编

基于自然的解决方案全球实践

——碳中和视角下的协同路径探索

中国环境出版集团・北京

图书在版编目（CIP）数据

基于自然的解决方案全球实践 ： 碳中和视角下的协同路径探索 / 李政，王彬彬主编. -- 北京 ： 中国环境出版集团，2022.8
ISBN 978-7-5111-5151-3

Ⅰ. ①基… Ⅱ. ①李… ②王… Ⅲ. ①生态环境保护－案例－世界 Ⅳ. ①X171.4

中国版本图书馆CIP数据核字(2022)第080240号

出 版 人　武德凯
责任编辑　丁莞歆
责任校对　薄军霞
装帧设计　金　山

出版发行　中国环境出版集团
（100062　北京市东城区广渠门内大街16号）
网　　址：http：//www.cesp.com.cn
电子邮箱：bjgl@cesp.com.cn
联系电话：010-67112765（编辑管理部）
010-67147349（第四分社）
发行热线：010-67125803，010-67113405（传真）
印装质量热线：010-67113404
印　　刷　北京中科印刷有限公司
经　　销　各地新华书店
版　　次　2022年8月第1版
印　　次　2022年8月第1次印刷
开　　本　787×960　1/16
印　　张　11.75
字　　数　180千字
定　　价　188.00元（全两册）

编写委员会

总指导：解振华

总顾问：何建坤

主　编：李　政　王彬彬

成　员：张佳萱　霍　莉　常　江　杨　秀　董文娟
李　莹　李伟起　王裕祥　张尚辰　杜孟瑶
顾　戬　吕若平　汪丽君　林泽文　谭琭玥
胥辛未　宗贝贝　宫炳含　宋迈思　马海晶
付亚男　林　鹿　周靖蕾　洪　毅

序　言

自然既是气候变化影响的“受害者”，又是应对气候变化的“重要阵地”。基于自然的解决方案强调尊重自然规律，通过造林、加强农田管理、保护湿地与海洋等生态保护和生态修复、改善生态管理等实施路径，提升大自然的服务功能，控制温室气体排放，提高应对气候风险的能力，同时还能增加碳汇，是一种减缓和适应气候变化、提高气候韧性的综合手段。

2019年9月，在纽约召开的联合国气候行动峰会上，联合国邀请中国和新西兰共同牵头提出基于自然的解决方案。基于自然的解决方案要求人们更为系统地理解“人与自然和谐共生”的关系，更好地认识人类赖以生存的地球家园的生态价值，提倡依靠自然的力量应对气候风险，构建温室气体低排放和气候韧性社会，打造永续发展的“人与自然生命共同体”。过去一年多，我国与各方一道通过规划、政策、技术、投入等手段加强基于自然的解决方案的实施，积极推动该领域的国内行动和国际合作。

2020年9月，习近平主席在联合国大会发言中明确提出：“中国将提高国家自主贡献力度，采取更加有力的政策和措施，二氧化碳排放力争于2030年前达到峰值，努力争取2060年前实现碳中和。”这是我国统筹国内国际两个大局作出的重大战略决策，进一步彰显了中国深入贯彻习近平生态文明思想、坚定走绿色低碳循环发展道路的战略定力，以及坚定支持多边主义、积极推动构建人类命运共同体的大国担当。要实现碳中和，加速转型创新以实现减排是其中很重要的一个方面，通过基于自然的解决方案增加碳汇是另一个重要方面。

在碳中和的长期目标下，基于自然的解决方案在应对气候变化中的作用将越来越显著。在深度减排的情景中，技术的减排潜力不断收窄，减排成本不断增加。基于自然的解决方案能够在农业、林业、海洋和湿地等各领域依靠自然的生态功能增加碳汇，以抵消工业、交通和难减排部门的碳排放，最后实现碳中和的目标。因此，加强基于自然的解决方案与当前各项减排技术与政策的融合，将对我国更新、强化国家自主贡献，实现碳达峰、碳中和目标发挥积极的重要作用。

席卷全球的新冠肺炎疫情提醒我们，应对全球性挑战，不能就发展谈发展、就生多（生物多样性）谈生多、就气候谈气候、就环境谈环境，而是要将保护生物多样性、应对气候变化、改善生态环境质量的政策行动融入发展经济、改善民生、促进就业、保障健康、维护国家生态安全中，实现协同增效，促进高质量、可持续发展。从根本上看，就是要转变传统的生产方式、生活方式和消费模式，推动经济社会转型和技术、制度、机制创新，走绿色低碳循环的发展道路。基于自然的解决方案是一个协同增效的典范，将产生修复生态、保护生物多样性、增加碳汇、适应气候变化、促进经济发展、提高健康水平、保障民众（特别是妇女儿童）权益等一系列积极效果。

本书精选了全球范围内28个应对气候变化的基于自然的解决方案典型案例，有助于社会各界学习理解基于自然的解决方案的先进理念和经验，借鉴优秀实践，为我国推进生态文明建设提供创新思路；同时，书中收录的我国优秀实践案例也为全球推进基于自然的解决方案提供了中国智慧和中国方案。

2021—2030年是全球生态保护和气候治理进程的关键时期，全球保护生物多样性和应对气候变化两个多边进程呈现加强协同、相互促进的趋势。期待本书能起到抛砖引玉的作用，推进跨领域对话与合作，解锁更多自然潜力，激发创新动能，为推进全球生态文明建设、共建人与自然生命共同体、保护人类美好未来作出积极贡献。

中国气候变化事务特使

解振华

2021年10月6日

前　言

基于自然的解决方案（NbS）最早是在世界银行于2008年发布的《生物多样性、气候变化和适应：世界银行投资中基于自然的解决方案》中提出的，强调生物多样性保护对适应与减缓气候变化的重要性。2009年，世界自然保护联盟（IUCN）向《联合国气候变化框架公约》第15次缔约方大会（UNFCCC COP15）提交了建议报告，强调NbS对应对气候变化等一系列社会挑战的重要作用（IUCN，2009）。

NbS在区域层面的深化从欧盟开始。2014年，欧盟启动“地平线2020”议程，强调利用科技促进就业与经济增长。2015年，欧盟将NbS纳入该议程，以便更大规模地开展研究和试点。同年，欧盟发布《基于自然的解决方案和自然化城市》，将NbS定义为“受到自然启发和支撑的解决方案，在具有成本效益的同时，兼具环境、社会和经济效益，并有助于建立韧性的社会生态系统”，其本质在于强调把NbS视作将当前所面临的挑战转化为创新的机遇、将自然资本转化为绿色经济增长的源泉（European Union，2015）。

在2016年的世界自然保护大会（WCC）上，世界自然保护联盟通过了NbS的明确定义，即“通过保护、可持续管理和修复自然或人工生态系统，从而有效和适应性地应对社会挑战，并为人类福祉和生物多样性带来益处的行动”。同年，该机构发布研究报告，系统梳理了NbS的概念、内涵、社会价值、实施方案与实践案例（Cohen-Shacham E et al.，2016）。

为了更科学地预估NbS在减缓气候变化领域的潜力，大自然保护协会（TNC）联合15个研究机构的专家团队于2017年10月发表文章，分析了20个基于自然的气候减缓路径可能带来的减排效益，定量阐释了NbS的巨大潜力，即能够提供为将全球升温幅度控制在2℃以内而需要在2030年之前达到的减排量的37%（Griscom B W et al.，2017）。

2019年9月，在纽约召开的联合国气候行动峰会是NbS发展历程中的里程碑。在此次峰会上，NbS被列入全球加速气候行动的九大领域之一。峰会设立了全球NbS联盟，由中国和新西兰联合领导。两国牵头发布了该联盟的最新成果——《基于自然的解决方案促进气候宣言》（以下简称《气候宣言》）与《基于自然的解决方案最佳实践案例汇编》（以下简称《案例汇编》）。其中，《气候宣言》得到了全球70多个国家政府、私营部门、民间社会和国际组织的支持，为NbS的进一步研究与实践提供了坚实的基础。《案例汇编》整理了全球林业、农业、海洋、水资源、生物多样性保护、荒漠化防治等领域的196个案例与倡议，为全球的气候行动提供了新思路。

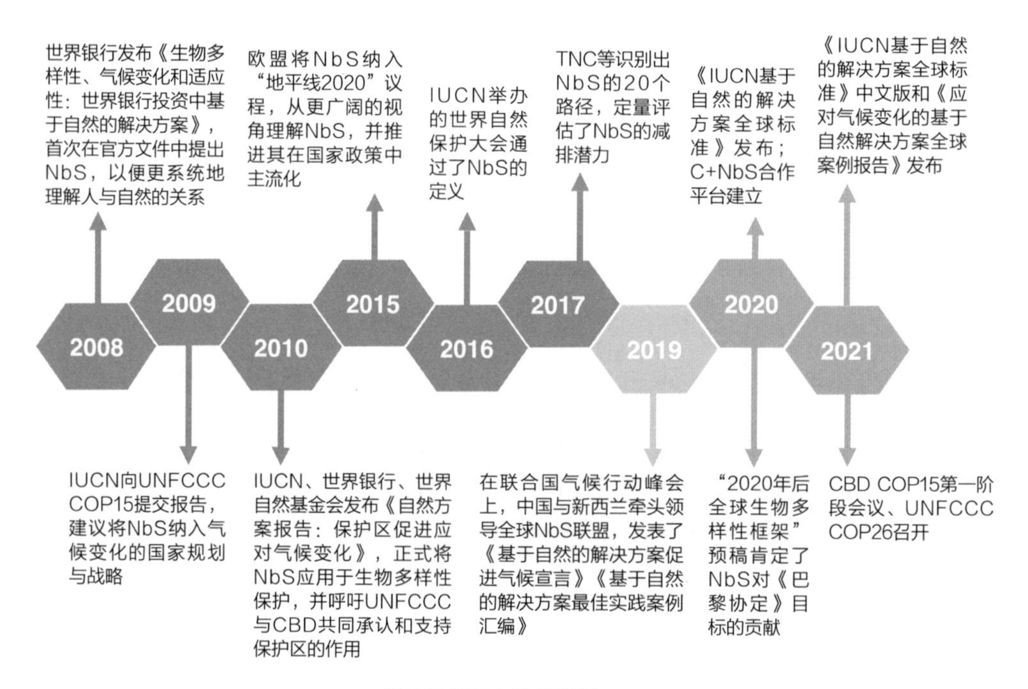

NbS的提出与发展进程

2021年，《生物多样性公约》第15次缔约方大会（CBD COP15）第一阶段会议在中国昆明召开，《联合国气候变化框架公约》第26次缔约方大会（UNFCCC COP26）在英国格拉斯哥召开。作为生物多样性保护与应对气候变化之间的重要联结，NbS成为这两次大会的重要议题。中国作为全球NbS联盟的共同领导者，积极学习与实践NbS相关理念，亟须了解先进的国际经验，尤其是符合中国国情、具有指导价值和实践意义、可推广实施的具体案例；同时，NbS正处于发展的关键期，中国作为生态文明理念的提出者，在国内已开展诸多生态保护与气候减缓协同并举的成功实践，可以为其他国家，尤其是发展中国家的政府、企业和组织，利用NbS应对气候挑战、造福人民提供创新且有前景的新思路。

本书基于《案例汇编》，结合2019年联合国气候行动峰会前后其他国际组织、地区政府、高校、智库开展的NbS案例研究，建立了由全球300个NbS案例组成的案例库，并根据案例的特点进行归类。除NbS涉及的林业、草地、农业、湿地 4 类主要生态系统外，本书还增加了城市、国家、平台与倡议、企业4个新的NbS应用领域，旨在帮助全球NbS项目设计者了解最新趋势、开拓创新思路。

为进一步提炼案例优势与闪光点，发掘符合中国国情的优秀案例，本书以习近平生态文明思想为指导，参考联合国可持续发展目标和《IUCN基于自然的解决方案全球标准》，结合中国国情和地方特色，制定了一套全新的案例筛选与评价标准，包括6个初级筛选指标和3个高级评估维度，希望为决策者、研究机构、非政府组织、企业、社区和个人提供借鉴。本书最终从300个案例中筛选出28个优秀案例，包括10个中国案例、18个国外案例，其中国外案例的实施地点覆盖欧洲、北美洲、南美洲、非洲、亚洲、大洋洲，具有广泛的代表性。本书还分析了每个案例的成功要素，总结出同一类型案例的特点，有望为全球，特别是发展中国家，在相关领域实施NbS项目、规划城市NbS发展、部署国家NbS战略、鼓励企业参与NbS发挥积极作用，并提供有针对性的参考。

在编写本书的过程中有三点观察。

第一，随着NbS 议题的演进，越来越多的案例突出了NbS在生态保护领域的贡献和潜力，而缺少了从气候变化视角的审视。本书的出发点是推荐以应对气候变化为主题的案例，以便与其他普适性的NbS案例报告有所区分，因此在筛选指标和维度设计上更加强调气候因素。但在实际筛选过程中发现，真正以应对气候变化为出发点进行设计的NbS案例相对较少，有的案例材料里虽提到了气候变化，但减缓和适应气候变化的措施、效益及后续改进等信息尚不足以支撑对案例的深入分析。

第二，现在国际上接受度比较高的是世界自然保护联盟对NbS的定义。该定义强调了对自然或人工生态系统的保护、管理和修复，如果从这3个角度来考虑，NbS能够提供为将全球升温幅度控制在2℃以内而需要在2030年之前达到的减排量的37%。习近平总书记在2021年3月召开的中央财经委员会第九次会议上指出："实现碳达峰、碳中和是一场广泛而深刻的经济社会系统性变革，要把碳达峰、碳中和纳入生态文明建设整体布局，拿出抓铁有痕的劲头，如期实现2030年前碳达峰、2060年前碳中和的目标。"实现碳中和需要更多创新，也就需要解锁更多的自然潜力。现在，越来越多的新能源是取自自然、源于自然的技术创新，是"受到自然启发、支撑并利用自然的解决方案"（European Union，2015）。本书选取了宝丰集团利用太阳能电解水以制取绿氢的案例，这个案例虽然暂时不符合世界自然保护联盟对NbS的定义，但从解锁自然潜力寻找创新解决方案的角度来审视，绿氢的制作流程是利用自然又不危害自然的技术创新。

第三，中国在NbS议题上的探索与世界领先水平同步。2019年，当纽约联合国气候行动峰会将NbS列为九大行动领域之一时，中国国内了解这个概念的人并不多，但在整理案例的过程中发现，这并没有影响国内利益相关方在实际工作中已经运用了这一视角。通过进一步的研究发现，NbS与我国倡导的生态文明理念异曲同工。本书精选了10个中国案例，每个案例与其他同类型的国际案例相比都毫不逊色。

鉴于以上三点观察，提出如下思考和建议：

首先，2021—2030年是全球生态保护和气候治理的关键时期，全球保护生物多样性和应对气候变化两个多边进程呈现加强协同、相互促进的趋势。NbS提出的初衷是应对气候变化，强调生物多样性保护对适应与减缓气候变化的重要性。在碳中和目标指引下，建议回到其初衷，从应对气候变化与生物多样性保护协同的角度深入研究并推进NbS相关工作，为未来10年全球治理的关键窗口期贡献更多应对气候变化的基于自然的解决方案优秀案例。

其次，实现碳中和目标需要系统性变革，建议以未来的视角看现在，从开放包容的角度重新思考NbS概念的内涵和外延，以适应新形势，使NbS在未来更具有包容性，留给各方更充分的探索空间。通过充分挖掘自然潜力，NbS对全局的贡献将超过目前的相关评估。

最后，我国在NbS方面的实践扎实、丰富，居世界领先水平。作为NbS工作方向的引领国之一，建议在接下来的国际推广和合作中展示更多的战略自信，主动推动更多的交流，帮助国际社会理解生态文明理念与NbS的共通性，在话语框架上推进与国际社会接轨，同时可以基于了解的情况和思考主动引导对NbS概念和议程设置的讨论，建设性地发挥引领作用，在展示真实、立体、全面的中国的同时，为解锁更大的自然潜力以实现碳中和目标贡献中国智慧和中国方案。

本书得到了中国气候变化事务特使解振华的悉心指导，由清华大学气候变化与可持续发展研究院（以下简称气候院）学术委员会主任何建坤教授担任总顾问并把握战略方向，生态环境部自然生态保护司副司长刘宁一直鼓励和关心相关工作的进展。本书的核心编写团队来自气候院，大自然保护协会的霍莉和中国环境科学研究院的常江研究员在不同的编写阶段分别作出了各自的贡献；自然资源部国土整治中心副主任罗明、世界经济论坛热带雨林及生态文明项目大中华区总负责人朱春全、世界自然基金会全球政策总监李琳博士为本书提供了审读意见，在此对所有参与和支持本书写作的专家和同人致以诚挚的感谢！

大自然保护协会作为气候院C+NbS合作平台的战略合作伙伴，在NbS议题的发展演进过程中发挥着关键作用，为气候院相关工作的开展提供了全方位的支持，在此对其中国首席代表马晋红及其团队成员王会东、Duncan Marsh、许进、靳彤、董珂等表示感谢。气候院C+NbS合作平台还得到了清华大学教育基金会全球气候变化与绿色发展专项基金、燕宝慈善基金会和汇丰银行的支持，在此一并表示感谢！

本书参考了《IUCN基于自然的解决方案全球标准》（以下简称《全球标准》），世界自然保护联盟中国代表处主任张琰在本书的写作过程中给予很多支持。2021年6月23日，世界自然保护联盟发布了《全球标准》的中文版，本书也成为《全球标准》中文版的首次应用，我们深感荣幸。

本书是气候院C+NbS合作平台系列成果的一部分。自2020年4月该合作平台启动以来，气候院组织了9场月度工作坊，与国内外NbS相关机构、同人建立了密切联系，累计影响人次超60万。接下来，气候院也将与各方一起继续跟进并参与NbS在全球范围内的理论研究与实践进展，继续为国内开展相关工作提供借鉴，为实现碳中和目标添砖加瓦，同时也希望能在中国与世界之间架起一座桥梁，为全球推进NbS、建设人与自然生命共同体贡献中国力量。

报告主编

清华大学气候变化与可持续发展研究院

常务副院长　李　政

清华大学气候变化与可持续发展研究院

院长助理　王彬彬

2021年7月1日

目　录

上篇　方法学

下篇　优秀案例

6 农业类

7 湿地类

8 城市类

9 国家类

C+NbS

上 篇

方法学

1 研究时间表

本书的研究工作于2020年6月启动，历时12个月完成内容的编写与修订，并于2021年7月以《应对气候变化的基于自然解决方案全球案例报告》的名称率先发布相关中英文报告（图1-1）。

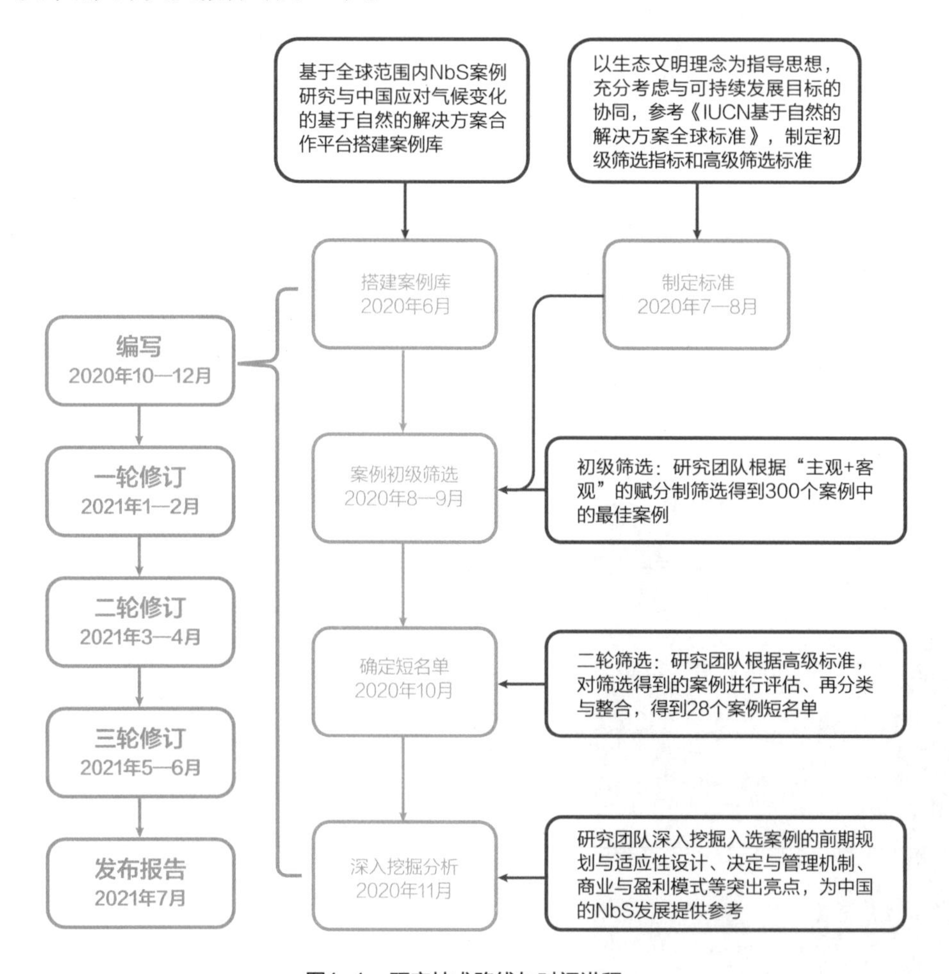

图1-1 研究技术路线与时间进程

2 案例库

本书基于现有的基于自然的解决方案（Nature-based Solution，NbS）全球案例研究和中国应对气候变化的基于自然的解决方案（C+NbS）合作平台搭建全球案例库，其中的案例总数达300个。以2019年联合国气候行动峰会（以下简称峰会）为界，案例来源可按合作平台启动与报告发布时间分为以下3类（表2-1）：

第一类案例来源主要包括欧盟“地平线2020”议程中启动的NbS案例研究平台，以及C40城市集团、大自然保护协会（TNC）在峰会前发布的NbS相关研究报告；

第二类案例来源即联合国环境规划署（UNEP）于峰会上发布的《基于自然的解决方案最佳实践案例汇编》及随之建立的NbS案例平台；

第三类案例来源主要包括峰会后世界自然基金会（WWF）和自然与气候联盟（N4C）发布的报告，以及大自然保护协会、世界自然联盟（IUCN）、牛津大学、清华大学气候变化与可持续发展研究院（ICCSD）启动的NbS案例研究平台。

表2-1　案例库参考报告与平台

序号	机构	名称	类型	发布时间
1	欧盟	欧盟城市NbS案例研究-Naturvation平台 https://naturvation.eu/cities	平台	2014年 （“地平线2020”议程）
2	欧盟	欧盟城市NbS案例研究-Oppla平台 https://oppla.eu/nbs/case-studies	平台	2014年 （“地平线2020”议程）
3	C40 城市集团	《可持续城市再生战略：通过绿色基础设施适应气候变化》	报告	2016年
4	TNC	《源头之外：水安全的自然解决方案》	报告	2018年

（续表）

序号	机构	名称	类型	发布时间
5	UNEP	UNEP-NbS案例库 www.unenvironment.org/nbs-contributions-platform	平台	2019年9月 （联合国气候行动峰会）
6	UNEP	《基于自然的解决方案最佳实践案例汇编》	报告	2019年9月 （联合国气候行动峰会）
7	WWF	《气候、自然与我们的 1.5℃ 未来》	报告	2019年12月
8	N4C	N4C全球NbS案例研究 http://nature4climate.org/nbs-case-studies/	平台	2019年启动
9	IUCN	《IUCN基于自然的解决方案全球标准》	报告	2020年7月
10	牛津大学	牛津大学NbS倡议平台NbS案例研究 http://www.naturebasedsolutionsinitiative.org/nbs-case-studies/	平台	2020年启动
11	TNC	TNC全球NbS案例库	平台	2020年启动
12	ICCSD	C+NbS合作平台	平台	2020年启动

根据案例中的行为主体、实施目标和主要用地类型，本书将300个案例划分为林业、草地、农业、湿地、城市、国家、平台与倡议、企业8类（图2-1），并对每一类别中的案例进行筛选、提炼、总结，以供借鉴。

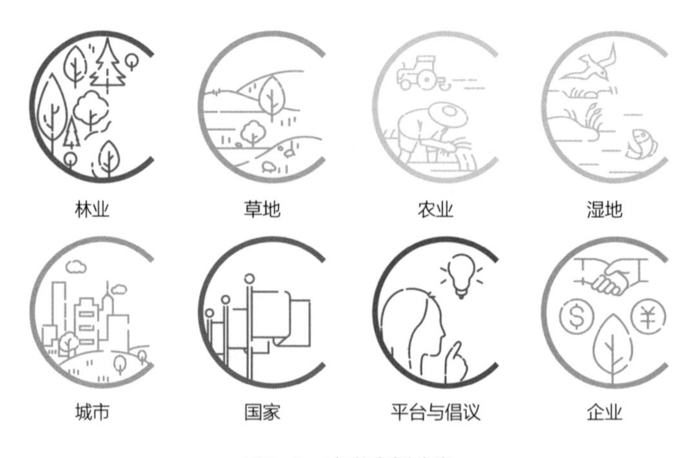

图2-1　本书案例分类

3 研究标准

本书以应对全球气候变化为主要目标，初步开发了适合在中国推进的NbS评价标准，并在此基础上综合评估了全球数百个优秀的基于自然的气候变化解决方案，并展开了深入的解读和分析。

本书以生态文明理念为指导思想，充分考虑与联合国17项可持续发展目标（SDGs）的协同，参考世界自然保护联盟制定并发布的《IUCN基于自然的解决方案全球标准》（以下简称《全球标准》，见附录1），并结合中国国情分别制定了服务于案例筛选的初级筛选指标和用于发掘案例潜力的高级评估维度。其中，初级筛选指标有6项，主要关注案例解决问题的能力和对多重利益的保障，旨在从数据库的数百个全球案例中快速筛选出能有效应对社会挑战的案例。在此基础上，本书又制定了包括治理模式、金融与市场机制、保障多元行为体及妇女参与在内的3个高级评估维度，对入选的初级案例进行二次筛选和深度剖析，以确保最终入选的案例对所有发展中国家和地区有深刻的借鉴和学习意义。

通过上述努力，这些标准可以促进NbS在中国更好地落地生根，在科学设计、实施并归纳NbS对应对气候变化的贡献和价值方面发挥了重要作用，同时还可以让国际社会更好地了解和认识生态文明理念，让更多的发展中国家可以在这一过程中学习和受益。

3.1 指导思想：生态文明理念

NbS与习近平生态文明思想高度重合，是生态文明理念在可持续发展领域的具体应用；生态文明为NbS在中国的推广提供了思想和文化的土壤，而NbS能够让世界更好地认识和接受生态文明的概念和内涵。世界自然保护联盟的《全球标准》要求“使NbS在合适的辖区成为主流，并发挥可持续性优势”，而中国提出的“建设生态文明、打造美丽中国”本身就是在国家范围内NbS主流化的最高体现。

本书正是以生态文明理念为重要指导思想。生态文明（Ecological Civilization）是人类文明的一种形式，是继原始文明、农业文明和工业文明之后人类向往的另一种文明，建立在人类对传统工业文明反思的基础上。生态文明的概念最早出现在20世纪80年代，由苏联环境学家在《在成熟的社会主义条件下培养个人生态文明的方法》一文中提出（Ye Q，1984）。2007 年，中国共产党第十七次全国代表大会首次提出要“建设生态文明”，自此生态文明作为一种重要的治国理念在中国开始实践并进入国际视野（刘静，2011）。2012年，中国共产党第十八次全国代表大会将生态文明建设纳入“五位一体”总体布局，并首次提出“美丽中国”的概念，指出“建设生态文明，是关系人民福祉、关乎民族未来的长远大计。面对资源约束趋紧、环境污染严重、生态系统退化的严峻形势，必须树立尊重自然、顺应自然、保护自然的生态文明理念，把生态文明建设放在突出地位，融入经济建设、政治建设、文化建设、社会建设各方面和全过程，努力建设美丽中国，实现中华民族永续发展。”2017年，党的十九大报告提出“建设生态文明是中华民族永续发展的千年大计”，把“建设美丽中国”作为现代化目标之一，强调“必须树立和践行绿水青山就是金山银山的理念，坚持节约资源和保护环境的基本国策，像对待生命一样对待生态环境，统筹山水林田湖草系统治理，实行最严格的生态环境保护制度，形成绿色发展方式和生活方式，坚定走生产发展、生活富裕、生态良好的文明发展道路，建设美丽中国，为人民创造良好生产生活环境，为全球生态安全作出贡献”（习近平，2017）。

生态文明的核心问题是如何正确处理人与自然的关系，人类对于自然应当获取有度，既要利用又要保护，努力保持经济发展、人口、资源、环境的动态平衡，实现人与自然和谐共处。习近平生态文明思想要求人类尊重自然、顺应自然和保护自然，其核心要义集中体现在以下“八个观”（李干杰，2018a）。

第一，生态兴则文明兴、生态衰则文明衰的深邃历史观。从世界文明的发展历史来看，生态环境的变化直接影响文明的兴衰演替。良好的生态环境和自然禀赋是人类文明形成和发展的基础。《联合国人类环境宣言》中写道：“环境给予人以维持生存的东西，并给他提供了在智力、道德、社会和精神等方面获得发展的机会。”尽管肥沃的土壤、温和的气候、丰沛的水源、茂密的森林给予文明繁荣兴盛的基础条件，但是无节制的放牧、伐木、垦荒、灌溉和污染排放最终将会导致文明的衰败。

第二，人与自然和谐共生的科学自然观。恩格斯曾在《自然辩证法》中提出：“我们每走一步都要记住：我们决不像征服者统治异族人那样支配自然界，决不像站在自然界之外的人似的去支配自然界——相反，我们连同我们的肉、血和头脑都是属于自然界和存在于自然之中的。”人是自然的一部分，人与自然是相互依存、和谐共生的命运共同体，人类对大自然的伤害最终会伤及人类自身，人类必须尊重自然、顺应自然、保护自然。

第三，绿水青山就是金山银山的绿色发展观。绿水青山既是自然财富、生态财富，又是社会财富、经济财富。绿水青山就是金山银山要求人类秉持可持续发展的理念，顺应自然规律，对自然开发索取有度，把不损害生态环境作为发展的底线，在保护自然生态价值的基础上创造经济价值，实现社会进步，不以牺牲生态环境为代价换取经济的一时发展。

第四，良好生态环境是最普惠的民生福祉的基本民生观。生态环境的质量密切影响着人类健康。当前，不同程度的重污染天气、黑臭水体、垃圾围城、农村环境问题依然是民心之痛、民生之患（李干杰，2019）。“良好生态环境是最公平的公共产品、最普惠的民生福祉。”这要求我们把对生态系统的保护和修复、生态环境质量的提升作为提高人民生活水平的重要内容，重点解决损害公众健康的突出环境问题，时刻坚持“生态惠民、生态利民、生态为民”。

第五，山水林田湖草是生命共同体的整体系统观。综合性和系统性是生态治理的重要特点，生态本身就是一个有机系统，生态系统中的各要素相互影响、相互依存，“人的命脉在田，田的命脉在水，水的命脉在山，山的命脉在土，土的命脉在树。”生态文明建设也要坚持“全方位、全地域、全过程”，在发展和治理的过程中以整体最优为目标，注意统筹兼顾各方利益，保障陆地与海洋、城市与农村、不同生物、不同排放之间的平衡。实现生态文明，不能仅仅局限于保护自然和修复生态环境，更应关注减少生物多样性丧失和环境退化背后的驱动因子，即我们的生产、生活、消费模式，财政、金融、投资方向，以及对发展模式的重新定义，对自然资源的内在和外在价值的赋值等。

第六，用最严格制度保护生态环境的严密法治观。从国家层面推进生态文明建设，必须依靠制度和法治，只有制定最严格的生态保护制度和法律法规，实现有法可依、有法必依，在此基础上强化制度执行，做到执法必严、违法必究，才能为生态文明建设提供可靠保障和适宜的政策环境。2018年3月，十三届全国人大一次会议审议通过了《中华人民共和国宪法修正案》，强调“推动物质文明、政治文明、精神文明、社会文明、生态文明协调发展，把我国建设成为富强民主文明和谐美丽的社会主义现代化强国，实现中华民族伟大复兴”，并将“生态文明”写入宪法，反映了新时代背景下对社会经济发展路径转变和优化升级的深刻理解。

第七，全社会共同建设美丽中国的全民行动观。生态文明建设是全人类共同的事业，既需要政府自上而下的制度设计，也需要自下而上的全民行动。这要求包括政府、企业、智库、国际组织、个人在内的多元行为体都成为生态文明的践行者和推动者，从而形成人人参与、人人共享的强大合力。

第八，共谋全球生态文明建设之路的共赢全球观。人类是命运共同体，建设绿色家园是人类的共同梦想。国际社会应当携手共进，在推动自身绿色改革的同时积极参与全球环境治理，通过积极合作形成世界环境保护和可持续发展的解决方案，保护好人类赖以生存的地球家园（李干杰，2018b）。

生态文明理念中的“尊重自然、顺应自然、保护自然”与NbS“保护自然生态底线、恢复自然生态本底和尊重自然规律”的基本原则一致（廖茂林，2020）。生态文明建设与NbS拥有一致的根本目标——谋求人与自然和谐共生。NbS以人与自然的共生关系为基础，推崇以可持续的方式利用生态服务价值，从而创造自然、社会及经济的协同效益，应对气候变化等社会挑战（贺庆棠，2019）。生态文明理念是指导人类发展的顶层设计与战略，NbS是生态文明在实践层面的方法学（卢风，2020）。优秀的NbS在设计和实践中也应与生态文明的各项理论保持一致。

3.2 初级筛选指标

在生态文明思想的指导下，结合可持续发展目标和世界自然保护联盟的《全球标准》，本书初步设立了6项初级指标，旨在从全球案例库中筛选出能够有效应对气候挑战、保障人民权益、实现多重效益的案例（表3-1）。

表3-1 案例初级筛选指标

<table>
<tr><th>序号</th><th>初级筛选指标</th><th>生态文明思想</th><th>联合国可持续发展目标</th><th>世界自然保护联盟的《全球标准》</th></tr>
<tr><td>1</td><td>气候变化的减缓与适应</td><td>共谋全球生态文明建设之路的共赢全球观</td><td>SDG13 气候行动</td><td>准则1：NbS应有效应对社会挑战</td></tr>
<tr><td rowspan="4">2</td><td rowspan="4">生物多样性保护</td><td>生态兴则文明兴、生态衰则文明衰的深邃历史观</td><td rowspan="2">SDG14 水下生物</td><td rowspan="2">准则1：NbS应有效应对社会挑战</td></tr>
<tr><td rowspan="2">人与自然和谐共生的科学自然观</td></tr>
<tr><td rowspan="2">SDG15 陆地生物</td><td rowspan="2">准则3：NbS应带来生物多样性净增长和生态系统完整性</td></tr>
<tr><td>绿水青山就是金山银山的绿色发展观</td></tr>
<tr><td rowspan="6">3</td><td rowspan="6">应对多重社会挑战</td><td rowspan="3">良好生态环境是最普惠的民生福祉的基本民生观</td><td>SDG2 零饥饿</td><td rowspan="3">准则1：NbS应有效应对社会挑战</td></tr>
<tr><td>SDG3 良好健康与福祉</td></tr>
<tr><td rowspan="2">SDG6 清洁饮水和卫生设施</td></tr>
<tr><td rowspan="3">共谋全球生态文明建设之路的共赢全球观</td><td rowspan="3">准则8：NbS应具有可持续性并在适当的辖区内主流化</td></tr>
<tr><td>SDG9 产业、创新和基础设施</td></tr>
<tr><td>SDG11 可持续城市和社区</td></tr>
</table>

（续表）

序号	初级筛选指标	生态文明思想	联合国可持续发展目标	世界自然保护联盟的《全球标准》
4	实现多重目标的协同	山水林田湖草是生命共同体的整体系统观	—	准则2：应根据尺度来设计NbS
				准则6：NbS应在首要目标和其他多种效益间公正地权衡
5	促进当地经济发展	绿水青山就是金山银山的绿色发展观	SDG1 无贫穷	准则4：NbS应具有经济可行性
			SDG8 体面工作和经济增长	
6	关怀弱势群体	良好生态环境是最普惠的民生福祉的基本民生观	SDG3 良好健康与福祉	准则5：NbS应基于包容、透明和赋权的治理过程
			SDG10 减少不平等	

1. 气候变化的减缓与适应

NbS在概念提出之初便以应对气候变化为目标，气候变化也是世界自然保护联盟《全球标准》中定义的七大社会挑战之首。习近平总书记在第八次全国生态环境保护大会上强调："要实施积极应对气候变化国家战略，推动和引导建立公平合理、合作共赢的全球气候治理体系，彰显我国负责任大国形象，推动构建人类命运共同体。"而坚持"共谋全球生态文明建设"更是习近平生态文明思想的六项原则之一。因此，旨在应对气候变化的NbS至少应在气候减缓与适应中的一个领域发挥积极作用。减缓指的是减少大气中的温室气体浓度，主要通过减少排放源与增加碳汇的方式实现；适应主要体现在减少暴露度和脆弱性，提高整个社会的适应能力，有效抵御极端气候事件及平均温度变化带来的影响和挑战（干旱、海平面上升等）。

2. 生物多样性保护

NbS本身源于生态系统所提供的商品和服务，在很大程度上依赖生态系统的完整性和多样性。"尊重自然、顺应自然、保护自然的生态文明理念"是推进生态文明建设的重要思想基础，"节约优先、保护优先、自然恢复为主"是推进生态文明建设必须坚持的方针。世界自然保护联盟的《全球标准》提出NbS应带来生物多样性净增长和生态系统完整性（准则3）。因此，NbS项目应以保护生态环境为前提，以保护生态系统、保护生物多样性为底线，致力于实现生物多样性和生态系统完整性的净增益，从而维持自身的稳定运行。

3. 应对多重社会挑战

生态文明建设不应局限于对生态环境的保护，而是要求实现资源利用、环境保护与社会经济的协同发展。世界自然保护联盟将人类当前面临的重要社会挑战定义为气候变化减缓与适应、防灾减灾、经济与社会发展、人类健康、粮食安全、水安全、生态环境退化与生物多样性丧失七项内容（图3-1）。除气候变化与生物多样性丧失外，NbS项目如果能够同时应对粮食安全、水安全等其他社会挑战，并发挥协同效应，将能达到事半功倍的效果。

图3-1　NbS应对的社会挑战
（图片来源：世界自然保护联盟）

4. 实现多重目标的协同

生态是统一的自然系统，是相互依存、紧密联系的有机链条，优秀的NbS项目应平衡好社会发展、经济进步、环境保护等多方面的利益，协同实现多重目标。世界自然保护联盟的《全球标准》要求NbS的设计应认识到经济、社会和生态系统之间的相互作用（准则2），在首要目标和其他多种效益间公正地权衡（准则6），并统筹考虑自然生态各要素、山上山下、地上地下、陆地海洋及流域上下游，进行整体保护、宏观管控、综合治理。因此，在NbS的设计和实施过程中，要坚持山水林田湖草是生命共同体的整体观，权衡取舍各方利益，以整体最优为目标，维护生态完整性。

5. 促进当地经济发展

世界自然保护联盟的《全球标准》要求 NbS应具有经济可行性（准则4），在应对气候变化、保护生态环境的同时实现盈利，增加经济收入，贡献市场和就业，逐渐脱离资金投入从而实现自身的可持续发展——这是决定NbS能否长期有效的关键。2008年印发的《全国生态脆弱区保护规划纲要》指出："我国生态脆弱区大多位于生态过渡区和植被交错区，处于农牧、林牧、农林等复合交错带，是我国目前生态问题突出、经济相对落后和人民生活贫困区。"生态脆弱和生态功能的重要性决定了中国在扶贫攻坚中应以"精准"和"绿色"为两个显著特点（刘解龙，2016），必须走绿色扶贫的道路。促进当地经济发展的NbS就是对"绿水青山就是金山银山"的解读。在农村地区，在生态文明理念指导下实施的NbS可以保护当地生态环境、拉动经济发展、实现绿色减贫，是推动乡村振兴的重要手段。

6. 关怀弱势群体

"社会弱势群体的生存状况直接决定着一个良序社会能否稳定和健康发展，也是一个社会是否公平正义的重要标志。"（邹海贵，2012）世界自然保护联盟的《全球标准》要求NbS 应基于包容、透明和赋权的治理过程（准则5），尊重和维护获得土地与自然资源、使用和管理土地与自然资源的法律权利和习惯权利，特别是要保护弱势和边缘化群体的相关权利。实施NbS项目应当保证社会公平，尤其是要照顾到妇女、儿童、残疾人、慢性病患者、老人、少数民族和偏远地区人民等弱势群体的特殊需求与利益。

本书针对6项初级筛选指标采取客观与主观相结合的办法进行打分——客观统计案例材料与报告对于给定指标的表述与量化数据，主观判断案例的突出亮点、本土适应性与可复制程度，各指标具体的打分标准见表3-2。通过比较林业、草地、农业、湿地、城市、国家、平台与倡议、企业某一类别内所有案例的总得分，可以得到该类别的4～6个最佳案例，再进入二次筛选与深入挖掘环节，最终入选案例在这一阶段的得分情况见附录2。

表3-2 案例初级筛选打分标准

分值	描述
0	案例未体现该指标对应的内容，或案例对指标的部分内容有所体现，但流于形式，没有实质成果
1	案例体现了该指标对应的内容，且描述较为详细，有一定的实质成果，但规模较小、可复制性不强
2	案例体现了该指标对应的内容，描述详细，有量化和实质成果，且案例具有一定的规模效应，对其他地区与全球有政策借鉴价值
3	案例体现了该指标对应的内容，描述详细，有量化和实质性的突出成果，且案例具有规模效应，对当地及地区甚至全球都有积极意义，还具有深远的政策借鉴价值

3.3 高级评估维度

为充分挖掘案例的管理和决策机制设计及市场化、规模化、效益化的亮点与优势，使选取的案例更好地服务于国内外政策制定者，并为利益相关方提供值得借鉴的理念与方法，在深入挖掘环节，本书设立了治理模式、金融与市场机制、保障多元行为体及妇女参与3个高级评估维度（表3-3），对入选的优秀案例进行二次筛选、分类与整合，最终获得由28个案例组成的短名单，再进行深度分析与潜力挖掘，并通过归纳总结得到案例的借鉴价值（附录3）。

表3-3 案例高级评估维度

序号	高级评估维度	生态文明思想	联合国可持续发展目标	世界自然保护联盟全球标准
1	治理模式	用最严格制度保护生态环境的严密法治观	SDG16 和平、正义与强大的机构	准则2：应根据尺度来设计NbS
				准则5：NbS应基于包容、透明和赋权的治理过程
			SDG17 促进目标实现的伙伴关系	准则7：NbS应基于证据进行适应性管理

（续表）

序号	高级评估维度	生态文明思想	联合国可持续发展目标	世界自然保护联盟全球标准
2	金融与市场机制	—	SDG8 体面工作和经济增长	准则2：应根据尺度来设计NbS
				准则4：NbS应具有经济可行性
3	保障多元行为体及妇女参与	全社会共同建设美丽中国的全民行动观	SDG5 性别平等	准则5：NbS应基于包容、透明和赋权的治理过程
			SDG17 促进目标实现的伙伴关系	

1. 治理模式

治理模式和决策管理机制是NbS项目成功的关键因素。优秀的治理模式能够通过多方参与，在保证项目有效运行的基础上实现适应性管理，并有所创新。有效的创新治理模式应基于当地地理与文化特点，包括自上而下和自下而上模式的联动、透明公开的监督模式、可靠的信息沟通和对话机制、创新的绩效评价系统等。适应性管理要求NbS基于科学理解、传统习俗和当地知识对项目进行定期监测和评估，并适时调整实施方案，实现“识别问题—合理（调整）规划—有效实施—科学评估”逐层上升的科学管理模式，将人力、资金等资源的浪费降到最低，提高NbS的成功率与持久性，有效提升资源的使用效率。

2. 金融与市场机制

融资是NbS项目实施的必要过程，NbS的资金来源包括政府、企业、国际组织、基金会、个人等。优秀的NbS应使用或设计合理有效的金融工具与市场机制来规避风险，以保证资金的稳定供给，从而增加NbS取得长期成功的可能性；同时，世界自然保护联盟的《全球标准》中的准则2（应根据尺度来设计NbS）要求NbS积极寻求与工程项目、信息技术、金融措施等其他类型项目的互补，实现NbS与社会发展的其他部门之间的协同管理。

3. 保障多元行为体及妇女参与

NbS的基本原则要求项目须以公平公正的方式产生社会效益，促进利益相关方的广泛参与。优秀的NbS应保障参与行为体的多元化，注重城乡平衡、代际平衡，实现政府、企业、国际组织、智库、社区、个人的全面参与，尤其应关注性别平衡，鼓励妇女参与NbS（专栏1）。这是实现公平公正的重要方式，也是习近平生态文明思想中“全社会共同建设美丽中国的全民行动观”的重要体现。

专栏1

政府是设计、实施和进一步推广NbS的主要力量，并可以为广泛的NbS项目提供有力的社会条件，包括制定相关政策并促进法治建设、推广生态系统付费市场和认证项目、促进自然资本投资、改善土地使用政策等，政府对NbS的重视与参与可以促进相关政策和法律法规的完善，并有利于NbS的主流化发展。

企业是NbS的重要利益相关方，也是助力将NbS从理论推广到实践的中坚力量。企业参与投资、设计NbS项目既是承担社会责任、应对社会挑战的表现，也是提高自身生产与经营可持续性、拓展业务和增加盈利的重要途径。

联合国、世界银行、世界自然保护联盟、世界自然基金会、大自然保护协会等国际组织是NbS理论最初的提出者，也是NbS理论最重要的前期推广者，参与了诸多NbS项目的设计、资助与具体实施。

研究机构、高校和民间智库可以为NbS提供重要的理论依据和支撑，也能以资助具体项目的形式推动NbS发展，还可以将NbS纳入其政策建议，推动相关法律法规的制定。

个人、社区和本土机构是项目最重要的实施者，也是NbS的直接受益者，充分考虑其利益并将其纳入NbS的实施与管理是NbS平衡利益相关方的重要内容。

保障妇女参与是NbS包容性的重要体现，妇女既是易受气候变化与生态破坏负面影响的对象，也是助力经济社会发展、参与NbS决策与管理的一支重要力量。这要求我们以相互尊重、性别平等为目标，鼓励女性在NbS项目中担任角色，提高妇女参与公共事务的积极性和参与度。

C+NbS

下 篇

优秀案例

4 林业类

森林等生态系统具有巨大的社会与经济价值，是全球3.5亿人的家园，为全世界超过15亿人（20%）的生计提供了支持，为5 000万人提供了就业（Sunderland T et al.，2013）。与此同时，森林也是人类应对气候变化的重要资源。对于减缓气候变化，一方面森林可以产生碳汇，吸收大气中的二氧化碳；另一方面开展管护、修复工作可以避免或减少森林在过去数十年甚至上百年积累的碳在短时间内被分解排放到大气中。对于适应气候变化，增加的森林植被可以起到减少水土流失的作用，有效地保持水土、提升社区收入以提高适应能力。林业是NbS发挥作用的典型部门之一，国内外有大量致力于保护、修复和可持续管理森林的案例。在林业类别中，本章选取了哥伦比亚国家级森林保护计划、尼泊尔社区林业运动和“植万亿棵树领军者”倡议3个案例，为林业领域NbS项目的设计规划与实施提供借鉴。

4.1 哥伦比亚国家级森林保护计划

案例亮点

通过创新的融资计划促进保护区的长期可持续发展。“哥伦比亚遗产计划”项目尝试建立一个可撤销的过渡基金，利用公共和私人领域的投资发挥经济杠杆作用，以确保20年内长期投资的可能性，有效帮助政府和国际发展机构、资助方、私营部门和非政府组织（NGO）之间建立合作，提升资金的使用效率。

4.1.1　项目背景

哥伦比亚位于南美洲北部，地处热带，气候因地势而异。地形大致分为西部山地区和东部平原区：西部主要为安第斯山区，另外有多处沿海平原；东部主要为奥里诺科平原和亚马孙平原两个部分。受地理条件的影响，哥伦比亚境内的地貌景观类型丰富，包括安第斯山脉、亚马孙森林、太平洋和加勒比海沿岸及奥里诺科平原等。哥伦比亚在历史上常因强烈的气候异常波动事件（主要与厄尔尼诺现象有关）和极端水文气象事件而发生灾难。气候变化会导致平均温度升高、降水模式变化，加剧局部地区干旱、洪涝和山体滑坡等现象的发生，对哥伦比亚境内不同区域都会产生不同的威胁。在这种情况下，哥伦比亚的森林资源显得尤为珍贵。因此，哥伦比亚的气候变化政策非常重视森林资源在改善地形地貌和景观管理方面的作用，同时强调国家保护区的森林管护工作，要求合理有效地利用保护区附近可开发的土地资源。

哥伦比亚的天然森林面积约为6 000万公顷（WWF，2019），储存了约260亿吨二氧化碳。然而由于资源匮乏，哥伦比亚的许多保护区管理效率低下。1990—2016年，该国约有600万公顷的森林被砍伐，其中一个重要的原因是周边居民对保护区及自然资源的侵占。2012年，森林砍伐导致的温室气体排放约占哥伦比亚总排放量的16%。此外，哥伦比亚还面临严峻的贫困问题，该国一半以上的人口生活在贫困线以下，部分农村地区的贫困率高达70%。解决以上问题需要充分认识气候变化对贫困和弱势人群造成的威胁。因此，改善土地利用、实现应对气候变化与减贫的平衡是该国气候友好型可持续发展战略的核心。

4.1.2　项目介绍

“哥伦比亚遗产计划”项目于2015年《联合国气候变化框架公约》第21次缔约方大会（UNFCCC COP21）期间正式启动。哥伦比亚环境和可持续发展部、国家公园管理局及戈登和贝蒂·摩尔基金会、世界自然基金会、国际野生生物保护学会（WCS）、保护国际基金会（CI）等机构共同参与了该项目。项目为期20年（2018—2038年），旨在通过减少毁林和再造林等方式规范并改进国家保护区管理工作，同时实现保护生物多样性、改善水源及提高适应气候变化的能力等多重效益。

“哥伦比亚遗产计划”项目不仅直接推进了哥伦比亚《生物多样性公约（2011—2020年）战略计划》及其目标的实施，也支持该国实现多项气候与发展政策，包括国家气候变化适应计划、控制毁林与森林管理的综合战略、国家公园气候变化战略和保护区生态恢复的国家战略。过去，哥伦比亚政府与利益相关方（民间社会、社区、私营部门、地方政府）的合作有限，限制了其对于联合国千年发展目标的实现。“哥伦比亚遗产计划”通过多元参与机制，有效帮助该国政府与国际发展机构、资助方、私营部门和非政府组织之间建立合作，提升了资金的使用效率。

项目资助方包括世界银行（WB）与美洲开发银行（IADB），力求通过创新的融资计划促进保护区的可持续发展。资金机制的基础是建立一个可撤销的过渡基金，利用公共和私人领域的投资在20年内发挥经济杠杆作用，以确保长期投资的可能性。具体来说，该项目建立了“项目永久融资机制”，允许所有参与方在合作框架下作出资金捐赠或政策承诺，以用于弥补保护区发展和可持续管理的长期资金缺口。资助方需要尽可能地调动资源来支持开展最需要支持的优先行动，而哥伦比亚政府则承诺逐步增加配套资金和政策支持，以确保获得长效资金。

资金来源主要有3种形式（图4-1）：①在项目初期设立一个可撤销的过渡基金，在特定时期内利用政府和私人资本弥补资金缺口；②哥伦比亚政府为保护区寻找并开发新的可持续融资模式，如环境使用者付费、生态补偿机制或其他融资渠道，政府决定将碳税的5%用于支持该项目的实施；③政府为项目提供财政担保，在保证于项目执行期间提供50%配套资金的基础上，承诺每年都将增加保护

资金的投入，直到项目结束时承担100%的资金投入。

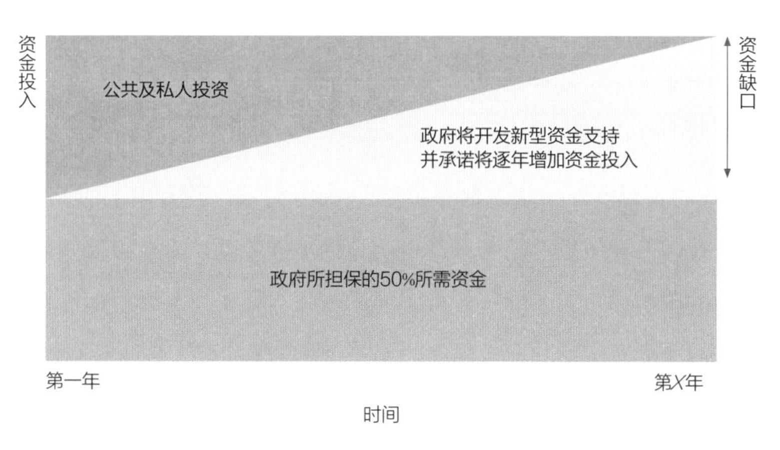

图4-1 “哥伦比亚遗产计划”项目利益相关方贡献

“哥伦比亚遗产计划”项目下设3个森林保护目标，即设立350万公顷的保护区；在现有保护区的基础上增加1 400万公顷的有效管理面积；在保护区周边建立9个“景观缓冲区”，用于保护生物多样性和土地可持续利用，总占地面积为3 580万公顷，占哥伦比亚国土面积的30%以上。该项目可以帮助保护区储存大量的碳，调节水资源平衡，并提高3 200多万人适应气候变化的能力。

该项目涉及多个利益参与方，包括国家及地方政府、民间组织、企业、高校和社区，其各自的贡献见表4-1。

表4-1 “哥伦比亚遗产计划”项目利益参与方的贡献

参与方	政府	捐助者和私营部门	民间社会和基金会	学术界和研究机构
贡献	制定规则，通过政策支持扩大保护区的目标和范围；设计和规划可持续的财政机制，给项目运行提供重要的资金和政策支持	提供重要的资金补充，支持开展优先行动，为国际社会通过混合金融和政企合作模式促进可持续发展提供新范式	通过在国际和该地区获得的专业知识、网络资源和经验，促进利益相关方的参与和互动，关注气候公平和性别平等议题，支持国际经验的传播	提供项目开展所需的技术和科学知识；促进研究与生物多样性保护和可持续发展目标协同的理论基础，支持最佳做法和经验教训的交流

4.1.3 项目成果

该项目预计在20年的实施期内，通过减少毁林、再造林和森林管护等手段减少约1亿吨二氧化碳当量的排放，并提供约500万吨的碳汇。项目前10年的工作聚焦于5个“核心景观区”，其占地面积为1 800万公顷，约占国土面积的15%。针对“核心景观区”开展的活动可以产生明显的早期收益，有效降低气候变化对生态系统造成的风险和压力。除气候减缓外，该项目还可以为贫困地区提供大量的就业机会，造福当地居民。项目规划的 9 个“景观缓冲区”将使约3 800个家庭直接受益，此外将有超过540个城镇、251个贫困社区、89个非裔哥伦比亚社区和超过2 500万名城市居民直接或间接受益。

4.1.4 项目信息

案例来源：UNEP-NbS案例库。

项目参与方：哥伦比亚环境和可持续发展部、国家公园管理局，戈登和贝蒂·摩尔基金会，自然遗产基金会，世界自然基金会，国际野生生物保护学会，保护国际基金会。

项目地点：哥伦比亚。

项目时间：2018—2038年。

项目链接：联合国环境规划署，https://wedocs.unep.org/bitstream/handle/20.500.11822/28927/Heritage_Colombia.pdf?sequence=1&isAllowed=y。

4.2　尼泊尔社区林业运动

案例亮点

尼泊尔的《森林法》为其开展社区造林运动提供了法律基础，允许成立森林社区合作社；法律规定将国有森林的使用权转移给社区合作社，进一步激发社区合作社参与；基层选举制度鼓励低收入人群、土著居民、妇女等参与社区治理，提升了弱势群体参与社区治理和发展的能力。

4.2.1　项目背景

尼泊尔是南亚内陆山国，地处喜马拉雅山南麓，北邻中国，东、西、南三面被印度包围，其森林覆盖面积（包括灌木）和被列入国家公园系统的面积约为5.83万平方千米，约占国土总面积的40%。尼泊尔社会经济发展水平较低，是世界上的贫穷国家之一，在186个国家的人类发展指数中排第157位。数年来，尼泊尔接受了大量的国际援助，2009—2013年官方发展援助占尼泊尔国民总收入的5%（World Bank，2014）。

4.2.2　项目介绍

尼泊尔政府在公共部门和自然资源管理方面具有丰富的管理和改革经验。尼泊尔森林管理部门将各级办事处的权力下放，任命“森林官”，并采用长官负责制。尼泊尔社区林业运动于20世纪70年代在民间兴起，此后在政府林业政策和立法的支持下不断发展，已覆盖尼泊尔大部分森林区域。尼泊尔政府于1993年通过了《森林法》，允许成立森林自治团体，为该国开展社区林业运动提供了法律基础。《森林法》提出要成立社区森林合作社（CFUGs），作为社区参与森林使用与管理的组织形式。1995年，尼泊尔政府出台了《森林规则》和《社区森林操作指南》，同年成立社区森林使用者联盟（FECOFUN），这一系列体制机制建设保障了森林使用者对自身权利的保护，同时使其有机会参政议政（Ministry of Forest and Soil Conservation，2013）。

尼泊尔现有的保护区、生态区和森林设置如图4-2所示。尼泊尔的社区林业将国有森林的使用权转移给社区，只要经地区政府部门批准，就能成立社区森林合作社（Pokharel R K et al.，2012）。林地的所有权仍属于政府，土地资源管理和控制权归社区森林合作社所有。社区森林合作社因此获得了使用和管理森林的权利，在此基础上产生的盈利、利润分配及森林权均可获得官方认可（Pardo R，1993）。社区森林合作社在森林管护方面采取的关键行动包括控制野火、管理露天放牧、控制非法侵占、恢复本地物种等。与此同时，尼泊尔政府也充分发挥行政引导和动员作用，帮助社区森林合作社建立起行之有效的治理模式和就业途径，主要行动如下：

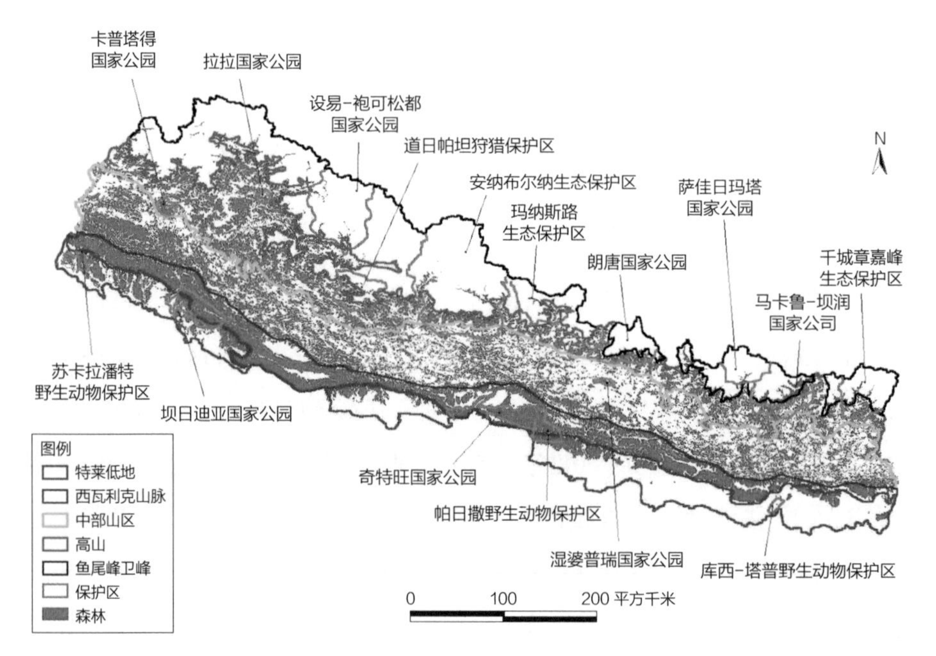

图4-2　尼泊尔现有的保护区、生态区和森林

（图片来源：Bhattarai B，2016）

●开展必要的能力建设，提供技术支持，帮助当地居民就业并提高收入水平；

●通过行政立法进一步提高妇女、青年、原住民、少数民族和弱势群体的能力，为其参与社区治理、获得社区帮扶创造基础和条件；

●增加天然林面积，保护本地物种，引导社区森林合作社以最有效和更可持续的方式保护森林资源；

●进行管理手段的创新，承认社区在气候行动中的价值，激励社区居民积极开展行动，形成自下而上的合力，提升基层治理效率，使能力建设公平有效地融入每一个基层社区。

尼泊尔政府森林管理部门、森林和水土保持部门、地方政府和非政府组织等为该项目的实施提供资金。获得森林使用和管理权后，森林使用者小组通过销售其管理的林业资源来增加收入。几乎所有的森林使用者小组都有自己的基金，用于执行森林管理计划和服务社区发展。尼泊尔政府还通过政策机制创新提供经济方面的激励措施，如将尼泊尔全国林业项目收入中至少25%的经费用于森林可持续管理，约35%的经费用于扶贫，其余收入则投入农村地区基础建设——建设太阳能和小型水电站、社区医院和学校，在洪涝灾害区建立预警系统，维护农村道路，实现社区一级的基本社会保障等。以上措施吸引了更多的林业企业和相关产业，促进了当地生态旅游的发展，为以林业为生的贫困家庭创造了就业机会，吸引了更多社区考虑将林业作为增收的主要方向，让社区林业运动成为一种可持续的项目。

除了有助于森林保护和可持续发展，社区森林合作社还在社会公平和能力建设上发挥了突出作用。作为自下而上以社区为核心的组织，社区森林合作社通过可持续森林管理和开发获得的收入可以进入专门的基金进行统一管理；社区统一管理的方式保证了社区对森林可持续发展的长期和稳定投入，也利于对居民形成正向激励。该项目在性别平等、帮助土著居民创造就业、发展社区文化等方面取得了显著的成绩。根据尼泊尔现有的林业立法，社区自下而上民主选举的执行委员会主席或秘书长必须是女性。因此，在2017年的选举后至少有11 000名女性在该项目的实施框架中担任领导职务（图4-3）。在促进社会公平方面，该项目模式除了创造性地在社区内部进行利润再分配和社区权利保障，还为单个社区的能力建设、提升社区的综合治理能力、缩小城乡差距等创造了条件。社区森林合作社是尼泊尔社区自治的典范。通过建立有效的自下而上的森林管理体系，尼泊尔将能力建设深入到每一个社区，保证了治理行动的高效运作和成效。

图4-3　尼泊尔女性参与社区治理

（图片来源：联合国环境规划署）

4.2.3　项目成果

尼泊尔社区林业运动以社区自发为基础，通过1993年《森林法》在全国的推广，迄今已有22 266个合作社参与，覆盖来自尼泊尔全国农村和城市约300万户家庭。据尼泊尔政府部门统计，截至2016年，尼泊尔全国的森林覆盖率接近50%，相比20世纪末增长了5%以上。今天，在尼泊尔境内几乎所有处于中等海拔地区的森林资源中，超过2 200万公顷的林区已被社区建设和管理成可持续森林，可提供应对气候变化和其他多种社会效益。尼泊尔社区林业运动的主要目标是减少毁林和森林退化，增加本地森林的覆盖面积，从而有效增加碳汇。同时，几乎所有的社区森林合作社都在开展以社区为基础的气候适应行动，这些行动考虑了社区环境和面对气候变化的脆弱性，增加了社区一年四季获得本地食物的机会，有效保护了植物、动物和水源地健康的生态系统，有助于提高全国各地适应气候变化的能力。在尼泊尔政府的积极推动下，原住民和少数民族也自发地保护当地生态环境。政府还开发了一系列的生态产品，发展生态产业以提高社区居民收入，吸引年轻人回乡就业，促进社区发展。上述一系列成果促进了城乡地区的平衡发展，也带动了经济和社会的良性循环，显示出该政策的长期成效。

尼泊尔社区林业运动不仅增加了森林碳汇，具备明显的生态和环境效益，而且在过去30多年中建立了良好的治理系统，社区林业也在国家立法中被明确承认。同时，基层选举制度鼓励低收入人群、土著居民、妇女等参与社区治理，以保证在公平、透明的前提下开展活动。这一举措使社区林业运动不仅在全国范围内促进了发展平衡，还有效实现了社区内部、不同群体间的公平和公正，提升了弱势群体参与社区治理和发展的能力。给其他国家有效开展社区层面的减缓和适应气候变化的行动提供了可长期借鉴的解决方案。

4.2.4 项目信息

案例来源：UNEP-NbS案例库。

项目参与方：尼泊尔政府与当地社区。

项目地点：尼泊尔。

项目时间：自1978年开始。

项目链接：联合国环境规划署，https://wedocs.unep.org/bitstream/handle/20.500.11822/28836/Forestry_Nepal.pdf?sequence=1&isAllowed=y。

4.3 “植万亿棵树领军者”倡议

案例亮点

通过更加高效的方式整合项目，并提供资金和政治支持平台；尝试利用无门槛的平台向全球有识之士征集促进项目发展的解决方案，最大限度地吸引全球伙伴的参与。

4.3.1 项目背景

2006年，联合国环境规划署首次提出“全球十亿棵树运动”，计划用一年的时间在世界范围内种下10亿棵树，以减缓全球气候变化的进程。截至2011年，在该倡议的资助下，全球一共新种植了120亿棵树。2018年3月，在联合国环境规划署等国际组织的推动下，摩纳哥亲王阿尔贝二世、《联合国气候变化框架公约》秘书处执行秘书帕特里西亚·埃斯皮诺萨等国际知名人士在摩纳哥共同签署了《万亿棵树宣言》，将“为地球种树”推上了新的高度。

4.3.2 项目介绍

在达沃斯世界经济论坛（WEF）2020年年会上，“一万亿棵树”的想法被列为大会的正式倡议，提交参会代表讨论（图4-4）。最终，包括中国、美国在内的多国政府代表与300位企业代表共同通过了“植万亿棵树领军者”倡议。该倡议旨在推动大规模再造林投资，提高全球应对气候变化及减缓生物多样性丧失等挑战的能力。该倡议为期10年，同时也是“2021—2030联合国生态系统恢复十年”决议的重要组成部分，为加快全球气候与自然领域的合作与行动提供了重要契机。“植万亿棵树领军者”倡议同步启动了1t.org平台，旨在进一步推动各组织在全球种植一万亿棵树以应对气候变化。

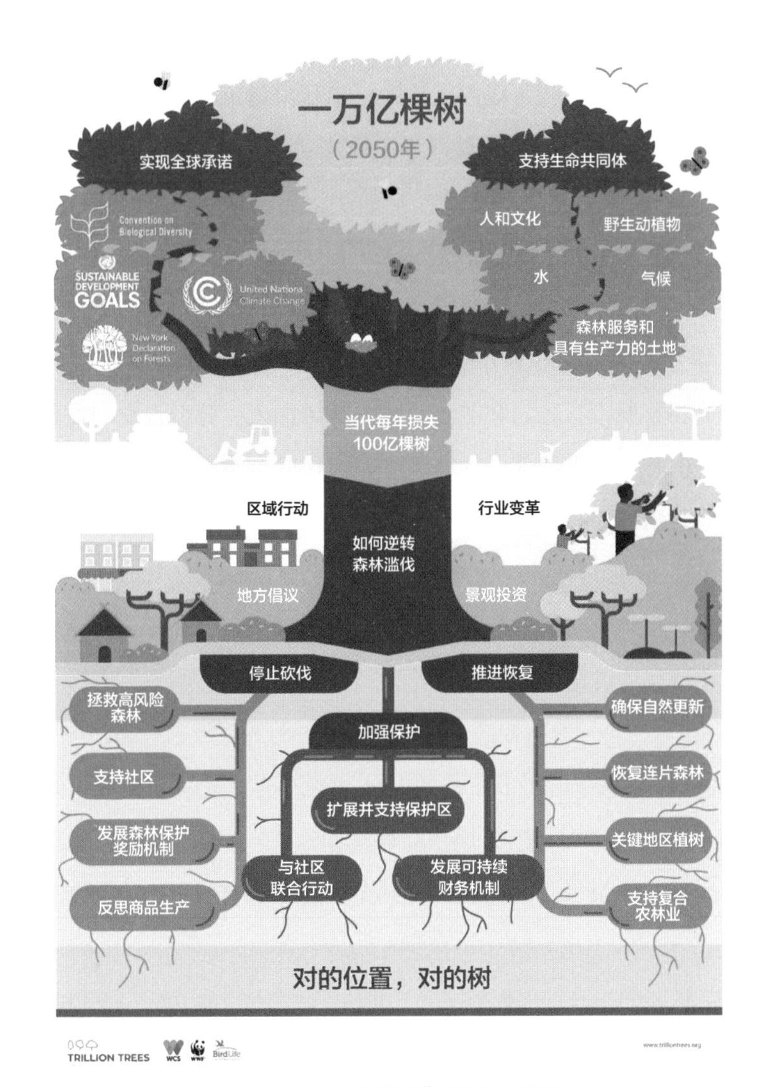

图4-4　“一万亿棵树”计划宣传海报

（图片来源：Trillion Trees）

为推动倡议的落地实施，世界经济论坛设立了秘书处，与世界资源研究所（WRI）、苏黎世联邦理工学院、联合国环境规划署等机构密切合作，为各利益相关方加入倡议并开展工作提供便利。能够依照科学的造林原则、秉持NbS理念，愿意公开展示其在森林保护、修复和生态恢复上的成果及进展，积极推进碳减排目标的各利益相关方均可以向世界经济论坛申请加入“植万亿棵树领军者”倡议，成为1t.org平台的一员，并获得相关的技术、资源支持（李鹏宇，2020）。

2020年7月，美国通过《伟大美国户外法案》，计划在国家森林中重新种植12亿棵树，并支持近5万个工作岗位。2020年8月，“植万亿棵树领军者”倡议在美国成立了第一个国家分会，由美国林学会和世界经济论坛共同主导，密歇根州底特律市、得克萨斯州达拉斯市、万事达卡公司、微软公司、美国银行、美国森林基金会等26家城市、公司和非政府组织承诺加入。分会设立利益相关方委员会，由来自美国政府和私营部门的高级代表担任联合组员，负责为整个分会提供战略规划建议，以确保有效完成“一万亿棵树”的目标。委员会设立秘书处，成员包括林业、社会动员、宣传、沟通等多个领域的专家。

与国家分会相对应，很多企业也纷纷推出行动计划：Salesforce[①]董事长宣布将在未来10年支持和动员超过1亿棵树的保护与修复，直接投资世界各地的森林，通过购买碳信用额度支持修复缅甸的红树林，保护巴西的热带森林和印度尼西亚的泥炭地；惠普公司将“森林零砍伐”作为企业战略，尝试通过与世界自然基金会合作，于2024年之前在中国和巴西修复并保护超过8万公顷的林地，实现企业造纸对森林“零砍伐”的目标，并与美国植树节基金会合作在2020年年底前种植100万棵树。

“植万亿棵树领军者”倡议通过在Uplink平台[②]上立项来保持其创新来源，通过无门槛的平台向全球有识之士征集促进项目发展的解决方案，从中形成案例合集以推广项目经验。多个参与方也应用Uplink平台寻找创新解决方案并吸引更多的民众参与。例如，欧洲的Ecotree组织利用Uplink平台吸引普通群众——个人可

①Salesforce是创建于1999年3月的一家客户关系管理（CRM）软件服务提供商，也是“植万亿棵树领军者”倡议的重要合作伙伴。

②Uplink平台是新兴的创新创意平台，通过多方参与的模式提出符合联合国可持续发展目标的创新项目。

以通过在初期投资150元认养一棵树来参与其中。Ecotree组织计划10年内在全球范围内种植1亿棵树，并有针对性地选择具备经济效益的本地树种，预计整个项目将产生32亿美元左右的利润。当所种植树木完成其生命周期之后，树干等部分将被制作成可循环利用的家具，其所得的经济收入用于保证项目的可持续进行。

4.3.3 项目成果

“植万亿棵树领军者”倡议计划在10年内（2030年以前）通过连接、动员和对开展植树造林项目的社区进行能力建设，在全球范围内保护、修复和种植一万亿棵树。如果美国分会的工作可以顺利实施的话，美国将会有1.31亿英亩[①]的林地被用于修复和再造林，4.3亿英亩的林地得到保护，5 500万英亩的林地树木密度将得以增加，而每棵树将贡献0.62吨的碳汇量。

4.3.4 项目信息

案例来源：UNEP-NbS案例库。

项目参与方：由世界经济论坛牵头，中国、美国等国家政府、组织和个人参与。

项目地点：全球范围。

项目时间：2020—2030年。

项目链接：“植万亿棵树领军者”倡议官网，https://www.1t.org。

---

①1英亩=0.004 046 9平方千米。

4.4 小结

在政府的引领与支持下，哥伦比亚提出“哥伦比亚遗产计划”项目，以保护境内丰富的森林资源。该项目为期20年，旨在通过减少毁林、森林退化和再造林等方式规范和改进国家保护区管理工作，同时实现保护生物多样性、改善水源和提高适应气候变化能力等多重效益。此外，该项目采用了“永久性项目融资”（PFP）的资金模式，有效帮助哥伦比亚政府与国际机构、资助方、私营部门和非政府组织建立合作，提高资金的使用效率，以促进保护区的可持续发展。该项目不仅直接有助于哥伦比亚多个与可持续发展相关的目标的推进，还为国际社会开展NbS提供了可借鉴的资金管理模式和土地规划利用等领域的经验。

尼泊尔社区林业运动起源于政府牵头的全国性林业管理改革，实施期超过30年。改革的重点是设置自下而上的治理模式，出台相应的法律法规，使森林使用者小组能够积极高效地配合政府在基层落实森林可持续发展项目。这让尼泊尔社区林业运动最大限度地动员了全国力量的参与，形成了长期有效且可持续的运行模式。同时，该项目通过较为完善的法律法规、选举和保障系统，保护当地居民的林地使用权，并给予当地居民长久参与林业项目的动力和权力；在选举和管理的过程中重点保障了多元行为体的参与，尤其是贫困人口、少数民族、原住民及妇女等弱势群体的参与。

“植万亿棵树领军者”倡议于2020年在达沃斯世界经济论坛上发起，旨在通过建立一个平台来实现“在2030年前种植一万亿棵树”的目标，包括中国、美国和欧洲多国在内的政府官员及300多家与会公司代表共同通过了该倡议，正式启动了 “植万亿棵树领军者”平台项目。该项目第一个分会设在美国，由美国林学会和世界经济论坛共同牵头。除了政府部门，包括惠普和Saleforce在内的很多企业也积极参与项目的推动，以便于利用互联网平台促进民众参与和创新解决方案。

5 草地类

草地是以多年生草本植物为主要生产者的陆地生态系统，是自然生态系统的重要组成部分，具有防风、固沙、保持水土、调节气候、净化空气、涵养水源等生态功能，对于发展畜牧养殖业、维护生物多样性、提升环境治理和维持生态平衡发挥着不可替代的作用。同时，草地也是面积最大的陆地生态系统，拥有巨大的储碳能力。草地生态系统碳库主要包括植被碳库（地上和地下生物量碳库）和土壤有机碳库两部分，据统计，全球草地生态系统的碳储量为2 663亿吨，占陆地生态系统总碳储量的12.7%，其中生物碳储量占世界的6.0%，土壤碳储量占世界的15.5%（Lieth H et al.，2012）。

荒漠化防治、气候变化和生物多样性保护被共同列为《21世纪议程》的优先行动领域。除了人类活动造成的影响外，气候变化所带来的气温升高、降水减少是荒漠化日趋严重的重要驱动因素。因此，NbS与荒漠化治理之间存在显著的协同效应，可以进行协同治理，恢复荒漠化土地可以提高碳储量、保持水土，进而减缓和适应气候变化。由于干旱与半干旱地区树木种植和管护的难度大、成本高且存活率低，在这类地区开展草地管理和修复是荒漠化治理的重要手段。本章选取了非洲“绿色长城”计划与中国毛乌素沙地治理两个生态修复案例，其采取的“林业+草地”修复策略可以在跨区域与国家两个层面为其他地区和国家实施基于自然的草地生态修复提供借鉴。

5.1 非洲“绿色长城”计划

案例亮点

倡导自上而下和自下而上两种模式的有机结合：在区域层面，通过非洲联盟（AU）协调各国政府在国家层面制定统一规划、设置远景目标；在地方层面，考虑到行政管理能力薄弱、难以通过行政管理手段来干预个体行为等局限因素，通过自下而上的社区治理模式变被动为主动，寻找适合当地情况的经济树种，或者以“林草结合”的模式激发当地民众参与的积极性。

5.1.1 项目背景

撒哈拉沙漠以南的非洲被认为是最易受气候变化影响的“热点地区”。当地生态环境脆弱，高度依赖雨养农业，持续受到干旱和荒漠化的挑战且适应能力有限。鉴于这些地区在一定程度上不可避免地会受到气候变化的影响，适应气候变化被列为撒哈拉沙漠以南的非洲地区气候政策中的优先领域。

5.1.2 项目介绍

受阿尔及利亚绿色大坝①和中国万里长城的启发，在世界银行、欧盟和联合国等的共同资助下，有22个国家（毛里塔尼亚、乍得、尼日尔、埃塞俄比亚、布基纳法索和尼日利亚等）参与了由非洲联盟牵头组织的“绿色长城”计划。该计划最初于20世纪80年代提出，在2000年前后重新受到关注，其主要目标是在整个

①绿色大坝又称绿色带，是北非正在建设的一项跨国林业项目，涉及摩洛哥、阿尔及利亚、突尼斯、利比亚、埃及五国，其基本内容就是通过造林种草建设一条横贯北非五国的绿色植物带，以阻止撒哈拉沙漠的入侵和土地沙漠化。

非洲范围内建造一条长达8 000千米的世界自然奇观。如果成功实施，“绿色长城”将成为地球上最大的活体绿色建筑，是大堡礁面积的3倍。“绿色长城”计划的具体规划见图5-1。

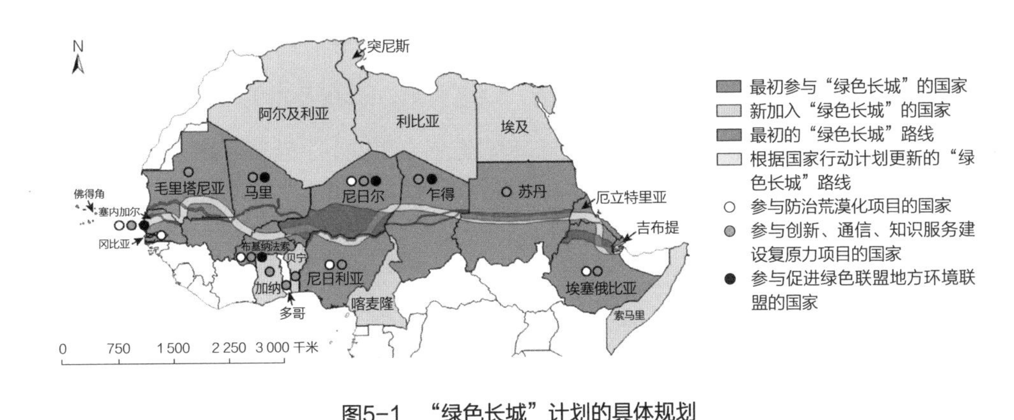

图5-1　“绿色长城”计划的具体规划

（图片来源：Goffner D et al.，2019）

“绿色长城”计划在设计之初就尝试建立一个横跨撒哈拉地区的连续森林地带，通过调节局地气候、温度、风速和减少土壤侵蚀，在减缓沙漠化进程的同时促进农业发展。经过一段时间的实施与评估，当前“绿色长城”计划不再倡导人为建造一组前后贯通的森林屏障，而是通过塑造不同的景观类型组成“马赛克式”的生态保护屏障，这将为改善区域环境、提高社区收入提供长期解决方案。“绿色长城”计划是一系列林业与非林业项目的集合，由位于撒哈拉地区的不同国家和国家间组织进行具体的设计和管理。该计划倡导自上而下和自下而上两种治理模式的有机结合：规划层面，自上而下由各国政府在国家层面因地制宜地制定统一规划、设置远景目标；区域层面，通过全球环境基金、常设委员会、监测站和国家联盟协调国家间的合作关系；地方层面，考虑到地区行政管理能力薄弱、难以通过行政管理手段来干预个体行为等局限因素，通过自下而上的社区治理模式变被动为主动，激发当地民众的参与积极性，同时保证项目设计和实施能够切实解决地方需求。

“绿色长城”计划中涉及多个国家的多个项目，通过政府和募捐两部分来筹措资金。2015年12月，《联合国气候变化框架公约》第21次缔约方大会承诺将为“绿色长城”计划提供40亿美元的额外资助；法国政府同意2020年之前提供10亿欧元的资助；世界银行总裁承诺为“绿色长城”计划及相关项目提供19亿美元的资助。“绿色长城”计划得到全球环境基金（GEF）、欧盟、联合国粮食及农业组织（FAO，后文简称联合国粮农组织）等的支持，通过世界银行新设立的PROGREEN多捐助方信托基金等对外筹资，为萨赫勒与西非项目协调国家行动并管理资金以保证维护费用。

“绿色长城”计划包含关于林业和草地两方面的具体任务。林业方面，以植树造林（如沙漠刺槐和猴面包树）为主。在沙漠地区造林后需要大量的人员管护费用，因此推广经济价值较高的树种有助于鼓励当地居民开展后续的森林管护工作。目前，塞内加尔正在大规模推广这种做法，该国的造林项目主要生产用于出口的阿拉伯树胶，已在约10万英亩的土地上种植了约1 800万棵树。此外，生态状况的改善使瞪羚、豺狼和鸣禽等多种动物开始回归。

需要注意的是，由于造林成本较大，在干旱地区种植树木的存活率只有20%左右，而在管理完善的种植园中树木的存活率能达到70%，因此“绿色长城”计划中的很多国家选择了修复退化的草原或草场。在马里、布基纳法索和尼日尔，人们用围栏围住了大片土地，指导当地农户在现有树木和芽苗周围种植，让土地从长期过度放牧的影响中获得修复和复苏。过去30年中，这种方式使尼日尔有近1 200万英亩的草地得到了修复。

草地修复项目中也不乏优秀的实践案例。例如，考虑到撒哈拉地区不适合开展大规模造林活动，局部地区开展了“生态围栏”项目，通过识别优先区、建设生态围栏等措施开展保护行动，以实现阻止动物破坏草场和恢复生态系统的目标。经验证，生态围栏中树木的数量是围栏外的3～4倍，平均树木高度超过围栏外20厘米。需要强调的是，“生态围栏”项目并非替换掉生态系统中的现有物种，而是通过减少外来物种侵袭、提升土壤肥力、降低火灾等提升树木生长和扩张的速度。“生态围栏”项目符合该地区适合造林的土地较为分散的特点，给社区居民提供了更多的就业机会，帮助当地实现了精准扶贫和生态保护的共赢。

5.1.3　项目成果

到2030年，“绿色长城”计划将恢复1亿公顷退化的土地，封存约2.5亿吨二氧化碳，给当地农村地区创造1 000万个就业机会。该计划于2007—2030年进行建设，目前已进入第二个规划期（建设进度15%）。除取得上述成绩外，该计划也面临着巨大的挑战：一方面，萨赫勒地区的荒漠化问题严重，每年都有大量林地因农业、建筑业、伐木业而无序发展或退化；另一方面，为该计划认捐的80亿美元只有一半到位，且种植和养护树木的成本巨大，给该地区较为贫穷的国家带来沉重的财政负担，部分地区的树木存活率也仅有20%（Aryn Baker et al., 2019）。尽管整体项目建设的进度相对缓慢，但是“绿色长城”计划取得的生态效益远超出预估。根据《联合国防治荒漠化公约》（UNCCD）公布的数据，“绿色长城”计划自2007年启动以来取得了重大进展（表5-1）。

表5-1　“绿色长城”计划进展

国家	成果
埃塞俄比亚	恢复了1 500万公顷的退化土地，改善了土地使用权保障
塞内加尔	种植了1 140万棵树，恢复了25 000公顷退化的土地
尼日利亚	恢复了500万公顷退化土地，创造了2万个就业机会
苏丹	恢复了2 000公顷的土地
布基纳法索、马里、尼日尔	约有120个社区参与，在2 500多公顷的退化地和旱地上建立了一条绿化带，用50种本地树种种植了200多万棵苗

信息来源：《联合国防治荒漠化公约》。

5.1.4 项目信息

案例来源：UNEP-NbS案例库。

项目参与方：非洲联盟牵头的22个国家和资助方（世界银行、欧盟、联合国等）。

项目地点：撒哈拉沙漠以南的非洲萨赫勒地区。

项目时间：2007—2030年。

项目链接："绿色长城"计划官网，https://www.greatgreenwall.org/。

5.2 中国毛乌素沙地治理

案例亮点

政府引领，落实系列生态治理政策；根据毛乌素沙地的不同条件在治理区域精准施策；创新防沙治沙模式，防用结合，推动林沙产业发展，使生态、经济和社会效益实现共赢。

5.2.1 项目背景

毛乌素沙地位于中国内蒙古自治区、陕西省和宁夏回族自治区三省（区）交界处，即鄂尔多斯高原向黄土高原的过渡区，是以草地放牧业为主的农、林、牧交错地区，同时也是具有特殊地理景观的生态过渡带。该区域为典型的温带大陆性半干旱气候，年均降水量为250～440 毫米，总体地势自西北向东南逐渐降低，主要的风沙地貌类型包括流动沙丘、半固定沙丘及固定沙丘。毛乌素沙地的植被地带性自西北向东南呈现荒漠草原—典型草原—草甸草原的过渡变化，各种隐域性植被[①]分布广泛，主要有沙地植被、湿地植被和盐生植被等。

毛乌素沙地被称为“最年轻的沙漠”及“人造沙漠”。历史上，这里曾是水草丰美、牛羊载道，有着“临广泽而带清流”的美景，但在唐代中叶以后，由于人类活动范围的扩大、过度开垦放牧及战乱等，再加上气候变化，毛乌素地区逐渐荒漠化，最终形成了连片的沙漠。随着地区生态环境的不断恶化，流沙侵袭压埋城镇村庄，草场退化，庄稼十耕九不收，当地群众生活愈加困难，毛乌素沙地一度形成了沙进人退的被动局面。

5.2.2 项目介绍

毛乌素沙地的治理措施主要有以下三个方面。

一是政府引领，构建绿色生态屏障。自1978年以来，中国先后实施了“三北”防护林、天然林保护、退耕还林还草等生态修复工程。地方政府在积极推进相关林业重点工程建设的同时，开展造林绿化、落实生态移民禁牧封育等生态治理政策，调整农牧产业结构，鼓励多种所有制参与生态建设，加大人力、财力的投入（图5-2）。得益于政府的引领作用，毛乌素沙地区域的沙漠化得到了明显的逆转。

（a）生态移民新村

（b）禁牧圈养

图5-2　地方政府采取的措施

（图片来源：中国治沙暨沙业学会）

①隐域性植被又称“非地带性植被”，是指受局部地形、土壤、地下水、地表水影响，出现两个以上植被带里的植被，如草甸植被、沼泽植被、水生植被等。

二是精准施策，创新防沙治沙模式。经过多年的努力探索与实践，一代又一代的治沙人总结出适合毛乌素沙地的治沙模式与经验，针对不同条件的治理区域提出了相应的对策。例如，建立以“带、片、网”相结合为主的防风沙体系；筛选出柠条、沙柳、花棒、樟子松等一批适宜生长的固沙植物；采取在流动半流动沙丘草方格内栽植固沙植物的治沙技术；对面积广、高低起伏、密集流动的沙丘地区采取“飞播+人工封育”的方法；在沙漠及活化沙丘的边缘，建立以灌木为主的防风固沙生物隔离带；实行谁造林谁受益，允许继承转让等造林绿化优惠政策和措施（图5-3）。

（a）草方格固沙

（b）飞播造林

图5-3　治沙模式

（图片来源：中国治沙暨沙业学会）

三是防用结合，推动林沙产业发展。以治理带开发，以开发促治理。毛乌素沙地利用沙区独特的光、热、土等资源优势，大力发展林沙产业，将防沙治沙与用沙相结合，使生态、经济和社会效益实现共赢。例如，发展以沙生植物资源为基础的加工型沙产业、以沙区光热土资源为主的绿色工业沙产业和以沙地旅游景观为主的沙产业等，具体包括利用沙生灌木需要定期平茬抚育的生物习性获得可再生的生物质原料（燃料），进行生物质发电；利用风积沙工业选矿，并将造矿后的产品作为生产原料广泛用于玻璃、陶瓷、冶金、电子、医药和化工等工业；开展螺旋藻养殖及深加工产业生产等（图5-4）。

（a）螺旋藻养殖园区

（b）沙漠中的生态公园

图5-4　林沙产业

（图片来源：中国治沙暨沙业学会）

5.2.3　项目成果

治沙半个世纪以来，毛乌素沙地荒漠化防沙治沙工程效益显著，其荒漠化程度呈明显逆转的趋势。

一是沙化土地面积减少、程度减轻。中国荒漠化土地和沙化土地面积自2004年开始出现缩减以来，已经连续10年保持了“双缩减”，并且荒漠化和沙化程度继续减轻。近5年来，毛乌素沙地重度和极重度沙化土地面积减少了628.2万亩[①]，内蒙古自治区乌审旗依托三北防护林、退耕还林还草等国家林业工程及地方生态修复政策和措施，2000—2017年沙化土地面积呈现明显的缩小趋势（图5-5）。

①1亩=1/15公顷。

（a）治理前

（b）改善后

图5-5　毛乌素沙地历史变迁

（图片来源：中国治沙暨沙业学会）

二是植被覆盖度提升。植被覆盖是防沙治沙的重要方式，也是荒漠化治理的主要目标之一（图5-6）。以代表旗县为例，乌审旗的植被覆盖度从20世纪70年代的28%提高到2018年的80%；榆林沙化土地治理率达93.24%，林木覆盖率由0.9%提高到34.8%；伊金霍洛旗2019年年底的森林面积达300万亩，森林覆盖率达36.85%，植被覆盖率达88%。

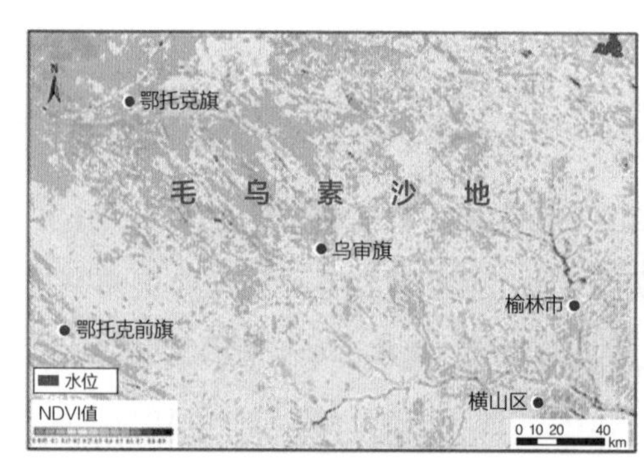

（a）2000年7月

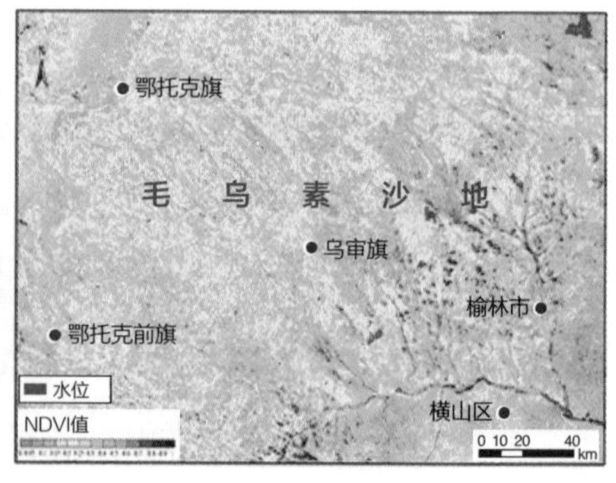

（b）2019年7月

图5-6　2000年与2019年毛乌素沙地植被指数对比

（图片来源：中国治沙暨沙业学会）

三是生态系统碳汇增加。宁夏盐池毛乌素沙地生态系统国家定位观测研究站对毛乌素沙地南缘典型荒漠生态系统的长期测算发现，以油蒿、杨柴、沙柳为典型植被的灌木生态系统在较长的时间尺度内发挥着重要的碳汇作用（每平方米每年吸收80克碳）。

四是人类居住环境和气候得到改善。气候改善是荒漠化治理的重要生态效益。以陕西省榆林市为例，扬尘天气由原来的每年100多天减少到每年10天以下。2017年，榆林市空气优良天数达到285天，气候得到了很大的改善；伊金霍洛旗的沙尘暴和沙尘天气明显减少，沙尘暴由2000年的每年23次降至2019年的每年13次，沙尘天气得到有效遏制。

5.2.4 项目信息

案例来源：C+NbS合作平台。

项目参与方：中国中央政府，内蒙古自治区、陕西省和宁夏回族自治区政府与当地群众。

项目地点：中国毛乌素沙地。

项目时间：1959—2020年。

5.3 小结

非洲“绿色长城”计划由22个国家主导且包含多个子项目，计划在整个撒哈拉沙漠以南的区域建造一条长达8 000千米的绿化带。在如此雄心勃勃的目标引导下，该计划倡导自上而下和自下而上两种模式的有机结合。在区域层面，通过非洲联盟的协调，各国政府在国家层面制定统一规划、设置远景目标；在地方层面，通过自下而上的社区治理模式有效提升社区参与的积极性，同时保证项目设计和实施能够切实解决地方需求；在国家层面，因“绿色长城”计划的强大号召力来源于其倡导的主人翁精神和区域认同感，故通过“建造新世界奇迹”的理念获得更多的支持与响应；在社区和个人层面，注重号召该区域青年共同参与。

毛乌素沙地的生态恢复是区域荒漠化治理的典型成功案例，是践行绿水青山就是金山银山理念的优秀实践案例。以毛乌素沙地为代表的中国荒漠化防治实践，首先得益于中国的林业保护与防沙治沙的法治体系、强有力的管理和有效的荒漠化监测与报告体系；其次，治沙学者在毛乌素沙地的治理过程中挖掘、创新了许多有效的固沙技术措施，包括飞机播种、封沙育林育草、机械沙障、风力治沙等，为中国和全球的荒漠化防治提供了重要的借鉴经验；最后，毛乌素地区治沙的成功更是当地人民一代代共同努力的结果，离不开基层人民的广泛参与。

6 农业类

农业是重要的温室气体排放源，据估计，温室气体排放总量中至少有1/5来自农业部门，主要来源是土地利用方式、反刍动物的肠胃发酵、秸秆焚烧、水稻生产、有机肥和化肥生产过程中释放的甲烷和氧化亚氮、林业和土地利用方式改变等。因此，粮食作物生产节能减排和土壤固碳减排潜力巨大。另外，农业生产需要高度依赖当地的气候条件，极易受到气候变化的冲击。实施NbS并发展可持续农业可以实现少投入、多产出，在减少碳排放的同时提高农业应对气候变化的适应能力，对于可持续发展意义非凡。可持续农业也可以带来更多的投资和就业机会，从而拉动区域和全球经济增长。商业与可持续发展委员会（BSDC）的研究表明，截至2030年，粮食和土地利用方式的转型每年可创造高达4.5万亿美元的新的商业机遇（BSDC，2016）。粮食与土地利用联盟（FOLU）分析发现，到2030年，每年在可持续粮食和土地利用系统上增加3 500亿美元的新投资可以创造1.2亿个新工作机会（FOLU，2019）。联合国粮农组织预测，向可持续农业和土地利用类型转变，可以为全球贡献2.3万亿的经济增长量，到2050年将创造2.0亿个就业岗位。本章聚焦印度、中国两个全球农业生产大国，选取了印度基于自然的“零预算农业”、中国气候智慧型主要粮食作物生产、中国杭州“三好农业”实践3个典型案例，旨在为农业领域的NbS实践提供可以学习和参考的经验。

6.1 印度基于自然的“零预算农业”

在生产过程中尊重作物的自然生长模式，能够在成本更低的条件下带来更高的净利润；关注小农户生产问题和性别平等，力求雇用同等数量的男性和女性领导，向女性提供创业激励等。

6.1.1 项目背景

印度作为农业大国，有超过一半的人口直接或间接地依赖农业生存。安得拉邦[①]是印度的农业中心，农业是其62%的人口的主要收入来源，占当地生产总值的1/3左右（Rythu Sadhikara Samstha, 2019）。过去安得拉邦多采用传统的农业种植模式，大量施用化肥农药，导致种植成本高昂、土壤肥力受到严重破坏。当地的自然条件恶劣，长时间面临干旱和缺水的问题，再加上由气候变化带来的极端气候事件和雨量分配不均等现象，农业发展受到严重威胁。随着城镇化进程加速、土壤退化、气候变化等问题的日益严峻，安得拉邦的农业用地迅速减少。除此之外，农产品市场的不确定性使原本就收入不高的农民雪上加霜，容易陷入债务危机。据统计，安得拉邦有高达90%的人口背负着还债压力，人均债务达1 500美元。这同时也是当今世界很多国家面临的困境，关乎全球是否能够真正实现可持续发展目标，为此联合国粮农组织于2018年开始倡导所有国家应用推广农业生态学与可再生农业，并联合印度政府在安得拉邦开展实践（图6-1）。

---

① “邦”是印度划分各个地区的行政单位，相当于中国的“省”。

图6-1 印度“零预算农业”的农田

（图片来源：自然与气候联盟）

6.1.2 项目介绍

为帮助农民修复已被破坏的土壤，同时提升小农户的生计状况，印度农业部和安得拉邦政府积极响应联合国粮农组织的号召，启动了名为“安得拉邦零预算自然农业”的可再生农业项目。“零预算农业”主要依靠低成本的自然投入来种植粮食，不使用昂贵的、具有污染性的化学产品。该项目旨在于2018—2024年在安得拉邦800万公顷的土地上推广可再生农业措施，将传统农业模式转变为基于自然的“零预算农业”（ZBNF），预计项目成果将惠及约600万农户。该项目在安得拉邦政府成立的农民协会和其他多家非政府组织的支持下进行。“零预算农业”技术的支持方包括世界混农林业中心、雷丁大学、印度科技学院和印度科学教育与研究学院，主要关注对农户进行能力建设和促进社区可持续发展。项目资金主要来源于联合国环境规划署推行的“可持续印度融资机制”（6年共计23亿美元）、慈善界捐赠及联合国粮农组织。

在项目初期的设计环节，安得拉邦秉承尊重自然的理念，在生产过程中尊重作物的自然生长模式，避免使用任何人工合成的化肥、农药等物质。这种农业生产方式能够在成本更低的条件下带来更高的净利润，因此被当地农民广泛采纳。作为日渐普及的再生农业策略之一，“零预算农业”除了能够提高农业的适应能力，还可以有效减少碳排放。该项目根据当地现有的农作方式，充分利用本地智慧推广气候智慧型农业，采取就地取材、利用天然肥料等做法有效增加了土壤的有机质含量，改良后的土壤提高了农业生产力、生物多样性、碳储存和水资源使用效率。该项目基于本地智慧的4种措施见表6-1。

表6-1 “安得拉邦零预算自然农业”项目基于本地智慧的4种措施

措施序号	措施名称	措施介绍
1	种子处理	运用牛粪、牛尿处理种子，为种子覆上微生物包衣，保护其幼根免受真菌、种子传播病或土壤传播病的侵害
2	零化学品	利用牛粪、牛尿、当地特产的椰糖与豆粉、未被污染的土壤等刺激微生物发酵，自制有机肥，使营养物质为植物所利用，在防病菌的同时增加土壤的固碳能力
3	土壤小气候	用作物和作物残渣来覆盖表层土壤，确保土壤中有适宜的气候，其好处在于可以产生腐败物质，保护表土，提高土壤保水性，控制杂草，营造土壤生物群，为土壤提供必需的养分
4	土壤保湿	通过第二项和第三项措施改善土壤结构及水分，采取本措施促进土壤通气，增加土壤含水量，提高用水效率，增强作物抗旱能力

信息来源：http://apzbnf.in。

当地奶牛的粪便、尿液和乳制品都可以作为“零预算农业”投入的原料，因此当地奶牛的数量对于该项目至关重要，一头奶牛足以满足30英亩以上土地的需求。除了就地取材，针对农民开展培训是提高项目参与度的一项重要手段。在项目执行过程中，首先选取约6 000名有影响力的农业大户开展培训和能力建设，通过他们的增收来带动其他村民。项目方有时会发现这些农业大户甚至会在夜间通过投影仪给其他村民展示视频内容，讲解并传授知识。不仅如此，通过培训农业大户使用智能手机及时跟踪项目进度，也在一定程度上减少了评估项目进展所需要的工作量。这样的培训每天都在不同的村庄发生，大大提升了项目的执行速度和普及率。

实施“安得拉邦零预算自然农业”项目有助于解决当地农户普遍面临的高负债问题，可以减少高达50%的农业转型成本，在减贫和性别平等方面有较强的借鉴意义。例如，该项目为符合条件的农民（拥有旱地不足2.5英亩、湿地面积不到1.25英亩的农民或单身女性农民）介绍特定的作物和牲畜模式，以提高收入和确保粮食安全。约有86 000个女性自助团体在非营利组织的帮助下参与规划、管理和监督项目的执行。该项目力求雇用同等数量的男性和女性领导，向女性提供创业激励，如支持她们在村里建商店、向农民销售天然肥料等有机肥料。

6.1.3　项目成果

2017年，已有58万名农民、3 000多个村庄和26万公顷的土地参与了该项目。其中，超过90%的农户收入和产量有所增加，尽管仍有10%的农民报告产量有所下降，但总收入是持续增加的。2018—2019年，印度经济社会研究中心针对“安得拉邦零预算自然农业”开展研究，结果表明在所有作物中基于自然的“零预算农业”每亩投入的生物成本和种植成本均低于传统农业下的化学投入成本（表6–2）。

表6–2　基于自然的“零预算农业”与普通农业措施之间的净利润、种植成本对比

作物	基于自然的“零预算农业”净利润/（美元/公顷）	普通农业净利润/（美元/公顷）	净利润增长率/%	基于自然的“零预算农业”种植成本/（美元/公顷）	普通农业种植成本/（美元/公顷）	种植成本降低率/%
稻田	667	603	11	526	604	13
花生	129	116	11	313	376	17
玉米	639	302	112	454	457	1
棉花	1 003	573	75	518	567	9
鹰嘴豆	769	655	17	398	464	14

信息来源：Galab S et al.，2019。

土壤的破坏及如何加强小农户适应气候变化的能力是全球面临的普遍问题，该项目利用本地智慧，一方面帮助农民修复受到破坏的土壤，另一方面也提升了经济收入，在印度甚至国际上有很强的借鉴意义。当地土壤测试的数据显示，83%的农户报告土壤肥力提高，而通过对土壤软化度、蚯蚓数量和田间植被覆盖率等数据的跟踪均证实了这一结论。从农民报告的谷物重量、茎秆强度和作物口味这三方面显示出产出质量的提高，42%的农民表示作物具有更强的抗旱和抗倒伏能力。可再生农业投入成本较低而产量较高，使农民从中获得了更高的收益。

6.1.4 项目信息

案例来源：N4C全球NbS案例研究。

项目参与方：印度农业部、安得拉邦政府。

项目地点：印度安得拉邦。

项目时间：2018—2024年。

项目链接："安得拉邦零预算自然农业"项目官网，http://apzbnf.in/。

6.2 中国气候智慧型主要粮食作物生产

自上而下的组织机构和运行机制的建立，不仅加强了项目的组织领导，还为项目正常运转提供了可靠的制度保证，使项目得以持续发展。

6.2.1 项目背景

中国安徽省怀远县是中国产粮百强县之一，冬种小麦、夏种水稻。2013年，该县水稻种植面积占总耕地面积的84.3%，小麦种植面积占总耕地面积的98.8%。

怀远县也是重要的劳务输出县，农村人口外出打工现象十分普遍。2013年，该县总劳动力为26 140人，外出打工的劳动力占总劳动力的49.4%，这些人在农忙期间基本不回家参与农业生产。在怀远县，村级合作社发展仍处于起步阶段，全县共有各类农民专业合作社9个，一般规模较小，部分已不再运行。其中，与作物生产相关的合作社仅有2个，基本服务于当地流转土地的农业公司。怀远县的农机服务现象比较普遍，主要涉及水稻和小麦机械收割两个环节，基本的组织形式是自发性的，即由本村或邻村拥有收割机械且具有服务能力的农户提供。但是，怀远县缺少统一的植保方面的服务内容，配方施肥也尚处于示范阶段。

中国河南省叶县是农业大县，玉米种植面积占总耕地面积的88.5%，小麦种植面积占总耕地面积的90.2%。叶县也是典型的劳务输出县，总劳动力占总人口数的56.7%，外出打工的劳动力占总劳动力的40.6%，打工地点分布全国，但主要集中在河南省内。由于距离较近，这些外出打工的劳动力在农忙季节（播种和收获）可以回家从事农业生产活动。在叶县，合作社发展同样处于起步阶段，共有各类农民专业合作社14个，最早的农民专业合作社成立于2008年，一般规模较小，部分已不再运行。大部分农民专业合作社围绕种植业开展，以提供农业技术服务为主。叶县的农机服务已经全面覆盖，包括玉米和小麦的机械播种、机械收割及秸秆还田三个环节，基本的组织形式也是自发性的。植保方面的服务和配方施肥与怀远县类似，尚处于示范阶段。①

安徽省怀远县和河南省叶县作为“气候智慧型主要粮食作物生产项目”的两个示范点，虽然主要粮食作物品类不同、环境各异、生产模式各有侧重，但是面临的挑战是相同的：人均耕地面积少、土地零散，农业基础设施落后，种植业生产管理模式不合理，温室气体排放多，作物秸秆还田及保护性耕作应用范围小，生产系统受气候条件影响稳定性差，化肥农药投入量大、耗能高、浪费严重，土

①主要数据来源：中国农业大学人文与发展学院于2014年发布的气候智慧型农业项目《社会影响评估报告》，详见http://www.reea.agri.cn/sttzgg/201906/P020190702531978504130.pdf。

壤固碳能力低且肥力有限。

6.2.2 项目介绍

“气候智慧型农业”这一概念由联合国粮农组织于2010年提出，旨在对农业系统进行改造并重新确定发展方向，帮助和支持其有效发展并确保气候变化下的粮食安全。气候智慧型农业有3个主要目标：持续增加农业产量和收入、建立和提高对气候变化的适应能力、在可能的情况下降低或避免温室气体排放。2014年，中国农业部和世界银行共同启动了由全球环境基金资助的“气候智慧型主要粮食作物生产项目”。该项目在全球环境基金510万美元赠款和参与的省、县2 500万美元的支持下开展，实施期为5年。安徽省怀远县和河南省叶县作为示范区，围绕水稻、小麦、玉米3种主要粮食作物，通过气候智慧型农业生产技术示范与应用、政策应用与创新及知识管理、公众知识的拓展与提升等活动，减少农业生产的温室气体排放，增加土壤有机碳含量，提高作物生产适应气候变化的能力。气候智慧型农业项目所提倡的发展模式可以概括为“固碳、减排、稳粮、增收”八个字。

6.2.3 项目成果

在“气候智慧型主要粮食作物生产项目”的实施过程中，国内外专家经过5年的共同努力，引进了一批位于国际前沿的新技术、新理念，实现了技术、理念的本土化，取得了丰硕成果。截至2019年年底，该项目在粮食主产区安徽省和河南省建立了10万亩示范区，项目区的粮食产量年均增加5%以上：小麦平均增产5.5%～6.9%，夏玉米平均增产2.1%～17.2%，水稻平均增产4.5%～8.5%。项目区单位面积的氮肥用量减少了约10%，农药用量减少了15%以上，土壤有机碳含量提高了10%。该项目5年累计实现固碳减排13.22万吨二氧化碳当量，超额完成了预期目标（6.5万吨二氧化碳当量）。项目监测与评价报告显示，在项目区开展的测土配方施肥、保护性耕作、秸秆还田和节水灌溉等技术活动对作物增产和固碳减排发挥了重要作用。同时，该项目提升了农民的科技意识和田间管理水平，制定了相关的技术规程与补贴制度，建立了资源高效、经济合理、固碳减排的生产模式，增强了作物生产对气候变化的适应能力，推动了中国农业生产的节能减排。2019年，中国水稻、玉米、小麦三大粮食作物的化肥利用率达到39.2%，农药利用率达到39.8%，化肥农药施用量连续三年负增长；规模化养殖污染防治有序推进，以农村能源和有机肥为主要方向的资源化利用产业日益壮大；以秸秆农用为主、多元发展的利用格局基本形成。

6.2.4 项目信息

案例来源：C+NbS合作平台。

项目参与方：中国农业农村部、世界银行、全球环境基金。

项目地点：中国安徽省怀远县、河南省叶县。

项目时间：2014—2019年。

项目链接：中国农业农村部，http://www.moa.gov.cn/xw/bmdt/202009/t20200922_6352773.htm。

6.3 中国杭州“三好农业”实践

案例亮点

创新地引入了慈善信托，依靠信托制度的优势为传统捐赠方式增加了更加多元化且具有影响力的投资渠道；通过开发“三好农业”，探索农户创收和保护环境之间的平衡，促进农户的持续参与；开发环保教育课程，促进社会共同参与环保事业。

6.3.1 项目背景

自2020年起，千岛湖已经成为中国杭州等城市约1 000万人的饮用水水源地，至少占杭州市饮用水总供给量的50%（图6-2）。千岛湖流域整体水质优良，在全国水环境状况排名中比较靠前，但仍然面临来自农业的面源污染、生态脆弱等压力。通过大自然保护协会与世界银行对该流域的分析，千岛湖流域保护的主要威胁来源于农村，这主要与集水区内的农户在种植中过量施用农药、化肥，畜禽养殖密度大等不合理的农业方式有关。面源污染又称非点源污染，主要包括土壤泥沙颗粒，氮、磷等营养物质，农村生活垃圾，各种大气颗粒物沉降等现象，通过地表径流、农田排水等形式进入水体环境，具有分散性、隐蔽性、潜伏性和模糊性等特点。此类污染源不易监测、难以量化，研究和控制的难度较大，目前缺乏有效的长期治理机制。

图6-2　千岛湖实景

（图片来源：大自然保护协会）

与此同时，气候变化对千岛湖周边水域的生态、农业和人民的生活都有重要影响。从气候方面来说，近年来由于高温、干旱天气增多，水源地缺水、断流的情况时有发生，严重影响周边居民的生活饮用水安全。过去，中国曾在各地重点发展工业和制造业，在提高人民福祉的同时也牺牲了许多自然资源，水源地的破坏逐渐成为显现的社会问题。工业发展带来的收益也让人们逐渐忘却了自然资源的重要性，一味地滥用资源只能让环境更加恶化，最终给当地人民群众的生产和生活带来威胁。以浙江省杭州市余杭区青山村的龙坞水库为例，2014年在大自然保护协会对该水库进行治理之前，其水质被检测出总氮、总磷和溶解氧不达标，主要是由于水库周围山上种植的毛竹林长期过度施肥。自20世纪八九十年代以来，当地村民因看到毛竹的经济价值而开始大量种植毛竹，并为提高产量而大量施肥。从政府层面来看，应大力呼吁水域附近的农户提高水源保护意识，改变对固有思维的认知，只有齐心协力才能实现保护和治理水源的终极目标。另外，千岛湖流域地处浙江省与安徽省交界处，因天然的自然地理条件极易受到气候变化带来的极端天气影响，容易引发山洪、滑坡、泥石流等自然灾害。从经济角度来看，千岛湖周边的农户多以种植茶叶为生，这类农产品的品质也因水源的损害而受到威胁，所以在保护水源的同时，还要结合生态的可持续发展，构建水源保护的长效机制，改变当前千岛湖流域农户的生产方式及环境意识。

6.3.2　项目介绍

生态保护不能仅靠一己之力，而需要全社会的共同参与。为探索水源地的长效保护机制并使之切实落地，2017年阿里巴巴公益基金会和民生通惠公益基金会联合万向信托股份公司设立了“中国水源地保护慈善信托”。2018年，“千岛湖水基金”项目作为“中国水源地保护慈善信托”资助的首个项目在杭州启动，由大自然保护协会担任科学顾问。

该项目致力于用科学的方法解决农业的面源污染问题，通过绿色消费帮助流域生态产业发展，最终实现千岛湖水源的长效保护机制，并改变当前千岛湖流域农户的生产方式及环境意识。“千岛湖水基金”创新地引入了慈善信托，依靠信托制度的优势为其助力，从而激发了环保事业的潜力。在机制的设计与构建中，信托以减少环境污染、保证经济利益为首要目标，通过灵活巧妙的金融设计，为传统捐赠方式增加了更加多元化且具有影响力的投资渠道，尽可能地拓展了公益慈善项目的募资范围、满足了各类公益慈善群体的需求，同时促进募集资金的增长，形成环保项目自身的良性循环，在决策机制与信息披露等方面保障了资金管理运行的公开与透明，实现了生态系统付费模式。

当前，千岛湖水源地的农业面源污染主要由水土流失和肥料、农药使用不当等直接原因造成，根本原因是容易造成水土流失的坡耕地自然条件与不可持续的农业生产模式。为实现对流域面源污染的控制，改革农业生产模式，“千岛湖水基金”项目采用绿肥覆盖、芒秆覆盖、自然落果和经济套种相结合的保护性农业实践措施，在安阳等地开展试点，推广“源头减量+过程拦截+末端处理”的流域面源污染治理综合示范。为充分发挥“绿水青山”的经济价值，“千岛湖水基金”项目提出了“三好农业”商业模式，所谓“三好”即“好品”“好卖”“好玩”。“好品”，即向公众宣传农产品种植过程中所使用的生态方法和所体现的对水源的保护。利用蚂蚁集团的区块链技术可以实现与农产品动态追溯的无缝对接，同时宣传绿色环保、有机等健康理念，让公众对产品产生放心食用的安全感，提升市场竞争力。“好卖”，即在“好品”的基础上建立完整的可持续销售渠道。通过电商直播、网店等交互式平台为千岛湖水源地优质农作物打造品牌，助力当地村民拓宽市场，保证经济收入来源。“好玩”，即提升当地村庄的知名度，大力发展特色旅游产品。推广生态旅游，开发深度体验活动，如亲子活动、音乐节、千岛湖马拉松赛、民宿与茶文化等。在保护环境的前提下，通过旅游业、服务业变现是保证生态系统可持续发展的方式之一。

在专注水源保护的同时，项目团队也致力于环保教育课程的研发和实践，如开发了“生态保护”主题的研学课程，吸引了来自北京、上海等地的中小学生及企业高管前来实践学习。“饮水思源——千岛湖水源保护项目研学路线”被联合国可持续发展教育专业区域中心评为“可持续发展教育精品课程”。环保研学基地被评为2019年“浙江农林大学生态环境保护志愿服务基地”。

“千岛湖水基金”项目是当时国内规模最大的水基金项目。该项目联合河南省安阳市打造了千岛湖流域第一个“源头减量、过程拦截、末端治理”的综合治理示范基地，成功地将公益、科研、金融、商业有机结合在一起，促成了吸纳农户、政府、企业、非政府组织及其他社会主体共同参与平台的建立，让环境治理、产业投资、业务合作以此为依托共同作用于生态之水的长期可持续发展。

6.3.3 项目成果

千岛湖周边散布着几百个小流域，经过世界银行的估算，最终决定投资8条流域。2020年，通过政府推荐农户、农户推荐农户等方式，千岛湖的生态护水农业已推广至5 157亩，覆盖水稻、茶叶、水果、山核桃等主要典型农产品种类，超额完成5 000亩的项目一期目标；2024年，经过“千岛湖水基金”项目的二期建设，生态护水农业有望在整个千岛湖流域推广到7.5万亩，相当于通过1%的关键土地保护削减了全流域10%的面源污染。

该项目团队选取杭州市淳安县的上梧溪与建德市的乌龙溪作为探索长效机制的试点。对当地茶叶种植治理的研究发现，通常地上肥料的流失率在70%，而坡地则会更高。农药的过度施用将加重周边水域的污染，最终进入千岛湖。因此，项目团队主要从技术与机制方面入手，对流域周边的试点农田进行精细管理，统一防治病虫害，统一采用无人机喷洒生物制剂与农药。其中，农药用量可以实现大幅减少，从而降低了施药成本。与此同时，生态茶园切实为茶农带来了每亩30%～40%的收入增加。此外，在水稻种植方面，农业面源污染防治也得到一定的突破。“千岛湖水基金”项目团队选取了约30亩的稻田作为试验田，通过测土配方施肥、精准施药、绿肥覆盖、稻鱼共养等生态护流措施提升了水稻产量，并实现了农户的创收。保护性的“三好农业”实践主要通过减少化肥的使用实现温室气体的减排，具体包括优化施肥项目与绿肥项目。据大自然保护协会相关团队测算，千岛湖优化施肥项目每亩可减排温室气体约39千克二氧化碳当量，以一期目标5 000亩计算，能实现195吨二氧化碳当量的减排，以二期目标75 000亩计算，能实现2 925吨二氧化碳当量的减排；千岛湖绿肥项目每亩可减排温室气体约47千克二氧化碳当量，以一期目标5 000亩计算，能够实现235吨二氧化碳当量的减排，以二期目标75 000亩计算，能够实现3 525吨二氧化碳当量的减排。

此外，综合千岛湖生态特色，项目团队同时开发了“千岛清泉茶”“千岛清泉米”等品牌，并获得“千岛湖茶叶”区域品牌的使用资格，实现了农业附加值的创新，提升了当地居民的生活水平。

6.3.4 项目信息

案例来源：TNC全球NbS案例库。

项目参与方：阿里巴巴公益基金会、民生通惠公益基金会、万向信托股份公司、大自然保护协会。

项目地点：中国浙江省杭州市。

项目时间：一期（2017—2020年）；二期（2021—2024年）；三期（自2025年起）。

6.4 小结

由于长期滥用化肥和农药，印度安得拉邦的土壤受到严重的破坏，加上极端天气事件频发，使农业领域非常脆弱。在此背景下，联合国粮农组织与安得拉邦政府和印度农业部积极开展了“安得拉邦零预算自然农业”项目，在生产过程中尊重作物的自然生长模式，避免使用人工化肥与农药。“零预算农业”的研究结果表明，在所有作物中，基于自然的“零预算农业”每亩投入的生物肥成本和种植成本均低于传统模式下化学肥料的投入成本。“零预算农业”尊重本地知识，就地取材，带动社区与当地农户广泛参与能力建设。此外，针对部分有影响力的农业大户开展培训，以一带多，通过其增收带动其他村民共同实施基于自然的“零预算农业”，提升了项目的执行速度和普及率。

“气候智慧型主要粮食作物生产项目”由中国农业农村部、世界银行和全球环境基金合作发起，在安徽省怀远县和河南省叶县进行实践。在中国农业农村部的领导下，项目管理办公室设立自上而下的项目指导委员会，涵盖国家、省级和县级代表参与的项目管理小组。 自上而下的组织机构和运行机制的建立，不仅加强了项目的组织领导，还为项目的正常运转提供了可靠的制度保证，保证了项目的持续发展。通过5年的实施，该项目取得了丰硕的“三赢”成果。在粮食主产区安徽省和河南省建立了10万亩示范区，项目区单位面积氮肥用量约减少了10%，农药用量减少了15%以上，土壤有机碳含量提高了10%，5年累计实现固碳减排约13万吨二氧化碳当量，产量年均增加5%以上。

千岛湖是中国杭州等城市约1 000万人的饮用水水源地，生态环境脆弱，面临着严峻的农业面源污染和气候变化带来的威胁与挑战。为此，由阿里巴巴公益基金会、民生通惠公益基金会共同发起，万向信托股份公司作为受托人，大自然保护协会作为科学顾问的“千岛湖水基金”项目于2018年正式启动。该项目致力于通过科学的方法减少农业面源污染，通过绿色消费帮助流域生态产业的发展，探索流域面源污染治理的长效治理机制。“千岛湖水基金”项目创新地引入了慈善信托，依靠信托制度的优势为其助力，从而激发环保事业的潜力，为传统捐赠方式增加了更加多元化且具有影响力的投资渠道。此外，意识到生态价值实现是保证生态系统可持续发展的“源头活水”，“千岛湖水基金”项目正在通过“三好农业”商业模式提升农产品的市场竞争力，拓宽市场渠道，增加农民收入，努力探索各种途径以充分发挥“绿水青山”的经济价值。

7 湿地类

湿地是应对气候变化的关键，拥有卓越的碳汇能力，是重要的“储碳库”和“吸碳器”，是气候变化的“缓冲器”。湿地具备强大的蓄水、净化水质的功能，被称为“地球之肾”。此外，湿地还是多种水生动物、两栖动物、鸟类和其他野生生物的重要栖息地。因此，湿地保护是NbS的重要领域之一，可以协同应对气候变化、生物多样性丧失、水源危机、极端气候事件等多种社会挑战。根据1971年签订的《关于特别是作为水禽栖息地的国际重要湿地公约》（RAMSAR，简称《湿地公约》），湿地是指包括天然或人工、长久或暂时性的沼泽地、泥炭地或水域地带，带有静止或流动的淡水、半咸水及咸水体，包括低潮时水深不超过6米的海域。广义的湿地分为自然湿地和人工湿地两大类：自然湿地包括泥炭地、海滩、河口、沼泽地、湖泊、河流、盐沼等，人工湿地主要有水稻田、水库、池塘等。据统计，全世界共有自然湿地855.8万平方千米，占陆地面积的6.4%。

本章重点探讨了天然湿地对于减缓与适应气候变化的重要作用，选取了泥炭地、沿海滩涂、河口湿地、红树林这 4 种典型的湿地生态系统，分享了印度尼西亚泥炭地保护、中国盐城黄海湿地生态修复、中国东营湿地城市建设、中国—东盟红树林保护与修复、南太平洋小岛国海洋保护5个典型案例经验，展示了国家、城市、地区和小岛国等不同行为主体设计与实施的湿地NbS项目。

7.1 印度尼西亚泥炭地保护

案例亮点

在泥炭地火灾后，印度尼西亚设立了国家泥炭地修复管理局，统筹管理和推进泥炭地保护和修复工作；在国际社会和多方机构的资助和支持下，通过开发在线监测管理平台来实现科学管理和加强预警监测，在管理部门的统筹协调下开展培训，通过科学保护等工作的开展有步骤地推进印度尼西亚泥炭地保护和修复工作。

7.1.1 项目背景

印度尼西亚拥有约2 400万公顷的泥炭地，约占世界热带泥炭地总面积的36%。泥炭地是从地方到全球范围的重要生态和气候保护屏障，其保护和修复对减缓和适应气候变化、发展循环经济均会产生至关重要的影响。保护好泥炭地有助于保护最大的天然陆地碳库，维护生物多样性和水循环。然而，农业发展、木材采伐、泥炭地排水[①]等人类活动正在威胁着这些脆弱的生态系统，在印度尼西亚苏门答腊岛有25%左右的泥炭地保持着良好状态[②]。2015年，发生在印度尼西亚的一场大火将260万公顷的土地化为焦土，燃烧释放了大量烟雾，导致上千人患病、6 900万人暴露在严重超标的污染物下。此外，泥炭燃烧所释放的二氧化碳比森林火灾多10倍，火灾使加里曼丹岛和苏门答腊岛的泥炭地释放了160亿吨左右的二氧化碳[③]。

①与森林不同，泥炭地没有保护自己免受水分流失到大气中的机制。排干水分会使泥炭密集的碳储存加速分解，并使它们的防火通道转变为火灾传播因素。

②数据来源：WWF Forest Solutions Platform，https://www.unops.org/news-and-stories/stories/restoring-indonesian-peatlands-protecting-our-planet。

③数据来源：UNEP，https://www.unenvironment.org/news-and-stories/story/fighting-fires-indonesias-peatlands。

7.1.2 项目介绍

2015年，火灾使泥炭地的保护受到了国际社会空前的重视。为减少火灾影响并保护泥炭地这一脆弱的生态系统，印度尼西亚总统宣布暂停对未开垦的泥炭地进行排水和转化，设立专门的政府部门负责推行该法令，修复印度尼西亚260万公顷左右退化的泥炭地，并与多个国际机构合作研发专门的信息公开与监测系统用于跟踪当前泥炭地的状况、监测修复活动的进展及其影响。同时，印度尼西亚政府通过多渠道筹措资金，开展“制定预防措施，提升热带泥炭地利用”项目及中加里曼丹省塞班高国家公园泥炭地修复工程，化被动为主动，通过加强监测评估系统提升泥炭地的安全管理工作。

为应对2015年的火灾并保护泥炭地这一脆弱的生态系统，印度尼西亚在2016年设立了国家泥炭地修复管理局，负责修复260万公顷因火灾退化的泥炭地。国家泥炭地修复管理局的主要职责是组织、管理、规划和协调各部门开展泥炭地保护和修复措施，包括对特许土地上的基础设施的建设、运营和维护实施监督，鼓励发展循环经济并开展宣传教育活动等（Dohong A，2017）。

在世界资源研究所印度尼西亚办事处、印度尼西亚技术评估与应用署、联合国粮农组织、联合国开发计划署（UNDP）和联合国项目事务署（UNOPS）等合作伙伴的帮助下，国家泥炭地修复管理局建立了“泥炭地修复信息和监测系统”。该系统是一个在线平台，用于跟踪当前泥炭地的状况，监测修复活动的进展及其影响。该平台显示的数据为官方公开信息，来源于相关部委和机构，具体包括自2017年起每年的泥炭地修复计划与实施数据、火灾热点数据和比较图表、参与修复工作的村庄数量，以及泥炭地修复研究活动的数量。不仅如此，“泥炭地修复信息和监测系统”平台上的互动式地图可以帮助用户按照自己的需求选择泥炭地分布、修复活动分布、火灾热点分布和修复相关指标等地图层（图7-1）。

修复活动

计划
BRG计划的泥炭地干预措施
7 个省份 ▼ 2019年 ▼
54 PHUs[①] 隐藏信息

活动	总计
1. 阻塞水渠	27 375 个
2. 深井（用于泥炭地复湿）	24 056个
3. 水渠回填	10个
4. 植被恢复面积	97 684公顷
5. 振兴生计区域	320个

实施
BRG计划的泥炭地干预措施
7 个省份 ▼ 2019年 ▼
56 PHUs 隐藏信息

活动	总计
1. 阻塞水渠	1 161 个
2. 深井（用于泥炭地复湿）	2 026个
3. 水渠回填	0个
4. 植被恢复面积	0公顷
5. 振兴生计区域	0个

（a）泥炭修复数据面板示例

（b）泥炭地分布（绿色）和修复区域分布（红色）

图7-1 泥炭地修复信息和监测系统

泥炭地失水会带来诸多负面影响，包括水位降低、泥炭面积减少、二氧化碳排放、火灾和干旱等。为防止泥炭地无节制地失水，国家泥炭地修复管理局利用信息技术创新开发了“泥炭地水监测系统”，监测水位变化，并报告土壤水分、降水量、空气温度和湿度、风向和速度等设备传输的实时数据。截至2018年12月，国家泥炭地修复管理局已经安装了142台水位监测工具设备，分布在7个优先修复的省份，对于防止灾难性的森林、泥炭地火灾及温室气体排放具有重要意义。

2018年，印度尼西亚政府在挪威政府的资助下与联合国项目事务署联合启动了为期两年的泥炭地项目，支持国家泥炭地修复管理局开发高效的综合泥炭地修复模式，针对7个优先省份的泥炭地开展修复行动。该项目注重意识提升和能力建设，支持当地居民合理利用泥炭地周边的自然资源，开发本地生产的环境友好产品，以提升收入水平。该项目重点加强了村与村之间的联系，从能力建设、社区预防等不同角度防范火灾的发生。

在此之前，印度尼西亚政府通过国际合作的方式在泥炭地保护实践和科学检

---

① PHUs是泥炭地水文单位。

测方面积累了大量的经验，如自2004年以来，世界自然基金会一直与当地社区和政府合作在塞班高国家公园开展泥炭地修复工程。此外，为了有效应对泥炭地退化问题，联合国项目事务署印度尼西亚办公室、美国国际开发署与联合国环境规划署于2015年联合启动了名为“制定预防措施，提升热带泥炭地利用”的项目，开展的主要活动包括建立可视化的火灾风险管理系统，以便提前1～3个月为高风险地区提供火灾预警信息，帮助其提升火灾的风险管理能力等。火灾风险管理系统综合考虑了天气、社会经济活动、脆弱和风险评估、厄尔尼诺现象和气候变化等不同维度的信息，通过计算机系统生成火灾风险预测图。在此基础上，项目组开展了各个层级的活动，包括促进部委之间制定联动的防火规划、共同开展泥炭地修复等。

7.1.3 项目成果

自成立以来，国家泥炭地修复管理局于2015—2018年修复了约70万公顷的泥炭地，到2020年修复了约315万公顷的泥炭地。暂停令和修复项目带来了显著的环境效益。在受保护的泥炭地地区，森林减少率在一年内急剧下降88%，达到了有历史记录以来的最低水平，预计15年内可减少55亿～78亿吨二氧化碳的排放。

2016年，“制定预防措施，提升热带泥炭地利用”项目开展一年后，火灾风险管理系统已经上线，数百名政府部门的工作人员接受了系统培训，并在所有项目覆盖区建立了防火论坛。现在，社区和地方政府已经意识到通过预防开展火灾管理的重要性，管理模式的转变使社区在火灾季到来前做好了充足的预防和准备。通过利用火灾风险管理系统开展脆弱性和风险评估（图7-1b），中加里曼丹省中部开展了以社区为主的泥炭地修复项目，如在Buntok镇建立了面积约为2万公顷的泥炭地修复项目，通过社区动员和社区参与的模式共同开展了泥炭森林火灾控制。此外，该项目还积极开发可持续的生计替代行动，如通过小额补助发展园艺、渔业和其他措施以改善收入，通过生计改善引导社区摒除非法砍伐和焚烧的陋习，更加积极地参与泥炭修复活动。

截至2017年，当地社区和政府已经在中加里曼丹省塞班高国家公园修建了1 400座水坝，灌溉了30多万公顷的土地，建立了70个社区苗圃，在泥炭地退化最严重的地区重新种植了1万多公顷的植被。此外，该项目还在当地建立了6个社区消防巡逻队（参与人数高达28 000多人）以维护、管理和监测水坝，同时开展植树造林活动。未来该项目计划再建造850座大坝，灌溉10万公顷的土地，建立100个社区苗圃，重新种植19 000公顷植被，建立涉及3万名当地居民的社区伙伴关系、15个社区消防巡逻队，包括联合执法和社区教育，预计在项目开展的30年内每年减少约150万吨温室气体排放。

7.1.4 项目信息

案例来源：N4C全球NbS案例研究、TNC全球NbS案例库。

项目参与方：印度尼西亚政府、地区政府与社区、国际组织与各资助方。

项目地点：印度尼西亚。

项目时间：印度尼西亚国家泥炭地修复项目（2016—2020年）；“制定预防措施，提升热带泥炭地利用”项目（2015年起）；中加里曼丹省塞班高国家公园泥炭地修复工程（2004—2033年）。

项目链接：自然与气候联盟，https://4fqbik2blqkb1nrebde8yxqj-wpengine.netdna-ssl.com/wp-content/uploads/2019/09/Peatland-restoration-in-Borneo-Indonesia_Sometimes-it-takes-a-disaster.pdf；联合国项目事务署，https://www.unops.org/news-and-stories/stories/restoring-indonesian-peatlands-protecting-our-planet；世界自然基金会，https://www.wwf.org.au/ArticleDocuments/360/pub-briefing-borneo-peat-restoration-sebangau-central-kalimantan-15jun17.pdf；泥炭地修复信息和监测系统，https://en.prims.brg.go.id/platform。

7.2　中国盐城黄海湿地生态修复

案例亮点

通过顶层设计，制定出台相关政策和条例，设立研究基地对科学保护进行尝试和分析，同时注意与国际机构、国外院校的沟通和交流，通过实践总结出科学保护的湿地NbS模式。

7.2.1　项目背景

中国江苏省盐城市黄海湿地是太平洋西岸和亚洲大陆边缘仅存的保存相对完好、面积最大的滨海湿地，也是中国最大的沿海滩涂，面积达45.53万公顷（图7-2）。同时，盐城黄海湿地作为中国黄（渤）海候鸟栖息地（第一期）的代表被列入《世界遗产名录》，范围主要包括江苏盐城湿地珍禽国家级自然保护区（以下简称盐城保护区）部分区域、江苏大丰麋鹿国家级自然保护区全境、盐城条子泥市级湿地公园和湿地保护小区，总面积为18.86公顷。遗产地有超过680种脊椎动物，包括415种鸟类、26种哺乳动物、9种两栖动物、14种爬行动物、216种鱼类及165种底栖动物，是濒危物种最多、受威胁程度最高的东亚—澳大利西

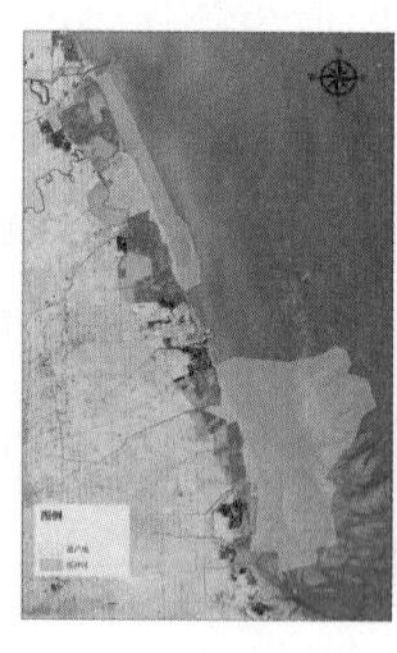

（a）地理位置

（b）自然景观

图7-2　中国盐城黄海湿地世界自然遗产地

亚候鸟迁徙路线的中心节点，也是全球数以百万迁徙候鸟的停歇地、换羽地和越冬地，为23种具有国际重要意义的鸟类提供了栖息地，支持了17种世界自然保护联盟濒危物种红色名录物种的生存，包括1种极危物种——勺嘴鹬、5种濒危物种——黑脸琵鹭、东方白鹳、丹顶鹤、小青脚鹬和大滨鹬。

7.2.2　项目介绍

多年来，黄海湿地坚持“从自然中来，到自然中去”，恢复湿地生态系统，发挥生态系统的生态功能，构建人与自然和谐共存、湿地保护和可持续发展的新模式，为遗产地生态保护功能的发挥提供了技术支撑。在政策制定方面，盐城市推动成立了盐城市湿地和世界自然遗产保护管理中心，研究出台了《盐城市黄海湿地保护条例》《盐城黄海自然遗产地保护管理及可持续发展三年行动纲要（2019—2021）》，为依法保护、科学保护提供了依据。

在学术研究领域，2020年9月，自然资源部国土整治中心批复在黄海湿地研究院设立“基于自然的黄海湿地生态保护修复科研基地”。该基地聚焦NbS的推广、实践与应用，重点研究湿地保护、生态修复、外来入侵物种防治（如互花米草治理）等问题，并开展相关培训活动。

为推动黄海湿地世界遗产保护管理与可持续发展工作，盐城市于2017年起推动设立黄海湿地研究院，加强与世界自然保护联盟、联合国湿地公约组织等有关国际组织及英国剑桥大学、澳大利亚昆士兰大学及韩国庆北国立大学交流互动，聘请了一支由国内外知名专家学者组成的特聘专家团队，开展滨海湿地保护、迁飞鸟类保育、生态修复、外来物种防治等专题研究，努力构建山水林田湖草生命共同体，为建设“美丽中国”和全球生态治理贡献力量。

盐城市在黄海湿地的生态修复中成功开展了一系列示范项目。

1. 川水港地块滨海湿地生态修复示范项目

川水港地块位于中国黄（渤）海候鸟栖息地（第一期）遗产地及遗产地缓冲

区，是盐城保护区南二实验区，位于江苏省整个滨海湿地生态系统的中间地带，是上海市崇明岛片区到盐城保护区核心区迁徙路径上的重要节点，能为迁徙候鸟提供必需的停歇和觅食场所。川水港地块紧邻东沙岛遗产地核心区，是鸻鹬类候鸟活动的主要区域，实现生态修复以后可为鸻鹬类候鸟提供广阔的高潮位栖息地和食物来源。此外，川水港地块也是丹顶鹤的历史活动区，但由于原生湿地面积减少、冬季食物不足，遗产地南区的丹顶鹤数量较少，修复后将成为遗产地南区丹顶鹤活动的重要空间（图7-3）。

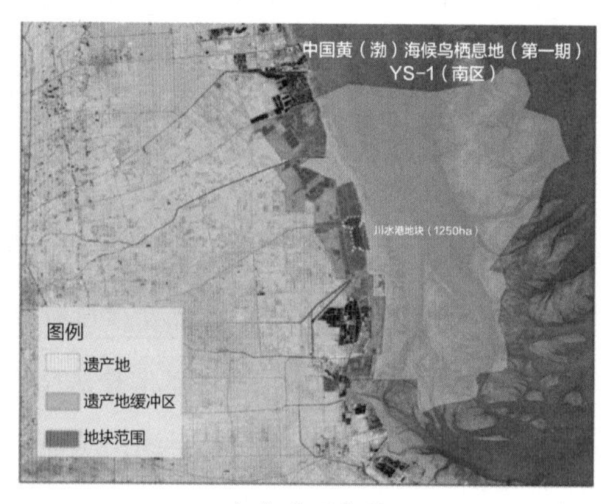

（a）地理位置

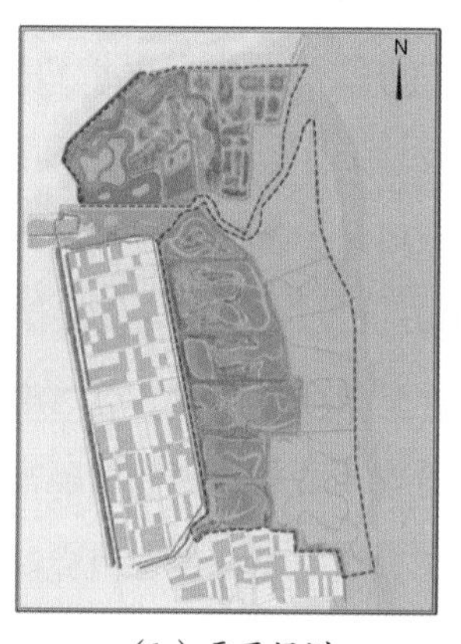

（b）平面规划

图7-3　川水港地块

几十年来，受自然因素和人为因素的影响，如海岸围垦、不合理的土地开发、地势淤高、湿地旱化、外来物种入侵等，川水港地块的滨海湿地生态系统已受到严重破坏，导致生境破碎化现象严重、生物多样性逐渐下降、生态系统日益脆弱、生态系统服务功能持续下降。川水港地块的滨海湿地分布呈总面积持续减少、自然湿地面积锐减的变化趋势。

川水港地块滨海湿地生态修复示范项目探索了滨海湿地在自然旱化背景下，结合自然保护区和遗产地主要保护目标，以NbS恢复湿地生态系统，发挥生态系统的生态功能，构建人与自然和谐共存、湿地保护和可持续发展的新模式，为遗产地生态保护功能的发挥提供了技术支撑；形成了基于多类型生境（海水生境、淡水生境、海水-淡水生境）营造、水位调控和多层次地形重塑的退化滨海湿地生态系统全方位综合治理修复集成模式，以及基于滨海潮汐引入调控、互花米草生态管控、滨海生态系统自然演化相结合的盐生沼泽湿地（碱蓬滩涂）与季节性海水滩涂湿地恢复模式。

2. 斗龙港地块生态修复项目

斗龙港地块位于中国黄（渤）海候鸟栖息地（第一期）遗产地缓冲区，在盐城市大丰区境内、斗龙港南侧，属盐城保护区南缓冲区，总面积为1 369公顷。斗龙港地块北部与盐城保护区核心区一河之隔，西部紧邻大丰区干河，南部为现有鱼塘养殖人工输水渠，东部隔河道与核心区相望。该区域原为高密度人工鱼塘养殖区，在2018年国家海洋督察时被列为必须整治、修复的区域。

斗龙港地块主要利用类型为已清退的淡水养殖塘，其生境较为单一，候鸟保护功能不足（图7-4）。由于长期的滩涂淤涨，在斗龙港地块海水能进入的区域有限，而淡水河道水位高程一般在0米左右，不通过泵站很难大规模进入项目区，养殖所使用的淡水由高水沟渠灌入，区域水资源难以自然到达，湿地旱化情况在枯水期特别严重。斗龙港地块由于长期开展养殖业，水体呈富营养化，湿地生态系统超负荷运转，养殖塘废弃后由于缺乏管理，鱼塘底部干涸，塘底生物量

图7-4　斗龙港地块已清退的养殖塘

较少，不利于鸟类觅食，形成了被芦苇群落占据的单一生境。该项目的地形改造工程包括生态岛塑造、浅滩塑造、浅滩驳岸塑造、开敞水面地形塑造、深水区地形塑造及堤岸地形塑造，考虑地形的多样化，充分满足了游禽、涉禽对栖息生境的需求。项目区的主要植物群落为沉水植物群落、挺水植物群落、盐地草本群落、盐生灌木群落，功能区分为潟湖盐沼生态修复区、咸水碟形湖生态修复区、树栖鸟栖息地修复区、淡水碟形湖生态修复区、鸟类友好型生态种植区。斗龙港地块采取NbS的技术理念，采用“以生态自然修复为主”和人工适度干预为辅的方针，尽可能地创造、保护或者修复纯生态系统要素，辅助结合生态系统要素和硬的工程性干预措施。

3. 野鹿荡野生动植物栖息地保护项目

野鹿荡位于中国黄（渤）海候鸟栖息地（第一期）遗产地缓冲区，长江以北的南黄海西岸，在国家级麋鹿保护区和丹顶鹤保护区之间，占地3 000亩，被两条海岸带相拥，直面黄海。整个区域有近5万亩的原生境海涂荒原，没有镇和村庄，夜晚没有灯光。每晚都有成群纯野生麋鹿出没，还有河麂等其他小野兽，夏季遍野都是萤火虫。自2009年起，盐城市大丰区政府和民间组织发起了野鹿荡保护工作，专门从事科研和人文自然保护。同时，与江苏省沿海滩涂和环境保护重点实验室联合建设了沿海野生植物种质库，目前拥有沿海地区485种野生植物中的近300种野草标本和种子，为沿海地区野生植物保护进行着扎实的基础工作（图7–5）。

图7–5　野鹿荡野生麋鹿种群

4. 互花米草入侵防治项目

互花米草于20世纪80年代前后引入中国，其根系发达、扩散能力强，在中国海岸带快速蔓延，对大部分沿海滩涂湿地的生物多样性等构成威胁，是造成盐城滨海湿地退化的主要原因之一（图7-6）。为加强遏制互花米草蔓延，加快海岸带生态系统保护与修复，寻找防护治理互花米草的“盐城方案”，黄海湿地开展了互花米草治理工作，以期为在中国更大范围内开展互花米草防治提供可借鉴的经验和建议。

图7-6　遗产地互花米草入侵现状

7.2.3　项目成果

以生物多样性保护为例，黄海湿地建立的珍禽保护区加大了缓冲区退渔还湿的力度，已退出近10万亩鱼塘，扩大了鸟类的生存空间。麋鹿保护区采取内部调节、青贮补饲、网格化管理、围网轮牧、水系改造、植被修复等措施，对麋鹿种群进行科学的调控管理，对麋鹿栖息地生境进行有效的修复改善。观测数据显示，一年来黄海湿地的鸟类种类及数量均有所上升，麋鹿种群数量达5 681头，占世界麋鹿总数的60%以上，其中野生麋鹿种群有1 820头。

2019年7月5日，盐城黄海湿地作为中国黄（渤）海候鸟栖息地（第一期）的代表，在第43届世界遗产大会上被成功列入《世界遗产名录》，成为中国第一块、全球第二块潮间带湿地世界遗产，填补了中国滨海湿地类型世界自然遗产空白，标志着中国世界遗产从陆地走向海洋。未来，盐城市将采取“三步走”战略：到2021年，盐城黄海湿地遗产实现全面保护，成立以盐城为核心的候鸟迁飞路线城市联盟，初步形成环黄海生态经济圈合作框架，成功创建国际湿地城市；到2025年，盐城黄海湿地遗产实现科学保护、活态传承、合理利用，环黄海生态经济圈合作框架成为地区间互惠互信的平台；到2035年，呈现“河海安澜、碧水畅流、鱼翔浅底、鹤舞鹿鸣、候鸟欢飞、游人如织”的美好画面。

7.2.4 项目信息

案例来源：C+NbS合作平台。

项目参与方：中国江苏省盐城市。

项目地点：中国江苏省盐城市黄海湿地。

项目时间：2016年起 。

7.3 中国东营湿地城市建设

案例亮点

积极动员各类社会条件，坚持规划引领，完善制度体系；广泛开展社会动员，通过“湿地学校”“小手拉大手”等各种活动让全体市民参与到湿地保护中来。

7.3.1 项目背景

黄河三角洲湿地是世界上少有的河口湿地生态系统，也是世界上暖温带保存最广阔、最完善、最年轻的湿地生态系统，位于中国山东省东北部的渤海之滨。山东省东营市（图7-7）是黄河三角洲的中心城市，受河海双重恩赐，境内湿地资源得天独厚，总面积为45.81万公顷，湿地率达41.58%。在这样的自然条件下，东营市在建市之初就被定位为“湿地城市”，在城市建设中注重发挥湿地资源的优势。

图7-7　鸟瞰东营市

受地形地势与城市基础管道的限制，东营市城市内涝问题突出，湿地缺水退化现象有所出现：一方面，黄河三角洲是典型的冲积平原，作为建立在黄河三角洲上的城市，东营市地势整体平缓，再加上排水通道相对单一、降水量集中在汛期、海水顶托严重及地下水位高等因素，容易产生内涝问题；另一方面，东营市的湿地用水主要依赖黄河水，近年来由于黄河径流量减小，加之降水量不均，雨洪水资源没有得到有效利用，东营市的湿地出现缺水退化现象。

7.3.2　项目介绍

为解决城市内涝问题，有效应对水资源匮乏、气候变化等挑战，东营市积极探索湿地保护与城市发展的兼顾性方案，主要采取了以下措施。

一是坚持规划引领，完善制度体系。编制了《东营市湿地保护修复总体规划（2018—2025年）》，制定了湿地保护任务目标，完善了湿地保护体系。出台了《关于加强规划引领提升城市品质的实施意见》，坚持以水为脉、以绿为衣、以湿地为特色，突出中心城“两带、三河、五片、多点”的湿地生态结构，彰显了“湿地在城中、城在湿地中”的城市特色，大幅提升了城市品质。启动了《东营市国土空间总体规划（2019—2035年）》编制工作，推进“多规合一”，科学划定“三区三线”，合理布局生态空间、农业空间、城镇空间。颁布了《山东省黄河三角洲国家级自然保护区条例》《东营市湿地保护条例》《东营市湿地城市建设条例》，依法保护湿地资源，推进湿地城市建设，并把湿地保护状况纳入生态文明考核体系。为落实湿地保护责任，山东省人民政府办公厅印发了《关于全面建立林长制的实施意见》，共设森林湿地长2 261名，其中市级森林湿地长5名、县级森林湿地长32名、乡级森林湿地长122名、村级森林湿地长2 102名。

二是建设国家公园，构建保护体系。近期，为推进黄河口国家公园的建设，东营市成立了“推进建设国家公园工作专班”，整合黄河口区域8个现有自然保护地作为国家公园创建范围，总面积为3 554.11平方千米，编制完成了《黄河口国家公园设立方案》《黄河口国家公园本底资源调查与评价报告》，设立了《黄河口国家公园社会影响评价报告》《黄河口国家公园综合研究和符合性认定报告》，全力推进国家公园建设。在全市域范围建立起6处水源地保护区、1处国家级城市湿地公园、1处国家级湿地公园、4处省级森林公园、9处省级湿地公园、1处省级风景名胜区、39处湿地保护小区，构建了以国家公园为主体的湿地保护体系，初步建立了海陆兼顾的生态保护框架体系。

三是提升蓄滞能力，建设无内涝城市。为提升城市雨洪蓄滞能力，东营市出台了《东营市中心城海绵城市专项规划（2016—2030）》《东营市中心城无内涝城市规划设计方案》，构建了“大水面、大绿地、大湿地、大空间”的生态系统，划定城市蓝线、绿线保护水系和绿地，实施积水点改造、水系贯通、蓄滞洪、泵站提升、管网完善、内涝严重小区应急排水、沿河小区就近入河等一系列工程建设，提升中心城防洪排涝能力。在工程的建设过程中，东营市以政府为主导，注重吸纳社会资本参与湿地城市建设，鼓励市场主体参与湿地城市建设多元营运，以商业收益反哺运营维护。

四是强化保护宣教，提升生态意识。近年来，东营市在黄河口湿地博物馆、黄河三角洲鸟类博物馆等生态教育基地的基础上，广泛开展“湿地学校”建设活动，通过“小手拉大手”活动，让全体市民参与到湿地保护中来。为此，东营市编写了湿地生态道德教育教材《美丽家乡——黄河口》，建立了生态道德教育讲师团，开设了“笔记大自然”“让燕子住我家”等深受学生和家长喜爱的教学科目，东营市的湿地学校得到了蓬勃发展（图7-8）。

图7-8　东营市湿地学校
（图片来源：东营市湿地保护中心）

7.3.3　项目成果

经过多年努力，东营市显著提高了城市的湿地率，通过提供供水、水产品、原材料等物质产品创造了巨大的经济价值，并通过水系贯通、蓄滞洪工程建设有效缓解了城市内涝问题。

一是提升湿地保护率，发挥湿地生态功能。东营市的湿地总面积为45.81万公顷，湿地率达41.58%，湿地保护面积为25.7万公顷，湿地保护率达到56.1%。由供水、水产品、原材料提供的物质产品价值达102.24亿元，带来的旅游休闲价值为

24.86亿元、文化科研价值为1.41亿元，而大气调节、洪水调蓄、水质净化、消浪护岸、生物多样性等间接生态价值为13.25亿元（表7-1）。

表7-1　东营市湿地生态系统服务功能价值（2015年评估结果）

单位：亿元

直接使用价值					间接使用价值+选择价值					合计
物质产品			旅游休闲	文化科研	大气调节	洪水调蓄	水质净化	消浪护岸	生物多样性	
供水	水产品	原材料								
5.41	59.56	37.27	24.86	1.41	0.29	2.70	0.84	0.27	9.15	**141.76**

信息来源：东营市政府“山东东营湿地生态系统服务功能价值评估与保护对策研究”项目。

二是解决内涝问题，改善生态环境。通过水系贯通、蓄滞洪工程建设，有效缓解了城市内涝问题，实现中心城内涝设防标准从不足“10年一遇”向“50年一遇”的转变；同时，也使雨洪水资源得到有效利用，保障了城市湿地生态用水需求。另外，区域内的生态环境得到了改善，生物多样性有所提升，城区鸟类由建市初期的不足100种增加到206种。

三是提升保护意识，守护湿地城市。目前，东营市40多所中小学加入了中国或东营市湿地学校网络，拥有国际生态学校1所、全国未成年人生态道德教育示范学校3所、东营市未成年人生态道德教育示范学校19所、中国湿地学校5所、东营市湿地学校7所，达到了教育一个学生、带动一个家庭、影响整个社会的效果，生态意识逐渐深入人心。

东营市湿地城市建设的成功离不开政府系统、科学的规划与管理，通过合理布局生态空间、农业空间、城镇空间，因地制宜地实施积水点改造、水系贯通、蓄滞洪等一系列工程建设，在解决城市内涝问题的同时有效应对水资源匮乏、气候变化等挑战。

7.3.4　项目信息

案例来源：C+NbS合作平台。

项目参与方：东营市人民政府。

项目地点：中国山东省东营市。

项目时间：1983年（建市）起。

7.4　中国—东盟红树林保护与修复

选取“社区协议”的创新治理模式，通过保护机制的形式规定了政府、农牧民、非政府组织等不同利益相关方在保护中的权、责、利；帮助当地乡村修建道路、安装路灯、改善交通环境，与村民共享红树林保护成果，使原住民变为共同保护红树林的“盟友”，减少了保护与社区发展的冲突，推动了政府、民间社会组织和企业等共同参与的生态环境多元治理。

7.4.1　项目背景

中国—东盟国家沿海分布着超过425万公顷、约占世界35%的红树林，是全球红树林生物多样性最丰富的地区。东盟成员国大多有较长的海岸线，形成了总长度为17.3万千米的海岸，同时拥有亚太地区最重要的海洋鱼类产卵地——海上亚马孙和著名的“珊瑚大三角区”，使中国—东盟地区成为地球上沿海和海洋生物多样性最丰富的地区。如果红树林消失和退化的趋势得不到改善，几乎所有未受到妥善保护的红树林将在百年之后消失。在整个中国—东盟地区，除中国因红树林种植修复工程使其面积得以增加外，其余东盟国家的红树林均以每年0.25%～20%不等的速率减少。

7.4.2 项目介绍

永续全球环境研究所（GEI）从2016年开始就以社区协议保护机制的模式，在中国—东盟红树林国家开展社区自主参与红树林保护的探索，在保护与发展之间寻找平衡。这种模式可以在实现对红树林生态系统有效保护的同时，让社区在参与保护的过程中持续享受自然所带来的福祉。永续全球环境研究所从国外引进了社区协议保护机制，其概念是在某个需要保护的区域，通过利益相关双方或几方（政府、企业、当地社区或个人等）签署协议的形式把保护权和有限开发权赋予不同的利益相关方，解决保护方和居民从自然中取得的经济利益有冲突的问题，缓解资源开发、环境保护和居民利益之间的矛盾。同时，由于当地社区的参与，提高了社区居民对自己土地的热爱和关注，改变了过去政府作为单一保护方的角色，形成了一套新的生物多样性保护模式。

在缅甸孟加拉湾沿岸，有绵延2 000千米海岸线的红树林。除了具有抵抗洪灾和风暴的作用，红树林生态系统还为众多野生动物和鱼类提供了重要的栖息地。对于当地人而言，红树林也为他们提供了各种生活必需品，包括食物、木材、薪柴和建筑材料等。然而近年来，缅甸的红树林面积正在急速缩减——从1980年的659 039公顷退化至2015年的462 963公顷，减少了近1/3，这主要与人们的生产生活和经济发展相关，如围塘养殖、水稻种植、棕榈油生产及薪材毁林等都对红树林构成了威胁。

图7-9　培育的红树幼苗
（图片来源：永续全球环境研究所）

永续全球环境研究所从2016年开始通过社区协议保护机制的模式与缅甸7个社区合作开展红树林生态系统保护（图7-9）。通过与当地政府、社区及缅甸本土非政府组织的合作，推动当地社区签署了为期5年的保护协议，约定各方保护社区周边红树林的责任与行动，并相继通过宣传与培训、日常巡护、培育红树幼苗、发放清洁能源设备、物种调查等，在提升社区对红树林认知的同时推动其自主开展保护行动。

中国广东湛江红树林国家级自然保护区分布着中国连片面积最大的红树林，约占全国红树林总面积的33%，分成了68个小区分散环绕着整个雷州半岛，守护着沿海800万居民的生产和生活安全（图7-10）。伴随人口增长和社区发展而产生的生产生活围垦、围塘养殖、城镇化建设占地、污水垃圾等问题，是雷州半岛红树林保护面临的主要压力。

图7-10 广东湛江红树林国家级自然保护区红树林景观

（图片来源：永续全球环境研究所）

永续全球环境研究所与广东湛江红树林国家级自然保护区合作，探索基于不同问题社区如何有效参与红树林保护的对策研究。永续全球环境研究所结合利益相关方访谈、文献收集整理、现场踏勘，对雷州半岛红树林保护面临的主要问题进行了分析，发现一般存在以下问题：①部分区域由于附近社区没有地方处理垃圾，直接将垃圾或废弃物倾倒入红树林中，对红树林生态系统的健康造成直接威胁；②随着地方发展的需求，依托红树林开展了休闲旅游，给红树林保护带来了风险，如高桥片区是整个雷州半岛最大、最完整的红树林片区，这里有雷州半岛上最壮观的红树林景观，因此该区域的自发型旅游也最多，而这些自发无序的旅游给红树林保护和当地政府管理带来极大的压力，如何平衡社区从旅游发展中获益的需求，同时积极调动其参与保护行动、实现社区共管将成为解决问题的关键。

通过引入社区协议保护机制的模式（图7-11），示范解决社区可持续发展与红树林保护的冲突，分别就减少垃圾对红树林生态系统的影响、无秩序旅游对红树林生态的影响提出基于社区合作的红树林保护方案，为实施中国典型红树林保护提供了研究基础。

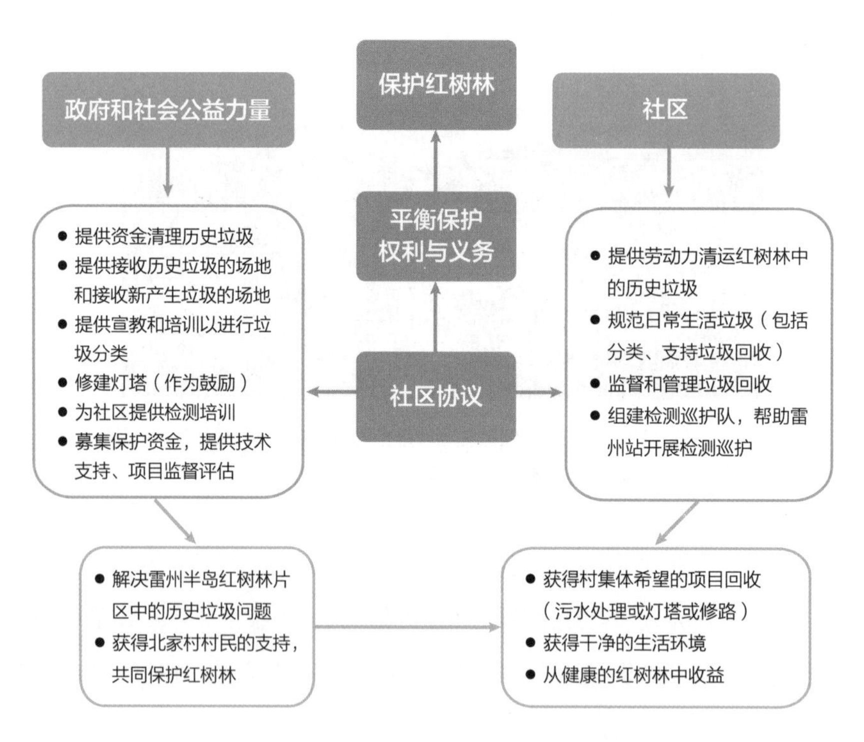

图7-11　试点研究社区协议保护模式

（图片来源：永续全球环境研究所）

7.4.3　项目成果

自2016年项目启动以来，中国—东盟地区的红树林保护行动成果颇丰。

一是社区试点研究，区域红树林保护成效良好。在缅甸和中国开设了2个试点开展社区参与红树林保护可行性研究，既完成了区域红树林保护目标，也提升了居民的实际收益。以缅甸为例，通过各方的努力，有超过5 000人、893户家庭通过参与项目受益，建立了总额为4 000美元的社区基金。在项目第二期结束后，社区基金平均使每户收益增加了17%。而在保护成效方面，建立了2 000英亩的社区保护地，调查发现当地红树物种有28种，其中有2种濒危物种，共培育了37 000棵红树幼苗。

二是合作开发红树林生态系统修复指南。在修复目标、原则、对象选择、树种选择、育苗、苗木运输及保存、种植、退塘还林、滩涂造林、退化红树林修复、监测与评估等方面形成了一套全面、细致、专业的红树林修复指导方案。同时，该指南提交给中国自然资源部国土空间生态修复司并译成英文，以支持日后的修复工作。

三是积极参与红树林保护的国际合作。永续全球环境研究所还与气候变化全球行动（GCAI）等合作，推动中国—东盟之间的红树林保护交流，包括于2019年8月在泰国召开“红树林保护——基于自然的气候解决方案圆桌会议”，2019年12月在第25届联合国气候变化大会上举办“NbS——以红树林保护及可持续农林产品为例官方边会”，各方共同探讨红树林保护在NbS的框架下如何更好地贡献于生态系统保护及气候变化的减缓与适应。

7.4.4　项目信息

案例来源：C+NbS合作平台。

项目参与方：永续全球环境研究所、缅甸当地非政府组织、项目地社区等。

项目地点：缅甸、中国广东省湛江市。

项目时间：2017年起。

项目链接：永续全球环境研究所，http://www.geichina.org/wp-content/uploads/2020/07/2019-Annual-Report_CN.pdf；中国—东盟环境信息共享平台，http://www.caeisp.org.cn/zh-hans。

7.5 南太平洋小岛国海洋保护

案例亮点

通过联合国家与当地政府、原住民村庄、国际组织、跨国公司、学者与智库等多方力量，为诸多小岛国提供海洋保护资金，同时还积极探索生态系统付费的项目，包括开发生态旅游、红树林保护与社区教育等丰富活动，尝试逐步建立起以应对气候变化、改善海洋生态环境质量、促进就业与经济发展的良性循环模式。

7.5.1 项目背景

小岛国所处的地理位置和经济发展能力十分脆弱，最容易受气候变化的直接威胁。萨摩亚独立国是一个由乌波卢、萨瓦伊2个主岛和附近8个小岛所组成的岛国，陆地面积为2 934平方千米，海洋专属经济区面积为12万平方千米，境内大部分地区被丛林覆盖，属热带雨林气候。斐济共和国位于西南太平洋中心，陆地面积为18 333平方千米，海洋专属经济区面积为129万平方千米，由332个岛屿组成，其中30%为有人居住的岛屿，属于热带海洋性气候，常年遭受飓风袭击。

7.5.2 项目介绍

世界团队组织（World Team Project）的“海洋机遇与小岛国”项目致力于建立国家、科研单位、企业和国际组织之间的合作网络，为小岛国的可持续发展提供技术、资金等资源，目前已与联合国教科文组织（UNESCO）、全球可再生能源倡议（Renewable Energy 100%，RE100）、萨摩亚与斐济政府及相关能源公司等合作并建立网络，为萨摩亚和斐济两国提供生态旅游、红树林修复、海洋保护等全方位的资助与服务（图7-12）。

图7-12 “海洋机遇与小岛国”项目开展的红树林保护
（图片来源：世界团队组织官网）

“海洋机遇与小岛国”项目致力于从不同领域积极践行生态友好海洋国家可持续发展的模式，通过“以点带面”的模式开发了一系列解决方案，创造就业机会，形成技术转让，启动新的和可替代的管理系统，帮助海洋和岛屿实现可持续转型。世界团队组织计划通过最先进的技术和方案，以及珊瑚种植、鳗鱼草种植、海龟保护等海洋修复或动物保护类项目保护海洋环境。其中，红树林修复是其解决气候危机的关键组成部分。“海洋机遇与小岛国”项目计划建立一系列集红树林保护、自然教育、生态监测和生态旅游于一体的行动，同时还计划联合网络伙伴开展当地居民的海洋能力建设活动，帮助人们从每一滴水到整个海洋生态系统层面重新认识和关爱海洋，让民众能够切身感受到如何利用新的技术和理念来合理开发利用海洋资源。

除红树林保护外，该项目还致力于在能源转型和环境整治方面为小岛国提供能力建设，如通过可再生能源微电网系统实现能源独立，构建清洁的用水和电力体系，发展可持续交通体系，利用潮汐能、氢能等新的可再生能源搭建保护区的能源系统，试点氢动力家庭，建立真空厕所，开展“限塑”活动，进行海洋塑料清理等工作。

在治理模式层面，世界团队组织通过联络利益相关各方，设立了区域性活动大本营，与国际志愿者一同支持项目的持续运行。此外，项目的治理模式体现出不同利益相关方的需求，由董事会和来自所有合作伙伴的两名代表组成，下设执行委员会（由董事会和主要组织组成），设立学术咨询委员会，每年召开利益相关方大会，参会的代表包括政府、项目组、参与新能源解决方法的研究机构、企业和投资银行等。

7.5.3 项目成果

该项目旨在通过对小岛国能力建设，设计因地制宜的可持续发展方案。以萨摩亚为例，2019年，世界团队组织连同其合作伙伴在萨摩亚自然资源部的帮助下，在埃莱帕塔群岛种植了超过25英亩（约合1万平方米）的近1 000棵树木，清理了6个海滩，为蛇、蜥蜴等当地物种提供了栖息地，有效保护了沿海社区和生态系统免受洪水侵害，并使其发挥净化水质和提供碳汇的作用。社区也可以通过此项目获得稳定的收入，进而带动了地方经济。相对于其他的海洋修复工作，针对红树林的修复成本效益较高，因为红树林的固碳效果比较明显。

基于红树林公园的生态旅游项目能有效带动当地经济发展，结合所在地的循环经济、可再生能源等方案，该项目为小岛国提供了优质的基础设施保障，在提升旅游产业的同时向世界展现土著文化，并建立起可持续的小岛海洋生态体系。以斐济为例，2019年，斐济旅游业对GDP的直接贡献约为10亿美元，与2017年项目开始时相比年均增长7.43%，年失业率也从2016年的4.32%下降至4.10%。旅游业是小岛国重要的收入来源，也是国家的经济命脉，开展生态旅游可以提供大量的就业机会，实现生态保护、社会进步和经济增长的协同发展。

在全球层面，“海洋机遇与小岛国”项目也在法制、技术等领域积极探索产生更广泛影响的途径，如通过专家小组影响联合国层面的海洋法等。在项目执行过程中，生态观测站网络提供了大量的生物保护本地数据。在该项目的推动下，将当地废物转化为能源的污水处理技术彻底改变了当地的水处理模式局面。目前，该项目已建立了多个小规模的集成系统以提供可持续的解决方案，展示了文化、教育、交通、农业、能源、政策、法律平等方面的相互配合和协同。

7.5.4　项目信息

案例来源：UNEP-NbS案例库。

项目参与方：世界团队组织及萨摩亚、斐济当地政府与社区。

项目地点：萨摩亚、斐济。

项目时间：2017年起。

项目链接：世界团队组织官网，https://worldteamnow.org/blogs/sos-is/；联合国海洋大会，https://oceanconference.un.org/commitments/?id=21714。

7.6 小结

泥炭地是最重要的碳汇之一，全球范围内的泥炭地仅占地球陆地面积的3%，却储存了陆地上1/3的碳，是全球森林碳储量的2倍。2015年发生在印度尼西亚的火灾使泥炭地的保护得到了国际社会空前的重视，为减少火灾影响并保护泥炭地这一脆弱的生态系统，印度尼西亚总统宣布暂停对未开垦的泥炭地进行排水和转化，并设立专门的政府部门负责推行法令并修复印度尼西亚260万公顷左右退化的泥炭地，还与多个国际机构合作研发了专门的信息公开与监测系统，用于跟踪当前泥炭地的状况、监测修复活动进展及其影响。同时，印度尼西亚政府通过多渠道筹措资金，开展“制定预防措施，提升热带泥炭地利用”项目及中加里曼丹省塞班高国家公园泥炭地修复工程，化被动为主动，通过加强监测评估系统提升泥炭地的安全管理工作。

沿海滩涂是指沿海大潮高潮位与低潮位之间的潮浸地带，在地貌学上称为“潮间带”。中国江苏省盐城市的黄海湿地是中国最大的沿海滩涂，于2019年在第43届世界遗产大会上被成功列入《世界遗产名录》，成为中国第一块、全球第二块潮间带湿地世界遗产。盐城黄海湿地除了具有涵养水源、净化水质、吸碳储碳的生态功能，其生物多样性资源也极为丰富，拥有丹顶鹤、白头鹤、麋鹿、中华鲟、白鲟等14种国家一级保护动物，是全球最大的丹顶鹤越冬地和全球最大的麋鹿基因库，具有巨大的旅游价值。成为世界自然遗产地进一步为盐城市的生态旅游提供了核心竞争力，也激励该市积极开展了一系列湿地生态修复项目。多年来，盐城市通过加强政策、支撑科研和试行自上而下的模式，为中国湿地生态修复提供了来自本土且对标国际的优秀案例。

黄河三角洲湿地是世界上少有的河口湿地生态系统，也是世界上暖温带保存最广阔、最完善、最年轻的湿地生态系统，位于中国山东省东北部的渤海之滨。

山东省东营市是黄河三角洲的中心城市，湿地资源丰富，湿地率达41.58%。受地形地势与城市基础管道的限制，东营市的城市内涝问题突出，湿地缺水退化现象有所出现。经过多年努力，东营市显著提高了城市的湿地率，通过提供水资源、水产品、原材料等物质产品创造了巨大的经济价值，并通过水系贯通、蓄滞洪工程建设有效缓解了城市内涝问题。东营市湿地城市建设的成功离不开政府系统、科学的规划与管理，通过合理布局生态空间、农业空间、城镇空间，因地制宜地实施积水点改造、水系贯通、蓄滞洪等一系列工程建设，在解决城市内涝问题的同时有效应对了水资源匮乏、气候变化等挑战。

红树林之所以被称为“海岸带卫士”，主要是因为其浓密的根系有助于提升地表粗糙度、增加摩擦阻力，使水流减慢并促进沉积物沉积，从而减少海岸侵蚀。复杂的红树林根系还可以从水中过滤硝酸盐、磷酸盐和其他污染物，从而改善流入河口和海洋环境的水质。红树林还是重要的碳汇，能够从大气中捕获大量的二氧化碳和其他温室气体，并在富含碳的水淹土壤中储存数千年。这种在水下被存储在红树林、海草床和盐沼等沿海生态系统中的碳被称为“蓝碳”（blue carbon）。海洋是地球上最大、最活跃的碳库，其容量约是大气碳库的50倍、陆地碳库的20倍。然而，全球的蓝碳却在急剧减少。自20世纪40年代以来，近海富营养化、填海造陆、海岸工程及海岸城市化使地球上大部分的蓝碳消失。全球约1/3的海草区域已消失，约25%的盐沼区域已不复存在，约35%的红树林区域遭受破坏，消失速率为每年1%～3%（Nellemann C et al.，2009）。由于全球对生物多样性、气候变化问题的重视度在不断增强，包括红树林在内的蓝碳资源在不同区域的重视度也在逐步提高。美国科学促进会的研究显示，在全球范围内，20世纪末至21世纪初的红树林损失率降低了近一个数量级，从每年的3%降至0.3%～0.6%，这在很大程度上得益于全球范围内红树林的保护工作。永续全球环境研究所的“中国—东盟红树林保护”项目与世界团队组织的“海洋机遇与小岛国”项目分享了非政府机构在区域层面通过创新的保护模式探索红树林保护和社区发展“双赢”的经验。

8 城市类

城市是人口增长与经济发展的集中点，属于重要的社会生态系统，同时也是经济发展与生态环境矛盾突出的重要区域，面临着气候变化、粮食安全、水安全、灾害风险等多重社会挑战。就规模而言，城市只占世界陆地面积的2%，却聚集了世界上80%的人口，消耗着全球2/3以上的能源，占全球二氧化碳排放量的70%以上[①]。而作为温室气体排放的密集地区，城市也可以成为提供解决方案的重要参与者和推动者。目前，全球关于NbS在城市中的应用一方面聚焦于减缓，如城市绿化等改善居民生活环境、促进城市资源和能源可持续利用以推进城市碳中和进程的项目；另一方面聚焦于适应，欧盟于2015年发布的《2020地平线报告：NbS使城市重返自然》特别强调了可以通过改造湿地或水生生态系统来提升城市的生态和观赏价值，同时也强调从保险的视角来探讨如何利用NbS降低与气候变化有关的灾害风险。

不同城市的地理位置、城市规模、人口密度、经济发展水平、城市功能等各不相同，需要结合自身特点设计和实施适宜的NbS。因此，本章选取了全球范围内处于不同发展阶段的6个典型城市案例进行梳理：从强化城市绿色基础设施建设、减缓气候变化的角度出发，选取了意大利米兰、英国伦敦与中国四川成都3个案例，简要介绍了如何利用NbS实现温室气体减排、提升能源使用效率、改善城市生态环境等多重目标；从完善城市适应性基础设施建设、提高气候韧性的角度出发，选取了荷兰鹿特丹、美国纽约曼哈顿和旧金山湾区的案例，展示了利用NbS提升城市防灾减灾能力的路径。

---

①数据来源：C40城市集团，https://www.c40.org/why_cities。

8.1 意大利米兰基于自然的旧城改造

案例亮点

通过顶层设计，结合区域发展规划，积极利用民间资本，鼓励民众参与到城市花园的改造和实施过程中；通过与建筑物的集合，塑造了以“垂直森林”为代表的绿色建筑项目，开创了在城市中践行NbS的地标性建筑。

8.1.1 项目背景

米兰是意大利的第二大城市及主要经济和工业区，也是欧洲经济最为发达和活跃的城市之一。然而，米兰同样面临着与大部分典型商业城市类似的挑战，如由城市化进程加快带来的交通和污染等问题。

8.1.2 项目介绍

自2006年起，米兰开始改造大都市区主要的旧城部分，涉及的人口超过410万人。米兰首先发布了针对城市环境问题的战略性方案《米兰宪章》，确立了以环境保护、促进社会可持续发展与改善社会福利等为目标，以推进绿色基础设施为主要手段的改造策略。与此同时，米兰所在的伦巴第大区（Lombardia）也推行了“地区生态网络”计划，通过绿色基础设施完善生态系统，同时提升阿尔卑斯山山区和波河河谷地区之间的生态系统完整性和连续性。该方案也包括为不同的城市和地区提供管理、提升生态系统的指导方针与筹资机制。

基于此，米兰将NbS应用于不同场景并作为其建筑和城市更新战略的一部分，实施了包括“垂直森林”、城南农业公园、水上公园、城市花园等系列项目。

1. “垂直森林”

“垂直森林”项目启动于2004年，旨在让城市街区规划和改造更多地运用

NbS措施，总投资超过20亿欧元。“垂直森林”项目由2座分别高110米和76米的住宅塔楼组成，种植了900棵树（每棵树高最多不超过10米）和2万多株其他植物（各种灌木和花卉植物），并根据外墙的阳光照射情况进行设计布局。据估计，这2座塔楼中的植物所提供的生态系统功效（碳汇、空气质量、生物多样性改善等）相当于2公顷森林所产生的碳汇（图8-1）。

图8-1　米兰的“垂直森林”

2. 城南农业公园

在《米兰宪章》的规划方案中，有一项建议是建立连接整个城区绿地与公园的绿色走廊系统。在这一建议下，城南农业公园项目建设了公园网络体系，为当地居民提供了农业、林业、文化和娱乐休闲活动场所，在保护景观的同时恢复了生态环境，并将外部地区与城市绿色廊道连接起来，改善了大都市地区的生态平衡。为了保护生物多样性，城南农业公园内设置了一些专门用于重建生态系统的区域，重新引入了日益稀少的本地动物物种（如伦巴德锹形蟾蜍、拉塔斯特蛙、沼泽龟和河虾）。同时，城南农业公园承载的城郊农业不但有利于水土保持，还可以满足米兰居民对本地粮食与蔬果日益增长的需求。

3. 水上公园

水上公园是一座多功能基础设施，内设的休闲娱乐区可以用于开展体育教育、骑行、跑步、野餐、观察动物等各种活动。场地内有各种关于本地动物（水鸟和小型两栖动物）、植物群（特别是具备净水功能的植物）的教育与宣传服务。

4. 城市花园

几十年来，在城市内推广城市花园的理念日趋流行。2012年，米兰市议会通过与非营利组织的合作，试图在较短的时间内在市区建立新的城市花园模式，减少维护成本。为此，市政府出台了激励机制，鼓励居民参与城市绿化，修复废弃绿地等活动。政府将园艺作为一种爱好来推广，将绿地打造为新型的社交空间；在操作层面，政府提供公共和市政用地专门用于种植供私人消费的水果和蔬菜。这些地块的使用者须遵守市政条例，每年支付少量的费用，由市政当局提供绝大多数地块的用水等资源。

米兰所在的伦巴第大区为旧城改造提供了400余万欧元的资金支持，用于改善和提升生态系统完整性。来自公共部门和私营企业的个人与实体最多可以申请每公顷30 000欧元的初始补贴和每年每公顷4 000欧元的维护费用补贴，最长期限为3年，补贴主要用于恢复森林生态系统及土壤修复。2007—2013年，伦巴第大区共投资6 120万欧元，其中20%左右来自欧盟基金（环境与气候行动LIFE计划、欧洲农村发展农业基金、欧洲区域发展基金），60%来自本地，此外还有部分民间的资金。米兰市支持与私人、私营或半私营公司合作维护绿地。政府推出了“领养一片绿地”项目，鼓励当地居民参与绿地管理，并寻求赞助以帮助城市财政。截至2016年4月，该项目已经促成了396个合作协议，管理超过米兰60%的领土面积。

8.1.3 项目成果

一是“垂直森林”内的植被能够降低周边空气中的颗粒物浓度，使可吸入颗粒物和总悬浮颗粒物平均减少20%～30%。这些结果证实了树木和绿色屏障（灌木和绿篱）在清除空气颗粒物方面的效果。此外，“垂直森林”的吸引力和知名度也让米兰市受益匪浅，屡获各项殊荣，包括“2014年高层建筑奖”和“2015年世界高层建筑与都市人学会奖”。事实证明，“垂直森林”项目是可以推广并复制的。自2017年起，负责开发该项目的Boeri公司就开始在中国江苏省南京市进行“垂直森林”项目的建设，这将是亚洲建成的第一个“垂直森林”，除了吸收二氧化碳，每天预计可产生约60千克的氧气。

二是在城南农业公园47 000公顷的总面积中，已有超过80%的区域被开发为农业用地，园内吸引了1 000多名村民。此外，将公园内的历史与建筑文化遗产联系起来是组成休闲和文化生活的重要途径，维护这些建筑文化遗产（公园内的小村庄、城堡、别墅和修道院）将成为增加公园旅游业的另一种方式。

三是水上公园是占地约3公顷的综合公园，包括一个防洪区（1公顷），一个污染物清理区（0.4公顷的短叶草芦苇床和0.3公顷的自然类多物种湿地）和一个休闲娱乐区（1.3公顷的公园）。园内建设了4个沙滤竖床用于处理联合污水溢流机制带来的第一道冲刷，同时建设了1个延伸的蓄水池用于第二道冲刷，并在河道中缓慢释放。绿色基础设施基本解决了污染防治和防洪问题。经过第三方评估，米兰的水上公园建设体现出欧盟的研究与创新资助战略对城市管理的服务作用，凸显了环境保护及社会参与的积极影响。

四是在米兰，大多数受访者希望获得更健康的食物与生活，这也是居民参与米兰城市花园项目、体会城市园艺的动机。城市花园为市政府开展本土化的城市园艺项目提供了许多政策建议，如鼓励园丁进行更多干预措施以提高土壤质量，为园丁提供土壤检测服务，开发城市实时土壤质量地图或研究花园附近潜在的污染源等。

8.1.4 项目信息

案例来源：欧盟城市NbS案例研究-Oppla平台。

项目参与方：米兰市政府及相关各方。

项目地点：意大利米兰。

项目时间：2006—2016年。

项目链接：欧盟城市NbS案例研究-Oppla平台，https://oppla.eu/casestudy/19446。

8.2 英国伦敦可持续城市绿色建设

案例亮点

通过设立可持续发展委员会对政策制定者提出建议和策略，给NbS等前瞻性理论提供最佳的实践场所；在实施可持续发展项目的初期应注意利用金融及市场模式来促进商业的参与，以保证项目的可持续性。

8.2.1 项目背景

英国伦敦是世界可持续发展的先锋城市之一。在2002年之前，伦敦就设立了可持续发展委员会，主要负责就伦敦的可持续发展问题向市长提出建议和策略。伦敦可持续发展委员会在该市的城市规划与建设领域具有极其重要的地位，涉及的领域包括城市艺术与文化、商业与经济、青年与教育、环境与健康、社区、资金、就业、交通等方面。伦敦可持续发展委员会通过开展节能减排和绿色低碳发展的相关研究，培养了多名在可持续商业和社区领域的领袖或活动家，为伦敦的可持续发展奠定了良好的基础。2018年，伦敦位居凯谛思“可持续性发展城市指数”榜首。

8.2.2 项目介绍

伦敦市长萨迪克·汗致力于将该市建设成为世界上最绿色、最健康的城市之一。为此，他启动了惠及所有市民的“伦敦可持续发展”项目，以改善伦敦的整体环境。该项目涵盖空气质量、绿色基础设施、气候变化、废弃物管理、城市韧性、环境噪声和循环经济在内的7个重点领域。同时，伦敦也制订了长期的可持续发展计划，以及到2050年实现碳中和的目标。此外，伦敦还决定于2040年前后实现全市区域范围内80%的出行方式由步行、自行车或公共交通等实现。

值得注意的是，“伦敦可持续发展”项目中包含了大量的NbS倡议和行动，能够有效减少城市内涝、改善空气质量、缓解热岛效应、增加人行道和自行车出行的便利、改善城市景观以增强生物多样性和韧性。这为助力伦敦成为绿色都市及实现可持续发展目标发挥着重要作用。伦敦在其新颁布的《伦敦2050远景基础设施规划》中强调了城市绿色基础设施的重要性，政府需要积极寻求商业的解决方案来助力规划的实施。伦敦计划利用两种金融及市场模式：一是向开发商收取费用的模式，以确保长期维护良好的绿色基础设施；二是共同出资的模式，用于泰晤士河、奥林匹克公园、绿化公园等大型公共区域绿色基础设施建设。同时，伦敦市政府于2019年成立了绿色基础设施工作委员会，其主要职责是采取更具战略性和更长期的方法投资推进绿色基础设施的建设工作。这将为伦敦带来新的经济机遇及就业机会，有助于提高社会绿色产业的发展及后疫情时代欧盟绿色复苏计划。在欧盟“城市韧性与可持续转型”（Transitioning towards Urban Resilience and Sustainability）项目的资助下，伦敦开展了如下NbS项目。

1. Quaggy河流域[①]天然节水项目

伦敦在位于东南部的Quaggy河流域推行修复天然河道、开展洪水管理等措施，以便改善天然河谷与洪泛区缩小等问题，降低洪涝风险。该项目对1990—2005 年实施的Quaggy河河道进行改造，对防洪设施的规划进行重新定位和升级。以上这些措施均有助于改善伦敦市的水质，降低城市污水处理成本，提升绿色基础设施覆盖区域，提升种群数量。除了可以起到固碳减排的作用，还可以降低干旱、洪涝风险，有助于城市水资源管理及生态循环。

2. 巴金河畔的棕地改造项目

棕地是指存在一定程度的污染、已经废弃或因污染而没有得到充分利用的土地及地上建筑物。巴金河畔是位于巴金与达根汉姆自治市的一块443英亩的棕地，该地区计划在20年内为2万名居民开发新的居住场所（图8-2）。项目在规划初期就考虑了需要充分发挥巴金河畔棕地的生态系统服务潜力，以确保项目开发的可持续性，具体包括保护该地块的生物多样性、保留该地块 40% 的绿地、开发一个全面的可持续城市排水系统、建设社区野生动物园等。在巴金河畔开发之前，政府已经确定了该棕地珍贵的无脊椎动物物种及其主要栖息地的特征，并将其纳入社区花园的设计中。建成后，该花园可以用来向居民介绍本土野生动物，并向居民展示通过园艺培养这些物种的方法。

①Quaggy河流域位于泰晤士河的东部区域。

图8-2　巴金河畔规划效果

3. 绿色屋顶项目

伦敦规定在全市40%的房屋上建设绿色屋顶，连同沼泽地、雨水花园和池塘等构建综合的可持续城市排水系统。欧盟“城市韧性与可持续转型”项目在巴金河畔开展了一系列绿色屋顶的建设工作，通过比较不同绿色屋顶系统在生态服务方面的性能（包括提供栖息地、节水和隔热等），实现了将传统以工业为主导的屋顶建设转化为由花草构建的生物多样性更丰富的屋顶绿化系统（图8-3）。在巴金河畔的建设中，“城市韧性与可持续转型 ”项目创建了少量的绿色湿地屋顶进行生态模拟实验：调查与该地域开发前棕地状态相关的关键生境再创造潜力，通过操纵排水、使用不同的骨料与改变基质深度在绿色屋顶上创造生境试验场。最后由项目组研究人员对屋顶进行监测，以评估栖息地对屋顶整体生物多样性的影响。此外，“城市韧性与可持续转型”项目还尝试对伊丽莎白女王奥林匹克公园现有的0.25公顷的绿色屋顶进行生物多样性价值评估，并在该屋顶上加装了太阳能电池板，利用生态位[1]方法来量化评估实施效果。研究发现，与伦敦其他绿色屋顶相比，该屋顶的无脊椎动物多样性很高，拥有多种本地特有物种。

图8-3　伦敦的绿色屋顶

4.“甲虫部落”项目

“甲虫部落”项目旨在拯救英国最稀有的昆虫之一——条纹甲虫。东伦敦大学与Buglife组织合作，使用生态仿生设计的原则，用65吨废旧材料（硬核、白垩、砖头和表土）制成了适合甲虫生活与繁殖的“甲虫部落”，并在该地区最高质量的开放生境中播种花卉，将“甲虫部落”植入其中，成为一个融合艺术、景观设计和生境保护的多功能空间。工作者把从建筑用地上救出的甲虫释放到校园，东伦敦大学逐渐形成了研究甲虫的行为和栖息地特征的露天实验室。

总体来看，位于伦敦东南部泰晤士米德镇的城市改造项目是“伦敦可持续发展”项目中值得借鉴的成功案例。该项目利用有效的政策规划和管理，将景观、植被种植和自然生态系统充分融合并集中在一个地区试点。泰晤士米德镇拥有一个超过45 000名居民的现代社区，其公寓和房屋周围拥有广泛的绿地、湖泊与运

①生态位是指一个种群在生态系统中在时间空间上所占据的位置及其与相关种群之间的功能关系与作用。

河网络。该社区始建于20世纪60年代，位于泰晤士河的洪泛平原上。为了防范大面积潮汐洪水的风险，泰晤士米德庄园的湖泊与运河在设计之初就拥有一定的蓄洪能力。但是多年来，由于进出路线设计和方向等问题，该地的湖泊与绿地没有得到充分利用，巨大的空白绿地功能单一，难以吸引游客和居民。泰晤士米德镇利用基于自然的城市改造项目，使城市融入更多的自然元素。

8.2.3 项目成果

伦敦的绿色城市建设已经在城市层面产生了多种积极效应。

一是坚实的理论研究基础为全球提供了学习分享的案例。伦敦的研究证明，绿色屋顶具有多重效益，包括减少径流、通过蒸腾作用缓解城市热岛效应、改善空气质量并提供新的休闲空间等。目前，伦敦已开发了评估城市绿色屋顶潜力的工具，可帮助政策制定者在规划过程确认哪些区域是可以开发为绿色屋顶的高潜力区域。需要注意的是，绿色屋顶战略也存在一些隐患，一般具备高潜力开发的区域都属于收入较高的地区，而低收入地区因为可开发的绿色屋顶区域有限，更加容易受到气候变化的影响，这可能会在一定程度上加剧贫富差距。以上这些研究成果为世界其他地区的绿色规划提供了有价值的参考和借鉴。

二是伦敦的NbS建设将会逐步实现多个目标，如市长发布的《伦敦规划》宣布，至2030年要种植200万棵树，增加5%的绿地，至2050年再增加5%。街道树木种植计划则针对已知的热岛区域要求在规划期内确保指定的重要自然保护地不出现净损失，至2016年将二氧化碳排放量减少到比1990年低23%的水平，功能性洪泛区不出现净损失，至2010年利用可再生能源生产945吉瓦时的能源等。

8.2.4 项目信息

案例来源：欧盟城市NbS案例研究-Oppla平台。

项目参与方：伦敦市政府及相关各方。

项目地点：英国伦敦。

项目时间：2018—2030年（第一阶段）、2030—2050年（第二阶段）。

项目链接：伦敦市政府，https://www.london.gov.uk/about-us/organisations-we-work/london-sustainable-development-commission。

8.3 中国成都公园城市建设

案例亮点

在习近平生态文明思想和理念的指引下，成都在推行公园城市的初期，一方面强调要突出公园城市的特点，另一方面把生态价值也考虑进去，保证了成都公园城市建设除了实现多重效益，还可以在经济和就业层面产生长期的可持续效益。

8.3.1 项目背景

成都地处中国西南地区、四川盆地西部，是四川省的省会城市，地势平坦、河网纵横、物产丰富、农业发达，属亚热带季风性湿润气候，自古有“天府之国”的美誉。2019年，其建成区面积达950平方千米，常住人口为1 600万左右，人口密度为1 180人/千米^2，城镇化率约为75%。成都是联合国人居署发布的国际可持续发展试点城市、国家第三批低碳试点城市，明确了“建立具有全球比较优势、全国速度优势、西部高端优势的西部经济核心增长极”的发展定位。然而近年来，成都的城市化与工业化进程快速发展，对周围城市的虹吸效应持续加强，人口总量不断增加，高楼大厦林立，小汽车保有量跃居全国第二，林地锐减、耕地面积减少、水质污染、生态板块碎片化、生物多样性降低等问题开始逐渐凸显。

8.3.2 项目介绍

2018年2月，习近平总书记在四川成都天府新区考察时指出，“要突出公园城市特点，把生态价值考虑进去。”作为首个提出“公园城市”概念的区域，成都正在加快“美丽宜居公园城市”的建设步伐，发布了《成都市美丽宜居公园城市规划（2018—2035年）》，塑造了“开窗见田、推门见绿”的田园风光和大美公园城市。成都通过保护、恢复和可持续管理生态系统，有效应对各类社会挑战，满足了人民对美好生活的向往，打造了可进入、可参与、可感知、可阅读、可欣赏、可消费的高品质全域公园体系，为NbS在城市的应用提供“成都方案”。

公园城市是生态文明时代主动协同保护自然和城市发展的高级形态，是新时代可持续发展城市建设的新模式，具有以下四大特征：一是突出了以生态文明引领的发展观，二是突出了以人民为中心的价值观，三是突出了构筑山水林田湖草沙生命共同体的生态观，四是突出了人城境业高度和谐统一的现代化城市形态。在公园城市的理念下，城市建设模式应实现三个转变。一是从“产、城、人”到“人、城、产”。从工业逻辑回归人本逻辑，依托良好的生态环境和公共服务，吸引人才聚集、企业汇聚，带动产业繁荣，实现“人、城、产”的和谐发展。二是从“城市中建公园”到“公园中建城市”。城市建设必须符合公园化环境的生态、美学、文化、经济与形态等要求，将公园形态与城市空间有机融合。三是从“空间建造”到“场景营造”。围绕人的需求，通过设施嵌入、功能融入、场景带入，全面营建城市生活场景、消费场景、创新场景等，增强空间归属感。公园城市作为回应新时代人居环境需求、塑造城市竞争优势的重要实践模式，具有一系列体现时代特点的重要价值，包括绿水青山的生态价值、诗意栖居的美学价值、以文化人的人文价值、绿色低碳的经济价值、简约健康的生活价值与美好生活的社会价值。成都的公园城市规划建设体系由森林公园、郊野公园群、城市公园绿地、天府绿道四部分构成（图8-4）。

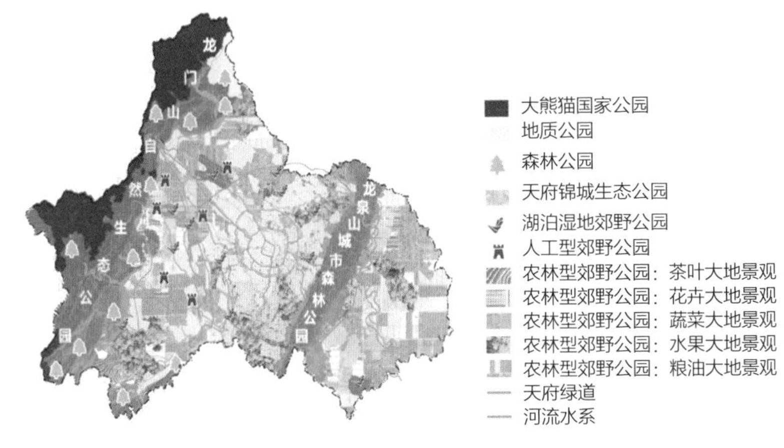

图8-4 成都公园城市建设规划

（图片来源：成都市天府公园城市研究院）

此外，成都正在实施以“碳惠天府”为品牌的碳普惠机制，旨在针对小微企业、社区、家庭和个人的节能减碳行为进行具体量化，并建立政策鼓励、商业奖励与碳减排量交易相结合的机制，正向引导人们的行为，从而调动全社会自下而上地践行绿色低碳发展行为的积极性，进而主动对接应对气候变化国家战略。根据《成都市“碳惠天府”机制管理办法（试行）》，成都的“碳惠天府”并不是照搬其他城市的经验模式，而是在国内首创提出建设面向公众和企事业单位的“双路径”激励机制，包括“公众碳减排积分奖励”和“项目碳减排量开发运营”两个路径。针对“公众碳减排积分奖励”路径，成都将围绕公众的衣食住行游，通过制定餐饮、商超、景区、酒店等低碳评价规范，引导企业创建低碳场景，实施低碳管理。运营实体可根据场景内的公众低碳行为数据，通过运营平台换算成碳积分并发放至个人账户，公众持有的碳积分在运营平台上可兑换普惠商品或服务。针对“项目碳减排量开发运营”路径，成都将重点引导龙泉山城市森林公园、天府绿道、川西林盘、湖泊湿地等重大生态工程开发碳汇，系统有序地制定碳减排潜力大、环境效益好的项目方法学。依据相应方法学开发的各类项目碳减排量可在四川联合环境交易所与有“碳中和”意愿的消纳方进行交易，从而将基于自然的生态建设工程、企事业单位节能降碳产生的环境效益以碳的属性呈现其经济价值。例如，根据碳惠天府方法学，锦城绿道每年可产生4 200吨二氧化碳当量的减排量用于碳中和交易，实现“生态有价”。

8.3.3 项目成果

自2018年以来，成都贯彻“园中建城、城中有园、城园相融、人城和谐”的规划理念，以绿道、水网串联森林公园、郊野公园和城市公园绿地，形成无缝衔接的全域公园体系。至2035年，规划建设1 275平方千米的龙泉山城市森林公园和1.69万千米的天府绿道，建成各级绿道3 689千米，新增绿地面积3 885万平方米，全市森林覆盖率达39.93%，建成区绿化覆盖率达43.5%。

以天府绿道为例，作为世界上最长的绿道，其在2020年已建成3 689千米，完成投资341亿元，目前正在以“绿道+”串联起“以道营城、以道兴业、以道怡人”的生态价值转化路径，将实现生态景观、慢行交通、休闲游览、城乡融合、文化创意、体育运动、景观农业及应急避难八大功能（图8-5）。

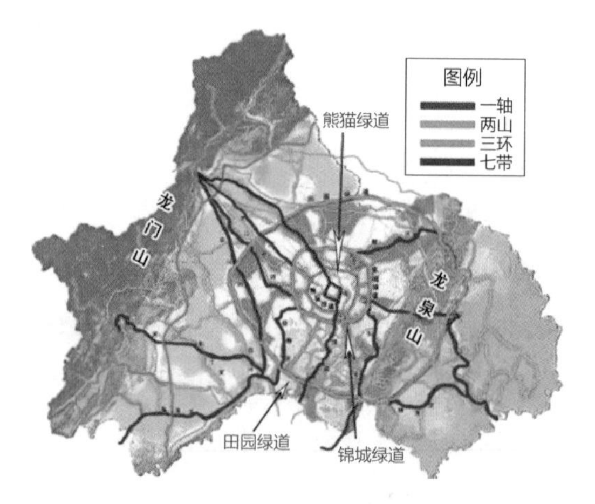

图8-5 成都市区域级绿道规划

（图片来源：成都市天府公园城市研究院）

天府绿道的前瞻性布局背后考验的是成都NbS长效运维和可持续发展的能力。天府绿道坚持政府主导、市场主体、商业化逻辑，以设施租赁、联合运营、资源参股等多种方式实施全球招商引资，天府绿道的社会投资占比达70%以上。以锦城绿道为例，能带动周边土地增值上千亿元，辐射周边千亿级现代服务业产业集群，每年吸引旅游者上亿人次，可创造10万个以上的就业岗位，吸引高素质人才，促进创新创业。此外，据中国科学院水利部成都山地灾害与环境研究所对气候调节、固碳释氧、土壤保持、水源涵养等18项生态服务价值指标的初步估算，锦城绿道每年的生态服务价值量约为269亿元，预计可产生40年以上的持续性效益，总价值达1万亿元以上。

8.3.4 项目信息

案例来源：C+NbS合作平台。

项目参与方：成都市政府及社会各界。

项目地点：中国四川省成都市。

项目时间：2018—2035年。

8.4 荷兰鹿特丹气候适应性基础设施

案例亮点

采取有效的创新治理模式，一方面意识到自身地理位置对于未来气候变化的脆弱性，另一方面积极主动地应对气候变化；将社区生活与防灾减灾和提升韧性有机结合，实现了资源的高效和集约化利用。

8.4.1 项目背景

低海平面是荷兰地形最突出的特点，全境1/4的土地海拔不到1米，其东北部的卫星城亚历山大斯塔德附近的区域为全境最低地，低于海平面6～7米，拥有居民17.5万人。类似的滨海城市在很大程度上会受到海潮的影响，如荷兰最大的城市阿姆斯特丹是一座名副其实的“水城”，河道纵横交错，海拔极低，雨洪灾害随时发生。因此，“与水共生”成为荷兰城市发展的重要理念。

鹿特丹是欧洲最大的港口，位于莱茵河与马斯河之间的三角洲。2008年，鹿特丹市议会批准了《鹿特丹气候防护规划》。2013年通过的《鹿特丹适应气候变化战略》旨在加强荷兰境内的洪水防御系统，通过综合规划提高城市利用空间，提升城市韧性，并探索应对气候变化带来的机遇，如发展经济、改善生活质量与增加生物多样性等。鹿特丹采取了一种量身定做的“内堤/外堤”方法，其大部分地区（包括主要港口）都位于外堤区（海平面以上3～5.5米），该区约有4万名市民居住，容易受到海平面上升或临时洪水的影响，将通过结合创新技术与传统方法对其进行改造；内堤区的大部分地区都在海平面以下，由排水口与水泵排干的堤坝系统组成，并由较小的二级堤防进行保护，如果该地区发生洪水，将是灾难性的。

8.4.2 项目介绍

《鹿特丹适应气候变化战略》中提出的目标是到2025年实现100%抵御气候风险。该战略强调至2025年，将采取措施确保每个特定地区在当前与未来几十年中受到气候变化的影响程度最低。此外，鹿特丹所有的城市规划都将考虑长期可预见的气候变化风险，主要从以下三个方面采取行动。

1. 蓄水

气候变化将导致极端降雨频发，会增加洪水造成的不利影响，特别是在蓄水能力不足、建筑物密集、路面透水能力不足的地区。需要对城市供水系统的排水与蓄水能力进行调整与改善，以应对未来可能发生的极端降雨事件。建议的措施包括雨水收集和存储、延迟排水，恢复城市的海绵功能等。具体工程包括在屋顶和外立面增加绿化面积、减少路面硬化、在公共街道与社区增加植被、在降水广场与渗透区建设绿色基础设施。这些措施在人口稠密、建筑密集、空间有限的地区尤其有效。

2. 三角洲规划

建立富有韧性的多层防洪建筑，如防洪建筑、防洪公共区域、浮动社区与“亲自然建筑”。外堤区重点关注港口与基础设施，保护其免受洪水侵袭；内堤区以预防为主，根据需要优化风暴潮屏障，加固堤坝，使其功能多元化，而这些设施平时则可以作为天然绿色堤防融入城市景观与休闲观赏路线。

3. 潮汐公园

气候适应行动需要与改善城市生态相结合。在潮汐公园方案（图8-6）中，

图8-6 鹿特丹的潮汐公园

正在开发若干外部堤防区，以替代固体建筑，防止高水位洪水。例如，在Esch和Mallegat的洪泛区域，创建绿色和蓝色走廊，创造高质量的自然环境，可以改善生物多样性与景观连通性，同时还可以增加碳存储，调节局地小气候。绿色和蓝色走廊的搭建可与新的娱乐区结合，使居民有可能利用这些走廊进行体育和休闲活动，实现强身健体的功效。同时，潮汐公园还设立了自行车道，鼓励民众绿色出行。潮汐公园的建立使与湿地相关的生态系统服务和功能得到修复与改善，可以发挥蓄水、调节水流与过滤水质的作用。城市的蓝绿韧性设施不仅有助于抵御气候风险，还使其更具吸引力且适宜居住。

8.4.3 项目成果

鹿特丹的气候变化适应方案不仅关注使用绿色基础设施，而且将“灰色”、“绿色”与“蓝色”方案相结合。例如，通过建设部分蓄水空间，包括容量达1万立方米的博物馆公园、停车场地下蓄水库、蓝绿走廊等项目降低强降雨事件带来的威胁。这些蓝绿走廊，包括水道与积水区都可以促进地下水补给，减少城市内涝，增加生物多样性，提高城市生活质量。仅2014年，鹿特丹就安装了超过185 000平方米的绿色屋顶，另外启动了100%韧性街区示范项目。

此外，鹿特丹还开发了创新的城市防洪模式，如2013年建成的Benthemplein水上广场能够在降雨高峰期发挥蓄水功能，在一定程度上缓解城市污水处理系统的压力。Benthemplein水上广场是世界上第一个全尺寸水上广场（Full-Scale Water Square），最多可容纳170万升降落在广场上的雨水，这些雨水会回渗到土壤中，或者被水泵抽送到城市其他地方的运河中。此外，水上广场有助于降低城市用于地下水网改造而产生的巨大成本，兼具体育和娱乐用途，为篮球、滑板和表演艺术提供了空间。

8.4.4 项目信息

案例来源：欧盟城市NbS案例研究-Oppla平台。

项目参与方：鹿特丹市政府。

项目地点：荷兰鹿特丹。

项目时间：2008—2025年。

项目链接：欧盟城市NbS案例研究-Oppla平台，https://oppla.eu/casestudy/19457；C40城市集团，https://www.c40.org/case_studies/benthemplein-water-square-an-innovative-way-to-prevent-urban-flooding-in-rotterdam。

8.5 美国曼哈顿下城区气候适应性计划

案例亮点

先识别高风险区域，根据科学适应的原则和社区情况制定合理的行动，再通过社区调研、风险分析并结合未来城市规划确立综合行动，以提升城市韧性。

8.5.1 项目背景

2012年10月29日，飓风“桑迪”登陆美国纽约，淹没了该市17%的土地，夺去了44人的生命，在全美造成约125人死亡，经济损失超过190亿美元。仅在纽约曼哈顿下城区，飓风“桑迪”就造成了毁灭性的影响，数以千计的房屋与多个重要的交通枢纽被摧毁。目前，科学界已经达成共识，气候变化会导致飓风或强降雨事件频率与强度的增加，海平面上升可能会淹没曼哈顿下城区的部分地区，同时危及纽约市的关键基础设施，包括地铁、轮渡网络和下水道系统，威胁城市10%的就业机会，以及许多历史、文化与社区资产。

8.5.2 项目介绍

基于上述因素，纽约决定采取行动，投资5亿美元用于提升城市的适应能力，以保护曼哈顿高风险区域，特别是下城区。适应行动制定的原则是切实考虑社区具体情况，通过社区调研、风险分析并结合未来城市规划确立综合行动以提升城市韧性，降低沿海暴风雨和海平面上升给曼哈顿下城区带来的风险，同时曼哈顿以此为契机利用全面创新的规划与设计手段来改善社区面貌。

《曼哈顿下城沿海气候韧性计划》包括以下内容：

一是炮台公园与炮台海岸适应行动计划。炮台公园城市管理局正在开发4个海岸带景观修复项目，以减少风暴潮与海平面上升给炮台公园及周边地区带来的风险。这些项目将分阶段实施。通过增加公园与绿地空间、修建部分固定防洪墙、布置临时防洪设施等手段降低沿海区域面临的风暴及洪涝风险。炮台公园南部适应行动包括建造景观护堤和防洪墙。炮台海岸计划重建年久失修的码头与长廊，并将其位置提升，以在2100年前防止炮台公园因海平面上升而出现每日潮水淤积的现象。作为独立项目的一部分，炮台公园城市管理局还将在公园后面建设景观护道与绿色墙体，以减轻沿海洪涝对该区域的侵袭，该项目于2021年开始施工建设。

二是金融区和海港地区适应气候变化行动方案。作为曼哈顿下城区适应气候变化整体计划的一部分，纽约市于2019年启动了金融区和海港地区适应气候变化行动方案。该方案的主要目的是确定开展具体行动来降低该区域的气候变化风险，这些风险包括海平面上升将导致曼哈顿下城区部分地区频繁发生4～6英尺[①]的洪水，气候变化带来的降水量变化可能会导致洪水泛滥与河道污染的加剧，沿海风暴潮带来的洪涝将严重威胁曼哈顿下城区。为此，纽约市将重点开发并设计防

①1英尺=30.48厘米。

洪基础设施以提升排水能力，开展雨水收集并管理废水，还为基础设施项目提供部分资金与筹资渠道。该方案的第一个阶段于2021年秋季完成。

三是布鲁克林大桥蒙哥马利海岸适应气候变化行动。该行动的具体方案包括建设由固定与可部署的移动屏障组成的综合防洪系统，改造排水设施以解决因降雨造成的内陆渗水问题。移动屏障位于滨海广场之上，在台风或强降雨来临之前可以用于保护两座桥梁附近的区域。它们将安装在一个升高的平台上，以解决海平面上升引起的潮水渗漏问题。移动屏障既保留了海岸带景观以开放给社区公众进行参观，又可以保护社区免受沿海风暴的侵袭。该工程于2021年开始建设。

此外，在金融区和海港地区的两座桥梁上还设置了临时防洪屏障，以保护这些地区免受更频繁、更严重的风暴影响。纽约市政府正在研究这项计划的潜在扩展区域，部分建设已于2019年完成。

8.5.3 项目成果

2019年3月，纽约完成并发布了《曼哈顿下城区气候韧性研究报告》，全面评估了当前与未来气候风险对下城区的潜在影响。该报告评估了与气候变化有关的危害，如慢性疾病、海平面上升、地下水位上升、潮汐淹没、风暴潮、极端降水和热浪等。研究发现，到2050年，曼哈顿下城区37%的建筑将面临风暴潮的威胁；到2100年，近50%的建筑物将面临风暴潮的威胁，20%的街道可能会因为海平面上升超过6英尺而每天遭受洪水的威胁，地下水位上升预计将威胁7%的建筑物的安全运营，39%的地下设施街道将遭受腐蚀与水渗透。

这些研究帮助纽约确定了长期的适应行动方案，包括曼哈顿下城区适应气候变化整体战略。相关项目将保护曼哈顿下城区70%的海岸线，并于2021年全面开工，主要通过加强海岸线防护、改善公共空间利用程度，将曼哈顿的双桥社区到炮台公园的5英里①左右的滨水区转变为综合防洪区域，以保护脆弱的沿海社区。利用创新的景观护堤和本土植物不仅能够保护曼哈顿海岸线免受洪水侵袭，还可以提供文化、娱乐和生态效益，并提升社区凝聚力。此外，项目的绿色基础设施，如生物洼地、雨水花园和行道树等，有助于增加当地的生物多样性，减少城市热岛效应。

8.5.4 项目信息

案例来源：C40城市集团。

项目参与方：纽约市政府及相关各方。

项目地点：美国纽约。

项目时间：2014年起。

项目链接：纽约市NYCEDC组织，https://edc.nyc/project/lower-manhattan-coastal-resiliency。

8.6 美国旧金山湾区“绿图计划”

案例亮点

“绿图计划”强调治理模式的创新，通过开发构建可视化的模型来支持决策，建立了一套基于自然资源保护的指导方针，实现“识别问题—合理（调整）规划—有效实施—科学评估”逐层上升的科学管理模式。

8.6.1 项目背景

旧金山湾区（简称湾区）位于美国加利福尼亚州北部，总面积为1.8 万平方

①1英里=1.609 344千米。

千米，人口约为770万人。作为全球三大湾区之一，它是世界上最重要的高科技研发和科教文化中心之一，也是美国西海岸最重要的金融中心。但与其他湾区一样，旧金山湾区面临着人口增长、城市基础设施发展、对自然资源需求的增加和气候变化带来的系列挑战。在快速城镇化的背景下，美国多数城市的土地与交通规划通常没有对自然资源予以充分考虑与重视。

8.6.2　项目介绍

意识到上述问题后，2017年旧金山湾区开展了“湾区绿图计划”（Bay Area Greenprint）项目（以下简称“绿图计划”），旨在支持建立一套基于自然资源保护的指导方针，从而在海湾地区促进可持续发展，平衡自然、农业资源和城市发展的需求。该项目的主要组织者包括大自然保护协会、绿地联盟、绿色信息网络、美国农田信托及湾区空间规划委员会。“绿图计划”主要关注自然和农业资源所发挥的 9 种自然价值与收益，包括粮食生产、水资源供应量、水质、降低水灾害风险、固碳、户外休憩、优先级高的栖息地、栖息地连通性及需要缓冲的物种与栖息地。此外，“绿图计划”也考虑了气候变化的情景，显示出气候变化对上述 9 种自然价值与收益潜在的威胁和机遇。“绿图计划”的构建流程见图8–7（冯艺佳等，2015）。

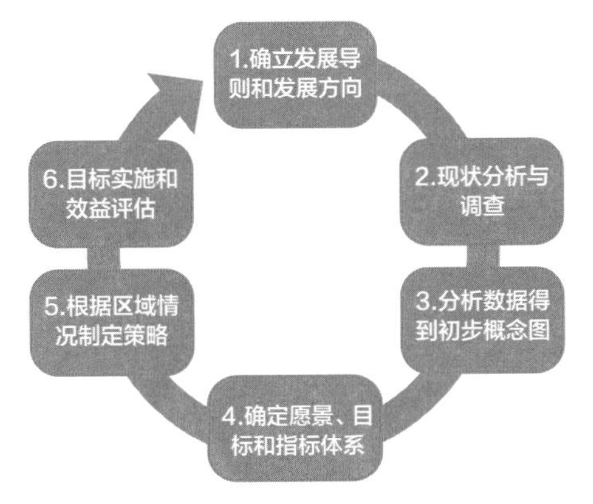

图8–7　“绿图计划”的构建流程

一是确立发展导则和发展方向。依据区域特点和计划目标、区域长期关注的问题及征集民意等方式，总结亟须提升改变的重点领域，进而确定“绿图计划”的发展方向。常见的方向为增加公园绿地、新建绿道、增强绿色空间可达性、改善环境与游憩设施状况等。在此基础上，针对区域存在的特殊问题再拟定相应导则。

二是现状分析与调查。通过一系列分析来了解区域状况，包括利益相关方调查、地理信息系统（GIS）数据分析、现有公园和绿色空间分析、资金来源分析。

三是分析数据得到初步概念图。利用地理信息系统将数据图示化为一系列现状图纸，将区域的主要自然特征，如较大的河流、起伏的山系等作为骨架，对一个区域的绿色基础设施进行识别并将其构建成绿色网络图。将绿色网络图作为基底，再根据要研究的问题（如人口、就业、住房、交通、文化资源等）进行相关因子的单独叠加，从而得到“绿图计划”的最初概念。

四是确定愿景、目标和指标体系。目标体系的确立基于“绿图计划”委员会、社区会议及实地调查。在目标体系确立之后，通过在社区发放调查问卷来了解民众的需求和优先级与认同度。综合考虑反馈意见，对城市资源进行分析并将指标叠加，并通过地理信息系统空间分析软件进行赋值，将对应目标进行图示。

五是根据区域情况制定策略。策略一般有多个，涵盖计划的方方面面。当区域较大或目标较复杂时，分目标下也同样会拟定实施策略，内容包括政策的变化、规章的制定、需要采取的行动等。

六是目标实施和效益评估。在“绿图计划”制订完成且经公示各方无异议之后，转而由地方政府协助“绿图计划”委员会进行实施与管理工作，主要工作内容包括估计预算并筹集资金、制定阶段性目标与行动计划并定期进行效益评估。

8.6.3 项目成果

“绿图计划”被开发为一个基于网站的应用工具（图8-8），其工具包可供政策制定者、城市管理者、环境保护组织和大众等各个利益相关方使用。这些工具利用可视化图集展示了各项自然资源的分布与土地利用，加上其个性化阅览分析土地的特点，为使用者提供了很大的便利，可以帮助其解决当地特有的挑战，并因地制宜地进行规划。例如，可以通过自然资源面板对湾区内的自然和农业资源进行详细介绍，展示湾区自然和农业资源分布、土地保护状况和发展风险，还可以对用户所关注的多种自然价值和收益进行整体评估，并在地图上展示这些价值和收益的重叠区域与重叠程度，支持管理者在规划中做出明智的协调或权衡决策。

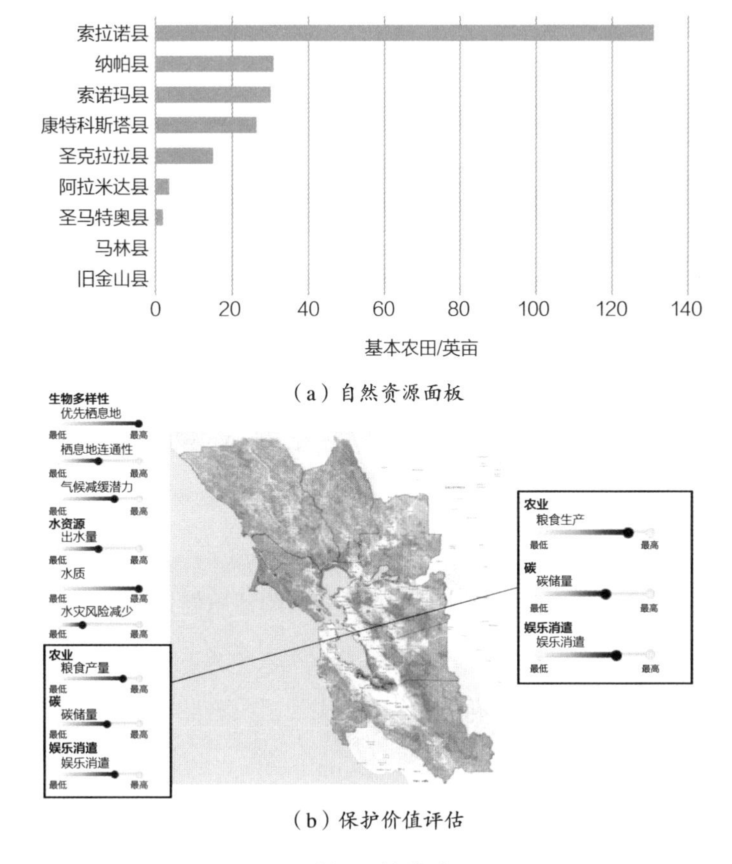

（a）自然资源面板

（b）保护价值评估

图8-8 “绿图计划”应用工具

“绿图计划”正在被湾区内的政府部门和自然保护组织使用，如负责制定湾区交通规划的部门正在尝试使用该工具对区域交通规划的项目生态足迹进行快速分析；加利福尼亚州海岸带保护协会正在尝试使用该工具指导湾区内的自然、休憩和农业资源保护工作；安提阿市正在将该工具所提供的自然资源价值信息考虑进当地沙溪重点区域的发展规划过程中，推动当地对区域内重要的自然和农业资源开展保护工作，使该区域更适宜人类居住。

8.6.4 项目信息

案例来源：TNC全球NbS案例库。

项目参与方：旧金山湾区政府及相关各方。

项目地点：美国加利福尼亚州旧金山湾区。

项目时间：2017年起。

项目链接：“湾区绿图计划”官网，https://www.bayareagreenprint.org/。

8.7 小结

在强调城市绿色基础设施建设与温室气体减排的3个案例中，米兰与伦敦都面临类似的问题，城镇化的进程使旧城区日渐衰落，亟须寻找新的契机对其进行改造和提升。

米兰通过旧城区改造项目开拓了NbS在城市的使用场景和空间，结合区域生态建设规划，利用多方筹资和鼓励市民参与等形式开发了“垂直森林”、城南农业公园、水上公园及城市花园等项目，基本实现了改善城市景观、提升空气质量及局部小气候、维持生物多样性、防灾减灾等多层次、全方位的需求。

伦敦将NbS与现有城市建设项目有机融合，优化老城区与城市棕地改造项目，积极保护本地物种和生物多样性，在项目实施中采用了大量生态仿生设计，如对绿色屋顶的研究、对本地特有条纹甲虫的保护、基于湿地公园的太阳能板建设等，将自然规律与现代技术充分结合，为其他城市的旧城改造、绿色创新带来了新的尝试。

中国四川省成都市是联合国人居署发布的国际可持续发展试点城市和中国第三批低碳试点城市，市政府首先从顶层设计上将公园城市作为新时代可持续发展城市建设的新模式，并在治理过程中通过系统规划和布局，旨在塑造“开窗见田、推门见绿”的田园风光和大美公园城市。

以上3个城市都通过尝试不同的形式与社区居民的生活进一步融合，米兰重视居民的生活需求，伦敦积极将本地的设计灵感和行动纳入改造项目中，而成都则通过以“碳惠天府”为品牌的碳普惠机制，调动全社会自下而上地践行绿色低碳发展行动，积极主动实施国家应对气候变化战略。

在强调城市适应性基础设施建设与提高气候韧性的3个案例中，荷兰鹿特丹作为地势低平的三角洲城市代表，长期以来一直是适应技术的最佳展示地和创新试验场。结合2013年颁布的《鹿特丹适应气候变化战略》，鹿特丹积极采取NbS提升城市防洪设施，以减少海潮、风暴和雨洪灾害带来的风险，通过“灰色”“绿色”“蓝色”基础设施的有机结合，加强了荷兰境内的洪水防御系统，探索了气候变化带来的经济与就业机遇。

受到飓风“桑迪”的影响，纽约选取了灾害风险较高的区域，即曼哈顿下城区的沿海地区开展试点，颁布了《曼哈顿下城沿海气候韧性计划》，针对城区内高风险地区制定了一系列因地制宜的政策，以提升适应能力。

NbS的成功实施离不开科学规划，以旧金山湾区为代表的“绿图计划”部分展示了如何利用数学模型和地理信息系统引导社区发展，推动自然、农田与区域扩张需求之间保持平衡的战略性规划工具成为解决城市发展与自然环境之间矛盾的媒介。

9 国家类

应对气候变化需要多方参与，其中国家是最重要的引领者和推动者之一，公共部门在NbS的地区与全球推广过程中起到至关重要的作用。实施NbS对于发展中国家意义重大。一方面，NbS可以突破传统的思路和模式，通过多元创新的模式推进减缓及适应行动；另一方面，NbS有利于将《巴黎协定》制定的2℃目标及各国提出的碳中和目标同多部门的业务实现有机融合。国家推动NbS发展的重要举措包括加强各部门对NbS的认识和了解、出台利于NbS发展的政策与法规、开展国家级试点项目、将NbS纳入国家自主贡献（NDCs）承诺，以及在相关的国家战略规划中体现NbS思想等。本章选取了哥斯达黎加、中国和巴西3个国家级NbS项目与战略，展示国家如何在设计和推进NbS的过程中发挥重要作用。

9.1 哥斯达黎加国家级生态保护计划

案例亮点

历任总统在国家层面给予了关注，同时得到了多部委的支持；建立了政企合作的新模式，开创性地使用了“永久性项目融资”的金融机制，集合了多家公益和企业的基金会设立信托基金开展保护工作。

9.1.1 项目背景

哥斯达黎加是全球著名的生物多样性热点国家（特有物种的数量多且受威胁的程度高），尽管国土面积和瑞士不相上下，但其物种数量却相当于整个北美洲的总和。哥斯达黎加共有26%的陆地面积和1%的国家海域（包括17%的领海）处于保护范围内，然而这些陆地和海洋保护区及其周围的生态系统却面临着经济发展带来的巨大挑战。过度捕捞、非法捕鱼、监管不足的旅游开发、城市化、伐木、水污染、珊瑚礁退化和渔业资源枯竭等都对保护区及其邻近的土地和水域造成了威胁。尽管付出了巨大努力，哥斯达黎加实际的保护区面积却距离《生物多样性公约》（CBD）的保护面积和目标有显著差距。

9.1.2 项目介绍

为解决上述保护差距，实现联合国《生物多样性公约》保护地目标，哥斯达黎加政府通过国家保护区系统确定了“永远的哥斯达黎加”国家战略。该战略的启动和实施离不开两位总统给予的关注和大力支持。奥斯卡·阿里亚斯总统在任时期提出了“与自然和谐共处”的理念，倡导提供持续的资金以开展自然保护工作。劳拉·钦奇利亚·米兰达总统在任期间对该战略也给予了大力支持，使该战略能够顺利在2010年启动。此外，包括该国环保部、能源和通信部及自然资源保护部、外交部等在内的多个部门的部长都为该战略提供了很多支持。除了政府部门的大力支持，该战略还开创性地设立了政企合作新模式，通过建立信托基金和契约管理等模式来确保参与方的责任和义务。通常来说，对保护区长期进行保护是实现生物多样性最直接的途径，但是发展中国家通常面临保护资源缺乏和保护力度不足等情况，目前还鲜有发展中国家完成《生物多样性公约》中规定的保护目标，“永远的哥斯达黎加”国家战略的成功实施使其有望成为全球第一个实现保护目标的发展中国家，将为诸多发展中国家提供宝贵的借鉴意义。

“永远的哥斯达黎加”国家战略在2010年《生物多样性公约》第10次缔约方大会（CBD COP10）上正式启动。在时任总统劳拉·钦奇利亚·米兰达女士的号召下，林登保护信托联合戈登和贝蒂·摩尔基金会、沃尔顿家族基金会和大自然保护协会共同建立了长效的资金机制，用于开展保护工作，同时在内阁层面设立了指导委员会以监督和评估项目的实施。值得一提的是，“永远的哥斯达黎加”国家战略对参与方的权责做出了明确的规定，设立了具体的阶段目标：确定通过国家保护区系统完成《生物多样性公约》的目标并预估费用；在项目初始时期确保资金到位，由政府明确其提供的资金比例和金额；实现5 000万美元融资并成立政企合作的信托基金，由该信托基金与哥斯达黎加保护管理局签署合作协议，明确资金拨付标准。

该战略的总体目标是“到2010年在陆地区域和到2012年在海洋地区建立并维持有效管理且具有生态代表性的国家和区域综合保护区系统”。哥斯达黎加是在全国范围内成功实施生态系统服务付费（PES）的先驱国家之一。国家融资基金于1997年启动了生态系统服务付费制度，以便惠及适合开展林业活动的中小土地所有者，促进国家保护和修复森林资源。哥斯达黎加通过投资1 400万美元用于支付生态系统服务，实现了再造林6 500公顷、可持续管理天然林10 000公顷、保护私人天然林79 000公顷的目标。在资金方面，“永远的哥斯达黎加”国家战略与林登保护信托、戈登和贝蒂·摩尔基金会、大自然保护协会等伙伴合作，设立信托基金，开创了“永久项目融资”方法，学习借鉴私营部门的融资模式，尊重项目参与方之间的相互依存性，并对其进行约束，使多个投资者能够同时进行大笔投资且以目标实现为支付依据。这是私营部门为发电厂等大型项目融资的常见做法，但在非营利领域却很少使用。

9.1.3 项目成果

“永远的哥斯达黎加”国家战略的成功实施离不开国际领域对生物多样性保护的重视、稳定的国内政局、超过40年的保护经验及政府强有力的实施。该战略设立了以下三个维度的目标。

一是提升保护区内不同生态系统的物种代表性。哥斯达黎加政府在大自然保护协会的支持下，确定该国具有重要保护意义的陆地、淡水和海洋生态系统，以及旗舰物种的保护目标，这些目标侧重于填补陆地保护区系统中被遗漏的区域，并大幅增加对海洋生态系统的保护。具体来说，哥斯达黎加的海洋保护区在原基础上增加了12个区域，使其保护面积翻倍。新增的海洋保护区优先考虑了受气候变化威胁（温度升高、海平面上升和酸化）较严重的区域。不仅如此，陆地保护区占国家总陆地面积的比例也由26%扩展至26.5%。

二是加强保护区管理。评估和改善保护区的管理有效性是政府与合作方共同关注的首要任务，包括对每个保护区的目标进行更新和管理，以及结合多重任务对目标进行评估。国家保护区系统制定了统一的《提升保护区管理规范导则：政策和评估指标工具》，共包含37个指标，涵盖社会、行政、文化和自然资源管理、政策法规及财务管理5个领域。这些管理工具和方法对于现有的和未来新设立的保护区都有重要意义。国家保护区系统还在改进机构效率方面做出承诺，将设立专门的海洋部门和评估保护区管理有效性的单位。

三是应对气候变化。当前，各方已经意识到气候变化将会严重威胁哥斯达黎加的生物多样性及安全。哥斯达黎加政府在战略实施过程中首次将气候变化对保护区的影响考虑在内，在项目初期就明确了气候变化对保护区生态系统和物种带来的威胁，以及可能的减缓和适应措施；在项目执行过程中持续地对项目实施开展气候变化影响监测评估，建立了动态的适应机制和措施；到项目后期，总结了保护区生物多样性及生态系统适应气候变化和极端天气事件的战略方案，并开展试点实施。

9.1.4　项目信息

案例来源：TNC全球NbS案例库。

项目参与方：哥斯达黎加政府、林登保护信托、戈登和贝蒂·摩尔基金会、沃尔顿家族基金会、大自然保护协会。

项目地点：哥斯达黎加。

项目时间：2010—2015年。

项目链接：“永远的哥斯达黎加”国家战略，https://www.geofunders.org/documents/534。

9.2 中国生态保护红线

案例亮点

“生态保护红线”是以国家层面为主导的协调经济发展和环境保护的重要手段，强调保护优先且因地制宜，从各地区的现实情况出发，用科学的方法对生态系统进行划分；通过出台系列法规、政策和规定，以及积极利用大数据和可视化的工具平台，可以提升决策的科学性；中国正在努力探索适合发展中国家的生态保护与应对气候变化的“良方”，以便让更多的欠发达地区、人群和濒危物种受益。

9.2.1 项目背景

随着工业化与城镇化的快速发展，中国的资源与生态环境形势严峻。尽管中国生态环境保护与建设的力度逐年加大，但总体而言，资源约束压力持续增大，环境污染仍在加重，生态系统退化依然严重，生态问题更加复杂，资源环境与生态恶化趋势尚未得到逆转。已建的各类保护区在空间上存在交叉重叠，布局不够合理，生态保护效率不高，生态环境缺乏整体性保护，且严格性不足，尚未形成保障国家与区域生态安全和经济社会协调发展的空间格局（李干杰，2014）。

9.2.2 项目介绍

生态保护红线是指“在生态空间范围内具有特殊重要生态功能、必须强制性严格保护的区域，是保障和维护国家生态安全的底线和生命线，通常包括具有重要水源涵养、生物多样性维护、水土保持、防风固沙、海岸生态稳定等功能的生态功能重要区域，以及水土流失、土地沙化、石漠化、盐渍化等生态环境敏感脆弱区域”[①]。生态保护红线是国土空间规划和管理的重要制度创新，是继“18亿亩耕地红线”后，又一条被提到国家层面的“生命线”。“生态红线”概念于2011年被首次提出，随后纳入《中华人民共和国环境保护法》（2015年）。2017年，中国政府发布《关于划定并严守生态保护红线的若干意见》，在全国范围内实施生态保护红线政策。截至2020年，生态保护红线划定面积占国土总面积的25%左右。该政策的目标是将几乎所有濒危物种及其栖息地都纳入保护范围，与此同时防止洪水和沙尘暴，并提供清洁的水和其他生态系统服务，实现应对气候变化、保护生物多样性的协同作用。

生态保护红线是协调经济发展和环境保护的重要手段。21世纪初，“全国生态环境调查与评估系统平台”的结果显示，中国林地、湿地、草地等生态空间被严重挤占。对于应该通过什么方式来改变这种现象，达到既能保护生态环境又能促进经济发展的目标这一问题，时任生态环境部南京环境科学研究所所长、全国生态保护红线划定专家委员会主任委员、首席专家高吉喜提出了“生态保护红线”这个概念——把一些重要的生态区域严格保护起来，保护区之外的地域可以为工业化、城镇化所利用。

2019年6月，中国发布了《关于建立以国家公园为主体的自然保护地体系的指导意见》，要求“将生态功能重要、生态环境敏感脆弱，以及其他有必要严格保护的各类自然保护地纳入生态保护红线管控范围”。同年11月，又发布了《关于在国土空间规划中统筹划定落实三条控制线的指导意见》，其中的“三区三线”是中国国土空间规划的核心内容。“三区”是城镇空间、农业空间、生态空间三种类型空间，“三线”分别对应划定的城镇开发边界、永久基本农田、生态

---

①定义来自中共中央办公厅、国务院办公厅印发的《关于划定并严守生态保护红线的若干意见》。

保护红线三条控制线（王应临等，2020）。

另外，中国于2020年出台了《全国重要生态系统保护和修复重大工程总体规划（2021—2035年）》，明确到2035年具体的保护目标，包括森林覆盖率达到26%，森林蓄积量达到210亿立方米，天然林面积保有量稳定在2亿公顷左右，草原综合植被盖度达到60%；确保湿地面积不减少，湿地保护率提高到60%；新增水土流失综合治理面积5 640万公顷，75%以上的可治理沙化土地得到治理；海洋生态恶化状况得到全面扭转，自然海岸线保有率不低于35%；以国家公园为主体的自然保护地占陆域国土面积的18%以上，濒危野生动植物及其栖息地得到全面保护。中国生态保护红线的发展演变思路如图9-1所示。

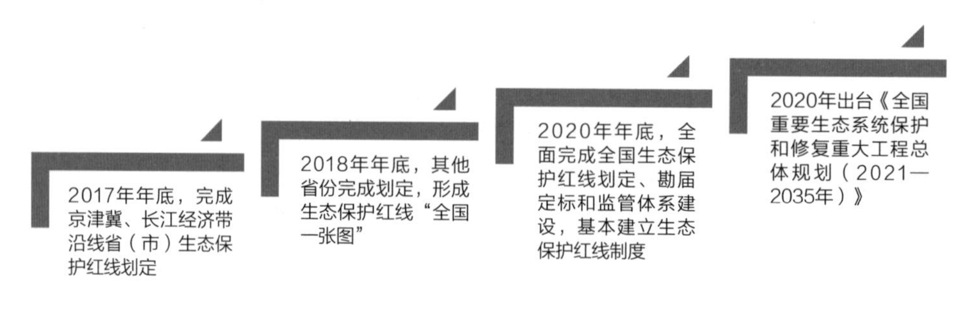

图9-1　中国生态保护红线发展演变思路

此外，中国政府对生态保护红线的落地和实施提供了巨大的支持。2017年10月，国家发展和改革委员会正式批复国家生态保护红线监管平台项目，总投资2.86亿元。国家生态保护红线监管数据库已完成设计，录入数据4类67种。该监管平台将依托卫星遥感手段和地面生态系统监测站点，形成天—空—地一体化监控网络，及时评估和预警生态风险，实时监控人类干扰活动，一旦发现破坏生态保护红线的行为立即依法依规处理，以确保生态功能不降低、面积不减少、性质不改变。

9.2.3 项目成果

经过中国生态环境部、国家发展和改革委员会、自然资源部等部门和相关省份的共同努力，通过采取国家顶层设计指导、地方组织的自上而下和自下而上两种模式相结合的方式，在科学评估、部门协调、规划协调、区域协调、陆海统筹的基础上，经专家论证和部际协调领导小组审核，目前中国已完成了包括京津冀3个省（市）、长江经济带11个省（市）和宁夏回族自治区在内的15个省（区、市）的生态保护红线划定工作。截至2018年，15个省（区、市）划定生态保护红线总面积约为61万平方千米，占其总面积的1/4左右，主要为生态功能极重要和生态环境极敏感脆弱的地区，涵盖了国家级和省级自然保护区、风景名胜区、森林公园、地质公园、世界文化自然遗产、湿地公园等各类保护地，基本实现了“应划尽划”。国家在划定的生态红线区域内实施生态修复等手段，区域内的固碳量占全国固碳量的45%，在气候减缓领域发挥了关键作用。

9.2.4 项目信息

案例来源：C+NbS合作平台。

项目参与方：中国各级政府部门。

项目地点：中国。

项目时间：2017年起。

项目链接：新闻报道 | 高吉喜：形成生态保护红线“全国一张图”，http://www.chinacses.org/hjkp_23038/hbkpxw/201811/t20181127_675124.shtml；新闻报道 | 15省份1/4面积划入生态保护红线，http://www.gov.cn/xinwen/2018-02/13/content_5266384.htm。

9.3 巴西保护水源地的“绿洲计划”

案例亮点

通过经济手段达成环境保护的目标，生态系统服务付费模式可用于生物多样性保护、应对气候变化、土壤防护、营养物质回收、水源保护等；尝试通过开发一套经济定价模型使农场主们获得切实的经济激励，以增加其土地上的森林面积，提高森林质量，并采取保护生态环境的最佳农业实践方式。

9.3.1 项目背景

气候变化与水资源安全是全球面临的重要社会挑战。在全球许多区域，气候变化带来的降水变化和冰雪融化正在显著影响水资源和水质，同时气候变化带来的生态系统破坏也给水资源安全带来了严峻的威胁。当前，全世界约有22亿人正在饮用没有经过安全处理的饮用水，42亿人没有安全管理的环境卫生服务意识，30亿人缺乏基本的洗手设施。联合国在《2018年世界水资源开发报告》中指出，全球对水资源的需求正在以每年1%的速度增长，未来30年这一需求会增加1/3。另外，自20世纪初以来，大约有2/3的森林和湿地已经消失或退化（联合国教科文组织，2018）。供需不足是水资源管理面临日益严峻的挑战的主要原因，通过保护和恢复与水有关的生态系统，包括山地、森林、湿地、河流、地下含水层和湖泊，可以提高水资源的安全性。以森林为例，据美国农业部统计，森林为美国68 000个社区的1.8亿人提供饮用水，是美国最大的饮用水来源。美国林业局管理的森林可以为美国33个州中的3 400个社区的6 600万人提供水资源，饮用水处理成本随着相关流域森林覆盖率的增加而降低。

9.3.2 项目介绍

地处亚马孙流域的巴西拥有丰富的水资源和生物多样性。然而，自20世纪70年代起，在经济高速增长的“巴西奇迹”中，巴西的自然环境为经济增长付出了沉重的代价。城市扩张，人口增长，重能源、重材料生产的模式导致巴西的环境资源面临严重的威胁，该国政府出台了一系列保护环境的法案和政策，包括很多通过经济手段达成环境保护目标的尝试。生态系统服务付费模式就是其中之一，这与NbS相关，旨在通过经济补偿的方式鼓励当地群众开展保护生态系统的工作。生态系统服务付费的逻辑非常简单：增加与环境保护相契合的经济活动的收入，以此来鼓励自然资源的可持续性使用，并惩罚掠夺性活动。生态系统服务付费模式可用于生物多样性保护、应对气候变化、土壤防护、营养物质回收、水源保护等。学术界近年来不断提倡通过经济手段在环境保护方面与现有的管制手段形成互补，而生态系统服务付费就是该提倡的产物。巴西的“绿洲计划”就是生态系统服务付费实践的成功典范。

巴西从2006年开始实施“绿洲计划”（The Oasis Project）。该计划是由非营利组织Boticario 集团自然保护基金会牵头，在地方政府的引导下联动企业（如当地自来水公司和三菱基金会）的力量共同实施的。该计划已先后在巴西阿普卡拉纳、圣保罗、南圣本图三地实施，其目标是通过保护私有土地上的原生森林来保护流域健康。

以巴西圣保罗2006年的“绿洲计划”为例，自来水公司将在圣保罗地区所获收入的1%捐助给当地政府的环境保护基金，政府再用该笔钱来奖励那些在保护原生森林工作上表现优秀的私有农场主。Boticario集团在其中起到技术支持的作用，通过他们独有的打分机制与评判标准给私有土地进行评级，并根据一定的金融换算公式决定应给予农场主多少奖励。具体来说，给予农场主的奖励可以表述为

$$奖励=Z\times X\times（1+G1+G2+G3）$$

式中，X——单位面积的补贴量，巴西雷亚尔/公顷；

Z——土地面积，公顷；

$G1$、$G2$、$G3$——三个加成项，分别代表淡水保护、自然生态系统保护及可持续农业行为，每个加成项有不同的权重：G1（0~1），G2（0~2.5），G3（0~1.5）。

除此之外，Boticario集团还通过具体的评级标准和调查问卷来衡量该项目的价值。通过这样一套经济定价模型，农场主们获得了切实的经济激励来增加他们土地上的森林面积，提高森林质量，并采取保护生态环境的最佳农业实践方式。

9.3.3 项目成果

这样的激励措施可以使整个地区的流域健康及水资源稳定，也直接让自来水公司获益，形成一个多赢的闭环。据Boticario集团估算，南圣本图通过“绿洲计划”的初步实施，增加了对1 620公顷森林的保护，并修复了3 239公顷的退化牧场。这些工作直接减少了54%的泥沙流失量、44%的土壤浊度，节省了13%~26%的水处理成本，并增加了2.8%的淡水产量。水处理成本的降低及产生的地区性的收益（减少洪灾的损失）、全球收益（碳汇）使该计划在6年之内就能收回投资成本。Boticario集团同时预测该计划在碳汇方面的价值将达到600万美元。“绿洲计划”在提升适应能力和社会与环境方面的贡献在于通过提高经济效益提升了适应能力、社会影响力（增加收入、减少贫困、气候公平等），保证了粮食安全，减少了物种灭绝和生态损失，保护了生物多样性。

9.3.4 项目信息

案例来源：UNEP-NbS案例库。

项目地点：巴西阿普卡拉纳市、圣保罗市、南圣本图市。

项目时间：2006年起。

项目链接：联合国环境规划署，https://wedocs.unep.org/bitstream/handle/20.500.11822/28898/NBS_water_security.pdf?sequence=1&isAllowed=y。

9.4 小结

哥斯达黎加是全球生物多样性与生态保护的先锋国家。通过政府前瞻性的战略规划和部署，该国于2010年成功实施了“永远的哥斯达黎加”国家战略，使哥斯达黎加成为世界上唯一一个扭转了森林退化的热带国家，并使其有可能成为第一个实现《生物多样性公约》保护地目标的发展中国家。“永远的哥斯达黎加”国家战略的启动和实施离不开历任总统在国家层面给予的关注和支持。该项目不仅获得了包括该国环保部、能源和通信部、自然资源保护部、外交部等多部门的支持，还建立了政企合作的新模式，开创性地使用了“永久性项目融资”的金融机制，集合了多家公益和企业的基金会设立信托基金开展保护工作。值得一提的是，“永远的哥斯达黎加”国家战略对各参与方的职责都做出了明确的规定，设立了明确的里程碑目标作为资金拨付依据。该战略的成功实施离不开国际领域对生物多样性保护的重视、稳定的国内政局、超过40年的保护经验及政府强有力的实施。该战略共设立了3个维度的目标，包括提升保护区内不同生态系统的物种代表性、加强保护区管理和应对气候变化。

中国致力于生态文明建设，在保护环境和应对气候变化的前提下协同发展。在2019年的联合国气候行动峰会上，中国受邀与新西兰共同推进全球NbS工作。在此期间，中国的生态保护红线政策已经被越来越多的国家关注和了解，而中国也在不断地探索和创新。原本以保护具有重要生态功能和脆弱的区域、恢复野生动植物种群为目标的生态保护红线成果，有可能被开发成为减排固碳的潜在资源，从而在科学保护自然、协同治理生态环境与应对气候变化工作之间搭建沟通合作的桥梁，为国际社会生态治理工作开创先河。生态保护红线强调保护优先且因地制宜，从各地区的现实情况出发，用科学的方法对生态系统进行划分。在管理措施上，提倡采取自上而下的设计，强调多个省份对生态保护红线实行区别化管理。在国家和部委的政策和资金支持下，中国已经在积极利用大数据和可视化的工具平台来提升决策的科学性。未来中国可以不断创新，探索出适合发展中国家的生态保护与应对气候变化的“良方”，以便让更多的欠发达地区、人群和濒危物种受益。

为保护当地宝贵的淡水资源与亚马孙森林生态系统，巴西开展了“绿洲计划”，并已在阿普卡拉纳、圣保罗、南圣本图市3个地区进行试点。“绿洲计划”以“生态系统服务付费”的概念为指导，通过经济补偿的方式鼓励当地人民开展保护生态系统的工作。该计划由非政府组织牵头，集合当地政府、基金会及自来水公司的力量，通过制定简单有效的激励机制鼓励当地社区参与，实现对森林、退化牧场、河流流域的生态保护，促进整个地区的流域健康及水流稳定。

10 平台与倡议类

NbS作为一项新兴的、全球性的、跨学科的议题，需要世界各国和利益相关方的跨界沟通与通力协作。聚集全球政府、国际组织、企业、智库、个人的力量传播NbS理念、促进NbS实践、推动NbS发展，离不开相关NbS平台与倡议在其中发挥的宣传、协调和整合作用。本章通过选取不同时期和不同区域范围内的NbS沟通宣传平台，包括在NbS发展初期发挥了基础性作用的自然与气候联盟，2019年NbS获得全球关注后兴起的“倒计时”全球倡议，以及由中国牵头发起成立的C+NbS合作平台，分析这些平台与全球性的气候行动倡议在促进NbS发展的过程中所扮演的积极角色。

10.1 自然与气候联盟

案例亮点

借助世界经济论坛、联合国气候大会、气候行动峰会等全球重大会议与平台，宣传自然在减缓与适应气候变化领域所发挥的巨大潜力；通过有效的沟通战略提高NbS在应对气候变化领域的话语权，并通过创建全球性的NbS行动社区激励更多的NbS实践；借助可视化的图集为各界直观地展示NbS在全球及各个区域可能发挥的减排潜力，进一步提高大众对NbS的认识和了解。

10.1.1 项目背景

自然与气候联盟（图10-1）是一个多方利益联盟，于2017年由联合国开发计划署、联合国环境规划署、《生物多样性公约》、世界自然保护联盟、大自然保护协会等组织联合发起成立。自然与气候联盟希望通过建立一个合作平台，召集更多的机构和组织宣传NbS的潜力和可能的机遇，从而增加NbS对于全球气候行动的贡献。自然与气候联盟的主要受众包括决策者、政策制定者、私营部门、意见领袖等，当前在全球已有超过20个合作伙伴（图10-2），旨在获得基金会、联合国机构、民间组织和私营部门的支持，增加对NbS的行动和投资，帮助其实现《巴黎协定》的目标。

图10-1　自然与气候联盟标志

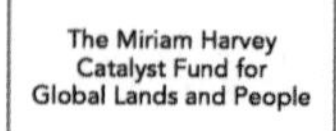

WE MEAN
BUSINESS

图10-2　自然与气候联盟的全球合作伙伴

（图片来源：自然与气候联盟）

10.1.2 项目介绍

自然与气候联盟指出，通过对生态系统的保护、恢复和可持续管理，到2030年前NbS能够为实现《巴黎协议》中规定的2℃或1.5℃温升目标贡献约 30%的减排潜力。自然与气候联盟希望通过在以下3个领域开展的活动来实现“到2030年通过NbS减排100亿吨二氧化碳当量”的目标：

①提升公众的意识，理解NbS的重要潜力和作用；

②提升和强化各国通过NbS实现国家自主减排承诺；

③提升气候融资，使更多的资金流向NbS。

自然与气候联盟的工作内容包括展示土地利用领域的气候减缓潜力、倡导优秀的土地管理科学方法、分享全球优秀案例、促进多方对话、传播科学知识等。自然与气候联盟主要通过高层对话与合作、战略动员、企业合作、青年动员及创新沟通等形式将NbS的故事尽可能多地带给受众。

10.1.3 项目成果

自2017年发起以来，自然与气候联盟每年都会在重大的国际事件或会议上发声，使与NbS相关的话题量比原来提升了3倍，NbS在应对气候变化领域的话语占有率从1%提升到了现在的12%，并形成了NbS社区，进一步扩大了NbS的影响范围；气候融资被更多地导向了与NbS相关的领域，NbS占净投资的比例从3%增长到8%；NbS的理论发展领域也突飞猛进，从2015年只有3篇学术论文提及NbS，到2019年已有100多篇NbS相关领域的论文。自然与气候联盟的沟通与合作战略取得了很大成效，通过建立一体化的沟通与交流平台把更多的企业与青年人吸纳到NbS的合作中。

10.1.4 项目信息

案例来源：C+NbS合作平台。

项目参与方：联合国开发计划署、联合国环境规划署、世界自然保护联盟、世界自然基金会、大自然保护协会等机构。

项目地点：全球（自然与气候联盟总部设于英国伦敦）。

项目时间：2017年成立。

项目链接：自然与气候联盟官网，https://nature4climate.org。

10.2 “倒计时”全球倡议

案例亮点

依托TED平台及其下属的TEDx项目，汇聚各地提出的NbS和本土案例经验并分享给全世界，扩大NbS在全球的声音和贡献，成功实现气候变化的“破圈”行动；聚焦不同层面的气候治理行为体，保障多元利益主体的广泛参与，利用TED的影响力，使气候变化议题获得了全球更多民众的广泛关注。

10.2.1 项目背景

要扭转气候变化的趋势，需要社会各界做出最积极的回应，包括坚强果断的政治领导、具有变革力量的商业愿景、每个公民的积极参与等。为此，TED平台总部与Future Stewards合作伙伴联盟共同发起“倒计时”（Countdown）全球倡议活动，旨在倡导和加速提出解决气候危机的方案，将绿色思想变为实际行动（图10-3）。该倡议活动的终极目标在于通过各方努力在2030年前至少减少一半的温室气体排放量，造就一个更安全、更清洁、更公平的世界。

图10-3 “倒计时”全球倡议标志

10.2.2　项目介绍

TEDx是由TED平台推出的一个项目，其最大的目标在于本着TED的使命，将“值得传播的想法”的精神带给全球各地的社区，倡导和帮助地方举办TED风格的分享活动。而“倒计时”全球倡议则是TEDx计划新打造的旗舰平台，旨在分享和传播各级行为体应对气候治理的解决方案和地方NbS经验，并在平台上汇聚来自全球各地解决气候危机的想法。“倒计时”全球倡议于2020年10月10日启动，为人类世界提出了5个相互关联的问题，试图通过倡导对这5个问题的讨论为清洁未来描绘蓝图，并由这5个问题引申出“倒计时”全球倡议中的5个分主题：能源、运输、材料、餐饮及自然。

该倡议活动邀请多方合作伙伴参与，地方组织、企业、城市、国家及各地公民等均可报名参加。“零碳征程”（Race to Zero）是“倒计时”全球倡议活动中提出的一项实质性全球运动，试图在《联合国气候变化框架公约》第26次缔约方大会（UNFCCC COP26）召开之前动员企业、城市、区域和投资者参与到碳减排的行动中。所有主办方都被邀请分享自己关于气候治理的案例、经验与想法，其中不同主办方的参与方式有所差异：企业主体被邀请作出切实的气候承诺（图

10-4），最大限度地减少对气候变化带来的影响；各城市和地区可以与平台相关的国际组织进行合作，展开碳减排实践；而非营利组织举办的活动则主要以一种公开分享的科普方式进行。

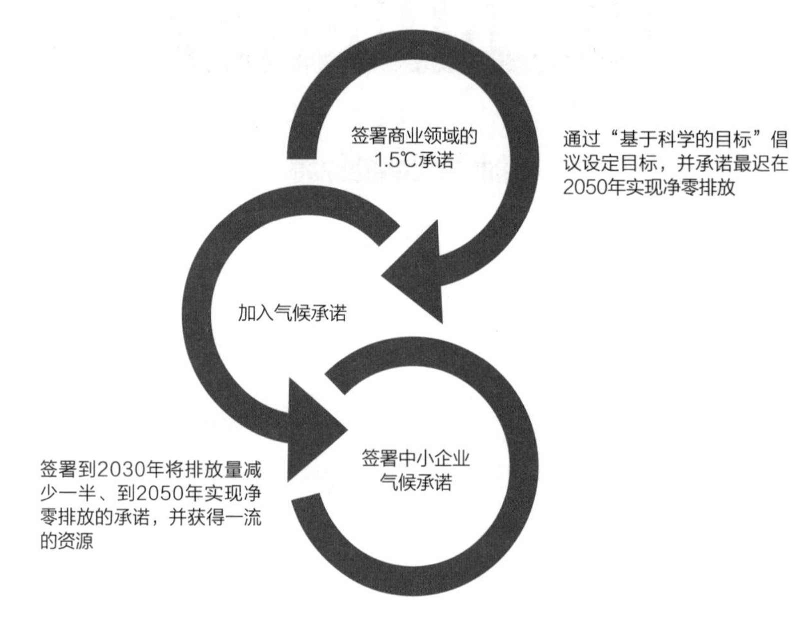

图10-4　企业参与“倒计时”全球倡议的路径

（图片来源：Countdown官网）

2020年10月10日，“倒计时”全球倡议在全球范围内通过线上发布会的方式正式启动。全球启动大会在视频平台YouTube上同步进行，有超过50位不同领域的思想家与行动者在5个分主题下分享了对气候变化的看法，呼吁各地的领导人和公民采取行动，也基于不同的视角提出了具有可操作性和研究价值的多元想法。自大会启动至2021年10月，“倒计时”全球倡议进入“加速发展”阶段，将接力棒交给各地方主办。各主办主体围绕解决气候危机的5个相互关联的问题，与当地实际情况相结合开展具有地方特色的具体活动，活动形式包括但不限于与听众一起观看分享“倒计时”全球倡议发布会的演讲和对话视频、主办当地气候对话与讲座、组织参与“倒计时”创意项目等。2021年10月12—15日，在苏格兰爱丁堡举办了为期四天的“倒计时峰会”，结合讲座、案例研究及研讨会的形式对2020—2021年的“倒计时”全球倡议活动进行整体总结。

10.2.3 项目成果

目前，在全球已有超过600个主体参与到“倒计时”全球倡议中。在中国，截至2020年12月，已经举办了近10次“倒计时”落地活动。举办主体包括一线城市、智库、非营利公益机构、企业等。上千人作为分享者抑或是听众参与到“倒计时”全球倡议活动中，这引发了中国社会媒体及学界的广泛报道和关注。

“倒计时”全球倡议预计将通过平台引领分享和传播对气候变化危机的广泛关注。在全球启动大会上对世界提出了5个相互关联的问题，并基于这5个分主题再次提出参与气候变化治理刻不容缓。在随后的一年中，气候专家、政策制定者、有远见的商业领袖及成千上万名当地组织者在当地讨论、辩论及制定策略时将这5个关键问题纳入考虑中[①]。此外，该倡议也通过汇聚各方治理力量，为解决方案的提出提供了集思广益的平台。“倒计时”全球倡议旨在宣传和改进气候领导者已经在做的工作，而非完全重新推翻再创造。到目前为止，已有几百个气候治理主体提出了有关气候问题的提案，并不断分享他们的想法和解决方案。科学家、活动家、企业家、城市规划人员、农民、首席执行官、投资者、艺术家、政府官员和其他人员聚集在一起，以找到最有效的、有据可循的想法。

“倒计时”全球倡议的终极目标是通过不断交流与分享来汇聚想法，提出解决方案，在未来通过气候峰会及后续行动进一步激活方案。该倡议得到了全球伙伴的通力合作，不断总结各治理主体提出的地方经验、案例与想法，支持了一系

---

①信息来源：TEDGoesFromIdeasToActionWithCountdown，EshaChhabra.https://www.forbes.com/sites/eshachhabra/2019/12/10/ted-goes-from-ideas-to-action-with-countdown/?sh=f4f24a96b707。

列旨在应对与气候相关挑战的跨部门项目。基于在启动大会上提出的5个相互关联的问题，该倡议制订了具体、大胆的目标，以期在2021年“倒计时峰会”上作出来自世界各地的解答，并试图在《联合国气候变化框架公约》第26次缔约方大会之前推动各治理主体作出新的承诺和切实的行动，努力达成在2030年前至少减少一半的温室气体排放量的终极目标。

10.2.4 项目信息

案例来源：C+NbS合作平台。

项目参与方：由TED与Future Stewards发起，全球伙伴响应。

项目地点：全球。

项目时间：2020—2021年。

项目链接：“倒计时”全球倡议官网，https://countdown.ted.com/。

10.3 C+NbS合作平台

案例亮点

中国首创的关注NbS这一新兴国际话题的平台，立足中国经验，面向世界需求，力争为国际进程贡献中国方案与智慧，致力于用更加强劲的NbS有效应对中国和全球可持续发展面临的挑战；本土实践与国际经验相得益彰，一流专家的学术研究和企业实际观察得以贯通；每期月度工作坊交流讨论得出的成果和共识都增长了各利益相关方对NbS这一议题在理论和实践层面的认识，使中国利益相关方快速与国际接轨。

10.3.1 项目背景

2019年9月，在纽约召开的联合国气候行动峰会将NbS确定为九大行动领域之一，邀请中国和新西兰联合牵头推进。清华大学气候变化与可持续发展研究院代表团在峰会现场学习了NbS的相关内容，了解到推进NbS需要跨学科、跨领域的交流与合作，而国内外关于NbS的系统研究还比较欠缺，国内也没有相关平台类组织把各方的力量聚合起来。

10.3.2 项目介绍

为推动国内外围绕气候变化与生物多样性的跨界交流及对NbS的系统研究，清华大学气候变化与可持续发展研究院于2020年4月开始组织搭建应对气候变化的基于自然的解决方案（C+NbS）合作平台（以下简称C+NbS合作平台），其中C既是Climate的首字母，也是China的首字母。C+NbS合作平台的标志和架构分别如图10-5和图10-6所示。

图10-5 C+NbS合作平台标志

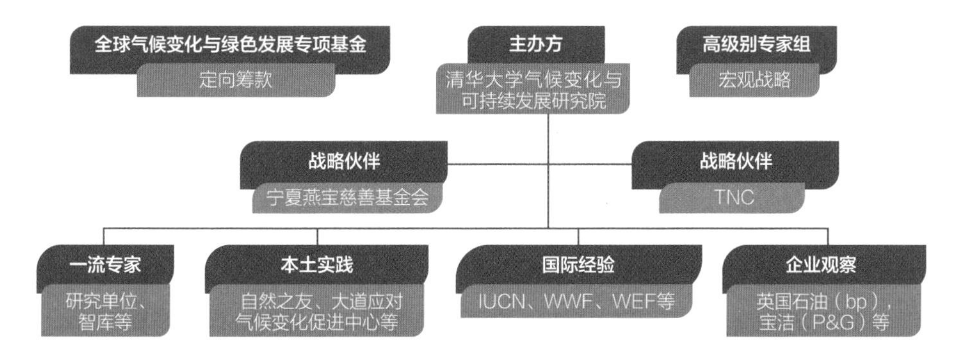

图10-6 C+NbS合作平台架构

C+NbS合作平台以“看清‘牌面’、理解内涵，了解NbS的‘前世今生’，形成初步的判断和建议”为目标，主要推进以下两方面的工作内容。

一是通过系统回顾NbS的现有研究基础，梳理“前世”，形成综合报告，产出气候变化视角下NbS的全面文献综述，同步推进综合研究与全球案例研究，相关成果于《生物多样性公约》第15次缔约方大会（COP 15）上正式发布。

二是通过组织月度工作坊，迅速聚拢NbS相关领域的国内外同行、伙伴，通过交流与互动，获取NbS领域的最新进展，从而了解 “今生”，整合资源，寻求合作。C+NbS合作平台在2020年共举办了9期月度工作坊，围绕NbS这一核心议题为每期工作坊设定了不同的讨论主题，包括NbS的概念、方法学、国际标准和经验、中国和全球实践等，旨在从理论与实践两个层面推进讨论合作，激发创新研究（表10–1）。根据统计，NbS月度工作坊的国际在线参与者覆盖四大洲，来自全球21个国家和地区，涉及多个领域和行业。

表10–1　C+NbS合作平台2020年月度工作坊内容

期数	时间	主题	简介
1	4月29日	NbS理论初探	举办线上闭门会议，邀请来自政府部门、科研院所、金融机构、国际组织、民间机构和私营部门的30余位代表探讨NbS议题及合作平台的价值、定位、内容，以及在国内国际的发展潜力
2	5月20日	NbS生物多样性专题	为迎接国际生物多样性日，邀请50余位代表参与线上闭门会议，深度解读生物多样性日提出的“以自然之道，养万物之生”（Our solutions are in nature）的内涵
3	6月24日	聚焦NbS方法学	举办线上闭门会议，邀请50余位代表参与，对生态系统和相关领域的碳排放现状、减源增汇潜力及评估方法学进行跨学科、多角度的深入研讨
4	7月20日	NbS国际经验Ⅰ	首次采用开放报名的方式，邀请中外专家聚焦分享国际经验，吸引来自政府部门、科研院所、国际组织、民间机构、全球高校、私营部门、媒体的200余位代表参与线上学习并展开热烈讨论
5	8月20日	NbS国际经验Ⅱ	线上公开活动，邀请中外专家就NbS的核心价值及在全球的先锋实践分享精彩观点，吸引了全球2 000余位代表参与线上直播，共同学习

（续表）

期数	时间	主题	简介
6	9月24日	NbS中国案例	线上公开活动，聚焦中国案例分享，邀请国内代表分享实践行动，交流NbS本土经验，吸引了全球1 600余位代表参与线上直播学习和讨论
7	10月27日	NbS与碳中和	线下与线上相结合，国家气候变化与专家委员会副主任、清华大学气候变化与可持续发展研究院学术委员会主任何建坤教授主讲中国长期低碳发展战略与转型路径，全球点击量超过12万人次
8	11月16日	TED青年专场	线下与线上相结合，举办TEDxICCSD“基于自然的解决方案”青年专场暨第二届世界大学气候变化联盟研究生论坛闭幕式，帮助青年学生开阔视野，了解NbS这一国际前沿议题，全球点击量超过32万人次
9	12月18日	NbS年度会议	线下与线上相结合，与世界经济论坛合办“2020回顾与展望”NbS年度会议，旨在加强各方合作，共同为《生物多样性公约》第15次缔约方大会助力

10.3.3 项目成果

C+NbS合作平台在NbS这一新兴热点领域逐渐发展出良好的国际化态势，引发了国内外的广泛关注与积极参与（图10–7）。目前，C+NbS合作平台已经形成了以月度工作坊为代表形式的定期国内外跨界对话机制，并通过适时调整参与形式，大大拓宽了国际化交流的程度。前三期工作坊均采取闭门定向邀请制，嘉宾都是与NbS领域相关的国内外研究机构和智库的学者、官员、专家。了解到还有其他许多想要参与交流的需求后，第四期工作坊首次开放了线上报名，吸引了来自国家相关部委、地方政府、相关研究机构、国际组织、新闻媒体和国内几十所高校等的代表报名并参与讨论。为了进一步扩大人群覆盖面，后续的沙龙活动通过中英文同步直播，为海外高校师生参与交流提供了便利。观看直播、回放和参与讨论的人数达2 000余人。第七期开启了全球直播的方式，点击量逾12万人次。第八期暨第二届世界大学气候变化联盟研究生论坛闭幕式的全球点击量达到了32万次。2020年12月，C+NbS合作平台被邀请与世界经济论坛合办主题为“基于自然的解决方案：2020回顾与展望”的年度会议，通过对话与交流进一步为2021年《生物多样性公约》第15次缔约方大会贡献智慧。此次月度工作坊在世界大学气候联盟官方Facebook账号和微博账号上同步直播，吸引了国内外人员的广泛关注。

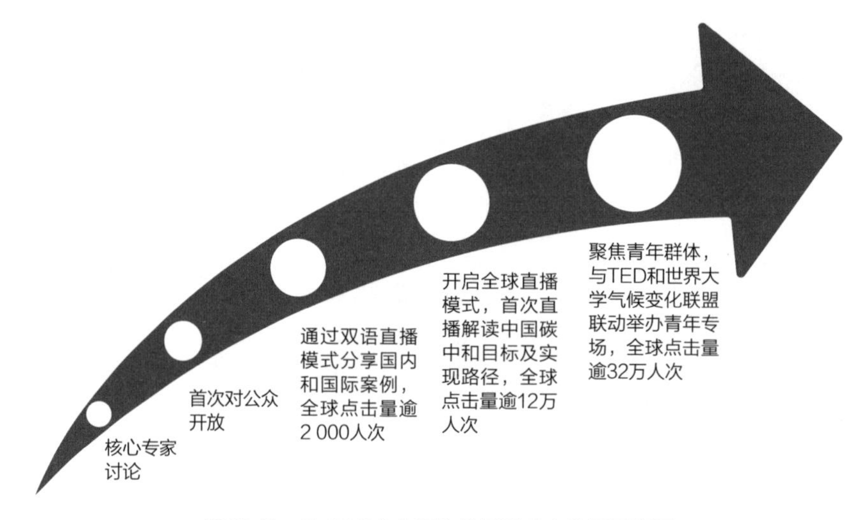

图10-7　C+NbS合作平台传播影响力发展示意图

截至2020年12月，C+NbS合作平台共组织了9期月度工作坊，并探索面向全球直播，累计收到400多家国内外机构报名参会，单场在线观众量最多达50万人次，累计参与人数超62万人。凭借迅速产生的国际影响力，C+NbS合作平台受邀与各大会议平台合作，推进在应对气候变化和保护生物多样性领域的观点传播与交流讨论。2020年10月，C+NbS合作平台受邀成为TEDx发起的“倒计时”全球倡议活动在中国的核心合作伙伴，与TED平台进行联动，两个月的时间里推出了“TEDxICCSD”系列活动，借助TED平台向全球传播中国应对气候变化的声音，在中国与世界之间架起了桥梁。

10.3.4　项目信息

案例来源：C+NbS合作平台。

项目参与方：由清华大学气候变化与可持续发展研究院主导。

项目地点：全球（C+NbS合作平台总部设于中国北京）。

项目时间：2020年起。

10.4　小结

自然与气候联盟成立于2017年，在NbS的理论与实践探索阶段发挥了重要作用。作为全球最早的专门致力于提升NbS应对气候变化潜力的国际平台之一，自然与气候联盟长期致力于提高NbS在决策者、私营部门、公众等群体中的重要程度，提倡用更多的气候资金支持开展NbS工作。

“倒计时”全球倡议于2020年10月10日发起，为交流和分享气候危机治理经验提供了一个全球平台，聚集气候多元治理主体，也为NbS项目关注度的扩大提供了重要途径。作为当前最具影响力的国际倡议，“倒计时”全球倡议在发起之时得到的关注度来源于其发起方——TED所积累的全球影响力：TED的YouTube账户有1 790万名订阅者，旗下的TEDxTalks有2 660万人订阅，Twitter账户有1 143.8万名关注者，TEDxTalks有55.3万人关注，发布的视频普遍有数百万至上千万的播放量。“倒计时”全球倡议借助TEDx项目的国际化多元大平台为NbS等气候解决方案发声，也为国际各方了解NbS提供了直达、专业、得到认证的路径。同时，还有很多国际组织、公益组织、大学智库及倡议联盟参与到该活动中。不同层次的治理主体可以通过TED平台和Future Stewards的伙伴网接触到更多企业、政府及非营利组织，为参与者与NbS领域的国际先锋增加联动的可能性。

C+NbS合作平台成立于2020年4月，以月度工作坊为代表的定期交流机制极大地促进了中国进行有效的NbS沟通与交流。C+NbS合作平台坚持立足中国国情，同时与联合国环境规划署、世界自然保护联盟、自然与气候联盟、大自然保护协会等国际机构和平台合作，向世界发出中国之声，提升中国的NbS全球胜任力。一方面，汇聚共识，为全球贡献中国智慧，在气候变化与全球治理领域提出中国方案；另一方面，积极学习国际先进的NbS科学理论与实践经验，内化以促进中国NbS的进步与发展，积极寻求与不同领域相关政策的协同。C+NbS合作平台通过每期的交流成果与战略共识在学术层面为全球智库、高校和研究机构未来开展相关联合研究奠定了思想基础，也为全球进一步在实践中落实NbS创造了条件。

11 企业类

企业是NbS重要的利益相关方，是助力将NbS从理论推广到实践的中坚力量。企业参与NbS项目的设计、投资与实施既是承担社会责任、应对社会挑战的表现，也是提高自身生产与经营可持续性、拓展业务和增加盈利的重要途径。本章选取宝洁、雀巢两家跨国公司与中国宝丰集团作为案例，为跨国公司和中国本土企业参与NbS、应对气候变化、推动绿色低碳发展提供借鉴。

11.1 宝洁公司利用NbS促进企业碳中和

案例亮点

作为全球日用消费品公司巨头，宝洁公司在可持续发展和企业社会责任方面做出了大量的尝试，除了通过创新的设计将可持续发展理念融入商品设计和包装，也成为在中国率先倡导企业参与NbS的行动者和引领者，并与多个国际组织联合开展实地项目，提升全球应对气候变化行动的信心。宝洁公司的整体战略与其NbS项目为更多的企业提供了宝贵经验。

11.1.1 项目背景

宝洁公司创始于1837年，是全世界最大的日用消费品公司之一，全球员工近11万人。宝洁公司在日用化学品市场上的知名度相当高，其产品包括洗发品、护发品、护肤品、化妆品、婴儿护理产品、妇女卫生用品、医药产品、织物、家居护理用品、个人清洁用品等。为推动全球加快应对气候变化进程，宝洁公司于2020年7月16日宣布了其2030年气候承诺，旨在通过企业内部改革和一系列保护、改善与恢复生态环境的措施推动宝洁公司在2030年前实现全球运营的碳中和。

11.1.2 项目介绍

宝洁公司承诺通过全面的能源转型减少温室气体的绝对排放量（图11-1）。当前，宝洁公司全球70%的运营地点都在使用清洁的可再生电力。到2030年，宝洁公司将实现生产过程100%的可再生电力覆盖，将温室气体的排放量减少50%。针对受技术限制在2030年前无法完全消除的排放（约3 000万吨）情况，宝洁公司将通过积极开展NbS等措施予以抵消。宝洁公司作为全球日用消费品巨头之一，通过跨界交流与国际对话传播NbS知识，推动全球的气候行动。2020年7月，宝洁公司与美国国家地理协会合作，主办名为“这是我们的家”圆桌会议，邀请企业领袖、非政府组织与青年代表共同探讨全球气候行动与NbS的贡献。宝洁公司还积极与C+NbS合作平台合作，在月度工作坊、TED青年专场为企业参与NbS发声。

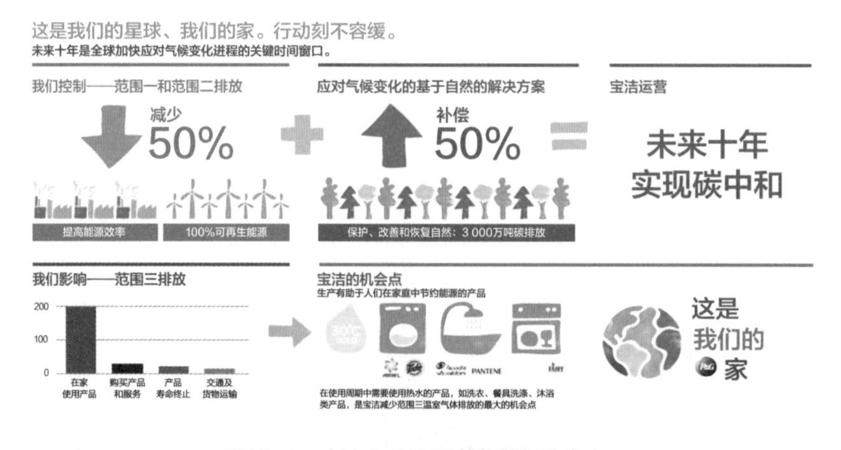

图11-1 宝洁公司2030年气候行动方案

宝洁公司宣布与保护国际基金会和世界自然基金会携手合作，共同确定并资助一系列旨在保护、改善与恢复森林、湿地、草原和泥炭地等重要生态系统的项目，在抵消碳排放之外，为环境和社会经济带来协同效益，进而保护自然环境并改善当地居民的生活。宝洁公司正在制订详细的项目资助计划，以为全球各地相关项目提供支持。已确定的项目如下：

一是与保护国际基金会合作开展菲律宾巴拉望保护项目——保护、改善和恢复巴拉望这一世界上第四大“不可替代”地区的红树林生态系统；

二是与世界自然基金会合作推进大西洋森林生态恢复计划——在巴西东海岸的大西洋森林为森林景观恢复奠定基础，并对生物多样性、水资源和包括粮食安全在内的其他协同领域产生重大的积极影响；

三是与美国植树节基金会合作建立常青树联盟——团结企业、社区和公民的力量，通过有效措施保护受气候变化影响的生活必需品，如通过植树来恢复加利福尼亚州被野火烧毁的地区，并改善美国的森林环境。

为解决家庭用水带来的能源消耗与温室气体排放问题，2020年在达沃斯世界经济论坛上，宝洁公司在世界经济论坛、世界可持续发展工商理事会、世界银行2030水资源小组的支持下牵头启动了“50升水家庭节水联盟”（50L Home Coalition），致力于以创新和技术实现将家庭每人每天的用水量限制在50升的目标（在部分国家，人均家庭用水量最高可达每天500升），开发和推广家庭用水系统创新，帮助解决城市水危机。该联盟计划在2020年内于1～2个城市开展试点，并在2020—2025年实现3～4个城市的创新试点，以确定规模化的路线并在全球范围内推广。如果“50升水家庭节水联盟”得以在中国城市地区推广，将每年节水约140亿立方米，并减少1 500万吨二氧化碳当量的温室气体排放。除此之外，宝洁公司也在企业生产与经营内部通过绿色信息公开、可持续供应链改革、推广可回收包装等方式减少资源消耗与温室气体排放。

11.1.3　项目信息

案例来源：C+NbS合作平台。

项目参与方：宝洁公司及保护国际基金会、世界自然基金会、美国植树节基金会等伙伴。

项目地点：宝洁公司全球工厂所在地及菲律宾、巴西、美国等地的自然保护区。

项目时间：2020年起。

项目链接：宝洁公司官网，https://www.pg.com.cn/environmental-sustainability/。

11.2 雀巢公司推广可再生农业与再造林以实现净零排放

案例亮点

作为世界上最大的食品饮料制造商，雀巢公司开展的企业可持续发展和气候变化领域的工作与NbS有天然的联系，其中绝大多数（95%）的温室气体排放来自其供应链的活动，这使NbS可以直接贡献于该公司的长期碳中和战略；雀巢公司对全球公布的“净零碳排放路线图”为国内外企业实现碳中和提供了借鉴。

11.2.1 项目背景

雀巢公司创始于1867年，是世界上最大的食品饮料制造商，也是最大的跨国公司之一，在全球拥有500多家工厂，最初以婴儿食品起家，以生产巧克力棒和速溶咖啡闻名，目前的主要产品有速溶咖啡、炼乳、奶粉、婴儿食品、奶酪、巧克力制品、糖果、速饮茶等数十种，公司销售额的98%来自国外，因此被称为“最国际化的跨国集团”。过去10年，雀巢公司致力于通过提高能效、使用更清洁能源及投资可再生资源的方式，来减少与食品、饮料生产配送相关的温室气体排放。2012年与2013年，雀巢公司在全球环境信息研究中心（CDP）①的“气候披

①全球环境信息研究中心（CDP）是国际性的非营利组织，其前身为“碳披露项目”，旨在为公司与城市提供测量、披露、管理和分享重要环境信息的系统。CDP指数衡量的是富时全球股票指数中的前500强公司在削减碳排放方面所做的努力，以及在信息披露的透明性方面的表现。

露领袖指数”和“气候绩效领袖指数”中获得了最高分，被评为“全球典范”。2019年9月，雀巢公司宣布“到2050年实现温室气体净零排放”的目标，并于纽约召开的联合国气候行动峰会上签署了联合国“企业1.5℃温升控制目标”。

11.2.2 项目介绍

2020年12月，雀巢公司发布了“净零碳排放路线图”，以2018年为基准年，承诺到2025年、2030年和2050年分别实现碳排放减少20%、减少50%和净零碳排放的目标。具体策略方面，雀巢公司着力于扶持农户和供应商以共同推动可再生农业发展，在未来10年内种植数亿棵树，并在2025年实现100%使用可再生电力，以实现净零碳排放目标。

雀巢公司的2050年净零碳排放承诺遵循“科学碳目标倡议”[①]（SBTi）的标准，与外部顾问机构合作计算了基准年的碳足迹（图11-2）。据统计，其基准年（2018年）的温室气体排放总量为1.13亿吨二氧化碳当量，包括范围1（燃烧煤、天然气和公司车队所用燃料产生的碳排放等直接排放）、范围2（电力生产等间接排放）和范围3（包括采购和已售产品在消费过程中产生的碳排放在内的所有其他间接排放）。其中，来源于直接运营的碳排放（范围1和范围2）仅占雀巢公司温室气体排放总量的5%，而绝大多数（95%）的温室气体排放来自其供应链的活动（范围3）。雀巢公司2050年净零碳排放承诺中包含原料采购、产品生产、产品包装、物流管理和差旅及员工通勤等部门所产生的碳排放[②]，这部分活动于目标的基准年（2018年）共排放温室气体9 200万吨二氧化碳当量。其中，来自供应

① “科学碳目标倡议”由全球环境信息研究中心、世界资源研究所、世界自然基金会和联合国全球契约组织合作发起，旨在为企业提供设定基于气候科学减排目标的清晰指导框架，以确保企业所设定的温室气体减排幅度和速度的目标与《巴黎协定》中控制全球温升幅度小于2℃的目标一致。

②目前，雀巢公司的净零碳排放承诺不包含下列碳排放：消费者使用的已售产品（1 270万吨二氧化碳当量）；购买的服务、租赁资产、资本品、投资（860万吨二氧化碳当量）。

链上游的原料采购共排放温室气体6 560万吨二氧化碳当量，占承诺范围内排放的71.4%。原料采购的总排放量中有90%来自奶业、畜禽业、土壤和森林，这为雀巢公司开展减缓气候变化的NbS提供了机遇。

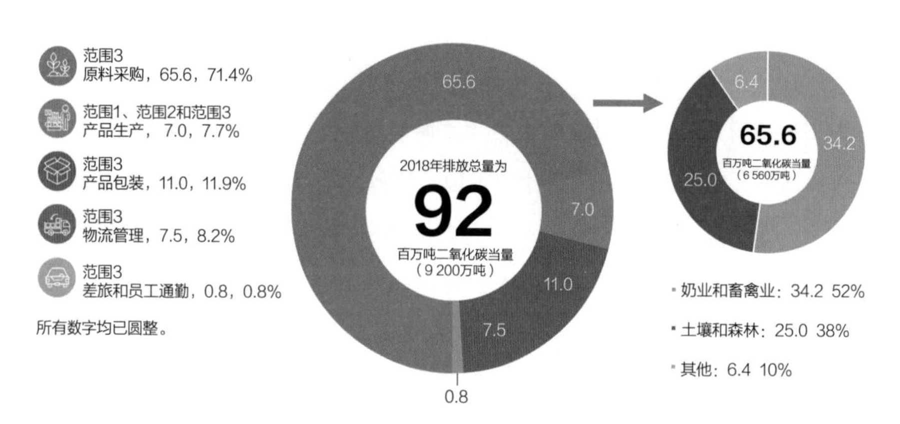

图11-2 雀巢公司承诺范围内的2018年排放总量（以百万吨二氧化碳当量计）

作为全球重要的食品和饮料生产商，雀巢公司近2/3的温室气体排放来自农业，因此实施再生农业和造林等NbS是该公司实现净零碳排放的重要策略。雀巢公司计划通过增加碳汇，到2030年从大气中移除或抵消1 300万吨二氧化碳当量的温室气体排放，具体措施如下：

①在水源和野生动物生态走廊周围种植植被，在改善水质的同时捕获碳；

②在牧场中栽种树木，使牧草长势更好，增加饲料产量，实现协同增效；

③尽可能采用有机肥料，使用本地堆肥（原料如咖啡果肉），积聚有机质并改善土壤结构及其储碳潜力；

④采取更加可持续的农业实践，实施免耕、轮作和覆盖作物等农业技术，从而避免氮耗竭，减少水土流失，控制病虫草害；

⑤种植乔木和灌木，形成自然保护屏障，保护作物免于恶劣天气和水土流失的危害；

⑥开展农林复合措施，利用遮阴树等保护咖啡类作物免受高温伤害，同时增加土壤中的有机质，提高土壤保持水分和储存碳的能力；

⑦修复森林和泥炭地，在增加碳汇的同时维持地下水位，降低火灾风险（雀巢公司投资了250万瑞士法郎用于保护和恢复科特迪瓦的重要森林资源）。

在再生农业领域，雀巢公司已经与超过50万农户及超过15万家供应商开展合作，支持其采取可再生农业实践，改善土壤健康，维持并恢复生态系统多样性。为此，雀巢公司采取了溢价、增加采购数量和共同投资等激励手段，预期到2030年将采购超过1 400万吨来自可再生农业的原料，从而推动市场需求，鼓励农户开展可再生农业。同时，雀巢公司还在扩大实施“造林计划”，未来10年每年将在其原料采购地植树2 000万棵。通过农林复合，树木将为农作物提供更多的庇荫以使其免于高温伤害，从而获得更高的作物产量；再造林还能显著提高碳汇，并改善生物多样性和土壤健康。雀巢公司承诺在其棕榈油、大豆等主要商品的供应链中到2022年实现“零毁林”。

在公司运营方面，雀巢公司预期在2025年前在其位于187个国家的800个工作场所中实现100%可再生电力，并努力实现低排放交通转型。雀巢公司还采取了水资源保护和再生措施，并努力解决运营中的食物浪费问题。此外，雀巢公司积极进行科技创新，推广新的植物基食品和饮料，并调整配方使产品对环境更加友好。

11.2.3　项目信息

案例来源：C+NbS合作平台。

项目参与方：雀巢公司及其全球工厂与供应商。

项目地点：雀巢公司全球工厂所在地与原料产地。

项目时间：2020年起。

相关链接：雀巢公司官网，https://www.nestle.com/sites/default/files/2020-12/nestle-net-zero-roadmap-cn.pdf。

11.3 宝丰集团的农光互补与绿氢创新实践

案例亮点

宝丰集团为企业参与NbS贡献了有益的创新实践，为碳中和目标的落实提供了可规模化推广的发展方向，“农光互补一体化项目”充分结合地方的资源和环境特色，在加快新能源替代化石能源的发展步伐外实现了促进就业等多重效益；“太阳能电解制氢储能及应用研究示范项目”为化工产业发展带来了广阔的绿色原料空间，打破了传统的原料限制，置换出更大的环境容量，也解锁了更多的自然潜力。

11.3.1 项目背景

宁夏回族自治区地处中国西北内陆地区，属于温带大陆性干旱、半干旱气候，水资源相对缺乏，是中国西部欠发达的省区之一。当地煤炭等矿产资源富集，同时该地区也是中国太阳辐射的高能区之一，其地势海拔高，日照时间长，光伏资源丰富。

宁夏宝丰能源集团股份有限公司（以下简称宝丰集团）以化工、新能源、现代农业为主要业务，积极在“新能源替代化石能源推动碳中和”领域开展创新实践。2020年，宝丰集团已建成中国当前规模最大的“甲醇、烯烃、聚乙烯、聚丙烯、精细化工、新能源”一体化循环经济产业集群和全球单体最大的集中式光伏电站。

11.3.2 项目介绍

当前，宝丰集团每年新增减排二氧化碳55.25万吨，同时承诺力争用10年的时间完成50%的碳减排，用20年的时间实现企业碳中和，以响应中国“在2060年前实现碳中和”的国家目标。宝丰集团重视以NbS解决环境问题，立足源头治理，开辟了“荒漠化土地可持续生态治理—创新建设新能源产业—利用新能源电解水制氢—制氢与现代化工融合协同发展”的碳中和新路径，从根本上建立了由末端治理转化为源头治理的技术路线，通过科学探索与技术创新开发得到经济可行的解决方案。

1. 农光互补一体化项目

宝丰集团利用宁夏地区丰富的光伏资源，将光伏发电与农业生产相结合，大规模推广“农光互补一体化项目”（图11-3），通过植被种植、土壤改良等措施，对宁夏银川市黄河东岸16万亩荒漠化土地进行生态治理，将植被覆盖率由原来的不足30%提高到85%，并在此基础上充分发挥宁夏枸杞特色产业优势，大力发展万亩优质枸杞基地，并因地制宜地种植经济林、经济草等其他经济作物。为综合利用土地和光照资源，2016年宝丰集团在枸杞上方建设了1 000兆瓦峰值太阳能发电项目，开创了“板上发电、板下种植”的“一地多用、农光互补”特色产业发展新模式，实现了经济、生态、社会效益“三赢”的目标。

图11-3 宝丰集团“光伏+农业”产业实景

技术优势方面，宝丰集团的太阳能发电项目全部采用华为智能光伏解决方案及最高效的单晶硅[①]组件，运用国际领先的带倾角平单轴跟踪技术，已建成全球单体最大的集中式光伏电站，转化率高，太阳能发电量较传统发电量提高了20%以上。该项目综合使用智能化监控管理后台和无人机巡检的新模式，能够及时发现故障并进行处理，有效保障了光伏电站的安全、高效、稳定运行。

环境效益方面，该项目用新能源发电替代火力发电，为企业直供“绿电”并解决部分社会用电。每年节约标准煤55.7万吨，减少二氧化碳排放169.3万吨，减少二氧化硫排放5.1万吨，减少氮氧化物排放2.6万吨，减少粉尘排放46.2万吨，相当于植树近9 000万棵，为宁夏地区传统能源的后续发展年增加约223万吨环境容量，有利于促进节能减排，加快新能源替代化石能源的发展步伐。同时，光伏板遮光减少约70%的蒸发量，有效提升了当地农业生态系统的碳汇能力。

社会意义方面，项目采用的“农光互补”新模式把生态效益转化为经济效益和产业优势，每年仅光伏发电组件清洁维护和枸杞采摘劳务用工就可以解决周边移民群众就业达8万人次，为每户增收4万多元，有效巩固拓展了脱贫攻坚成果，促进了乡村全面振兴。枸杞产业与新能源项目结合构建的新型特色产业发展模式（图11-4）加快了新技术、新产业的发展，催生了吸纳就业和创造财富的新业态，为社会带来了更多的有益价值。

2. 太阳能电解制氢储能及应用研究示范项目

在发展传统新能源转型与生态修复工作的同时，宝丰集团还积极科技创新，开发了以“绿电制绿氢”为核心的“太阳能电解制氢储能及应用研究示范项目”（图11-5）。在全球能源结构向清洁化、低碳化转型的背景下，氢能作为来源广

---

①光伏太阳能电池板分为单晶硅、多晶硅和非晶硅3种，现多数以单晶硅和多晶硅为材料，二者的原子结构排列不同，单晶硅是有序排列，多晶硅是无序排列，单晶硅的转化效率更高、使用寿命更长、效果更好。

图11-4 宁夏宝丰集团枸杞特色产业扶贫项目

图11-5 电解水制氢装置

泛、灵活高效、零排放的可再生能源，被视为最理想、最具发展潜力的清洁能源，是减少温室气体排放、应对气候变化、实现低碳工业的重要途径。宝丰集团应用全球先进技术，以“太阳能发电+电解水制氢”的最优组合，用太阳能生产绿色电能，再用绿色电能作为动力，通过电解水制取出绿氢和绿氧，替代化石原料和燃料，直供化工系统生产聚乙烯、聚丙烯等上百种高端化工产品，形成了一条完整的碳中和产业链条。“国家级太阳能电解水制氢综合示范项目”是目前已知的全球单厂最大、单台产能最大的电解水制氢项目，建成后每年可生产绿氢 2 亿立方米（标况），副产绿氧 1 亿立方米（标况）。

技术创新方面，太阳能电解水制氢是以太阳能为一次能源、以水为媒介生产二次能源氢气的过程。在“太阳能发电+电解水制氢”的最优组合下生产得到的氢气是真正意义上的“绿色氢气”。该项目引进了单套产能1 000米^3（标况）/时的电解槽，以及气化分离器、氢气纯化等装置系统，其先进性已达到国内先进水平，能耗低、转化率高，在技术上、工艺上、使用上保证了氢气的质量，生产的氢气质量高、纯度达到99.999%的国标高纯氢气标准，满足了精尖领域需求，对推广发展清洁能源具有重要的示范意义。

成本优势方面，该项目通过科技创新提高转化率，在实现制氢装置长周期高负荷运行、提高设备利用率的同时，有效降低了制氢的综合成本，制氢系统的电耗为4.5～5千瓦时/米^3（标况）氢气，绿氢成本可降低至0.7元/米^3（标况），实现可再生能源向高端化工新材料的有效转化。

环境效益方面，该项目全部建成后每年可减少煤炭资源消耗31.75万吨，减少二氧化碳排放约55.25万吨。此外，弃风弃光①是当前制约中国，尤其是西北地区可再生能源发展的重要因素，实施太阳能电解制氢储能项目可以有效消纳宁夏回族自治区内的弃光，其环境效益显著。

11.3.3 项目信息

案例来源：C+NbS合作平台。

项目参与方：宁夏宝丰能源集团股份有限公司。

项目地点：宝丰集团位于中国宁夏回族自治区的产业基地。

项目时间：农光互补一体化项目（2013年起）；太阳能电解制氢储能及应用

①弃风弃光是受限于某种原因被迫放弃风、水、光能，停止相应发电机组或减少其发电量，也可以是光伏电站的发电量大于电力系统最大传输电量+负荷消纳电量。弃风弃光的主要原因在于电源、电网、负荷3个系统要素。

研究示范项目（2020年起）。

11.4 小结

宝洁公司是全球日用消费品巨头，在全球80多个国家设有工厂及分公司，在家居护理、美发美容、婴儿及家庭护理、食品及饮料等领域经营了300多个品牌的产品，畅销160多个国家和地区。宝洁公司发布了2030年气候承诺，旨在通过企业内部改革及一系列保护、改善和恢复生态环境的措施，推动其在2030年前实现全球运营的碳中和；通过美国国家地理、C+NbS合作平台等传播碳中和与NbS理念，为企业参与NbS提供方案；通过与保护国际基金会、世界自然基金会合作，在菲律宾、巴西、美国、德国等地投资开展生态保护等项目。宝洁公司正通过企业承诺、平台发声与合作投资的方式，参与倡导全球的基于自然的气候行动。

雀巢公司是世界上最大的食品制造商，也是最大的跨国公司之一，以巧克力制品、速溶咖啡、婴儿食品、奶粉等产品闻名。受产品类型与原材料的影响，雀巢公司的温室气体排放主要源于农业生产。为加强气候行动以减缓气候变化，雀巢公司于2019年9月宣布了2050年净零碳排放目标，并于2020年12月公开了其“净零碳排放路线图”，利用科学工具详细评估了公司的温室气体排放情况，并以畜牧业、农业、林业等部门为重点，开展可再生农业和再造林等基于自然的减排行动。雀巢公司为企业科学评估碳足迹、提高信息公开与透明度、制定明确的碳中和路径提供了宝贵的参考。

宝丰集团立足于化石能源替代与生态修复的综合能源创新实践，有力地践行了“坚持人与自然和谐共生”的生态文明核心理念：其“农光互补一体化项目”依托现有的自然资源，推动新能源替代化石能源发展进程，不仅为节约资源、保护环境增添动力，同时也为当地贫困群众提供了良好的就业机会，具有重要的社会意义；其“太阳能电解制氢储能及应用研究示范项目”利用阳光与水两样简单易得的自然资源制取氢气，实现储能，顺应了国际、国内能源市场的未来发展趋势，将综合实现降本增效和节能减排，引领绿氢产业高质量发展，具有可操作、可示范和可推广的重要现实意义。

参考文献

- Aryn Baker, Mbar Toubab, Senegal. Can a 4,815-Mile Wall of Trees Help Curb Climate Change in Africa [EB]. Time, 2019.
- Bhattarai B. Community forest and forest management in Nepal[J]. American Journal of Environmental Protection, 2016，4(3)：79-91.
- BSDC. Valuing the SDG Prize in Food and Agriculture [R]. 2016.
- Cohen-Shacham E, Walters G, Janzen C，et al. Nature-based solutions to address global societal challenges [R]. IUCN: Gland, Switzerland, 2016.
- Dohong A. Strategy and Approach in Restoring Degraded Peatlands in Indonesia [A]// The Institute of Foresters of Australia (IFA) Biennial Conference [C]. Tropical Forestry: Innovation and Change in the Asia Pacific Region, 2017.
- European Union. Towards an EU research and innovation policy agenda for nature-based solutions & renaturing cities [R]. 2015.
- FOLU. Growing Better Global Report [R]. 2019.
- Galab S, Prudhvikar Reddy P, Sree Rama Raju D，et al. Impact Assessment of Zero Budget Natural Farming in Andhra Pradesh – Kharif 2018-19: A comprehensive approach using crop cutting experiments [R]. Telangana, India: Centre for Economic and Social Studies, 2019.
- Goffner D, Sinare H，Gordon L J. The Great Green Wall for the Sahara and the Sahel Initiative as an opportunity to enhance resilience in Sahelian landscapes and livelihoods[J]. Regional Environmental Change, 2019，19(5)：1417-1428.
- Griscom B W, Adams J, Ellis P W, et al. Natural climate solutions[J]. Proceedings of the National Academy of Sciences, 2017,114(44)：11645-11650.
- IUCN. No time to lose: make full use of nature-based solutions in the post-2012 climate change regime [R]. Gland, Switzerland: IUCN, 2009.
- IUCN. Global Standard for Nature-based Solutions: A user-friendly framework for the

verification, design and scaling up of NbS [R]. Gland, Switzerland: IUCN, 2020.

- Lieth H，Whittaker R H, et al. Primary productivity of the biosphere [J]. Springer Science & Business Media, 2012(14): 203-215.
- Ministry of Forest and Soil Conservation. Persistence and change: review of 30 years of community forestry in Nepal [R]. Kathmandu: Ministry of Forest and Soil Conservation, 2013.
- Nellemann C, Corcoran E, et al. Blue carbon: the role of healthy oceans in binding carbon: a rapid response assessment [R]. UNEP, 2009.
- Pardo R. Back to the future: Nepal's new forestry legislation [J]. Journal of Forestry, 1993, 91 (6): 22-26.
- Pokharel R K, Rayamajhi S, Tiwari K R. Nepal's community forestry: need of better governance [R]. Global Perspectives on Sustainable Forest Management, 2012.
- Rythu Sadhikara Samstha. Andhra Pradesh Zero Budget Natural Farming (APZBNF): A systemwide transformational programme [R]. Andhra Pradesh: Department of Agriculture, Government of Andhra Pradesh, 2019.
- Sunderland T, Abanda F, de Camino R V, et al. Sustainable forestry and food security and nutrition [R]. CFS-HLPE/FAO, Technical Report 11, 2013.
- World Bank. Nepal Overview [DB]. Washington, DC: World Bank, 2014.
- WWF. Concept Note: Heritage Colombia (HECO): Maximizing the contributions of sustainably managed landscapes in Colombia for achievement of climate goals [R]. Green Climate Fund, 2019.
- Ye Q. Ways of training individual ecological civilization under nature social conditions[R]. Scientific Communism, 1984.
- 冯艺佳，赵晶，王向荣. 绿图计划作为美国城市发展媒介的构建方法与推动力探究[J]. 风景园林，2015（9）：62-69.
- 贺庆棠. 生态文明建设与基于自然的解决方案[J]. 中国林业产业，2019（3）：77-80.
- 联合国教科文组织. 基于自然的水资源解决方案——联合国世界水发展报告[M]. 中国水资源战略研究会（全球水伙伴中国委员会），编译. 北京: 中国水利水电出版社，2018.

- 李干杰. 守护良好生态环境这个最普惠的民生福祉[N]. 人民日报，2019-06-03（009）.
- 李干杰. “生态保护红线”——确保国家生态安全的生命线[J]. 求是，2014（2）：44-46.
- 李干杰. 以习近平生态文明思想为指导 坚决打好污染防治攻坚战[J]. 行政管理改革，2018a（11）：4-11.
- 李干杰. 以习近平生态文明思想为指导，动员全社会力量建设美丽中国[J]. 中国人大，2018b（15）：45-50.
- 廖茂林. 共谋全球生态文明建设之路的理论认知及实践路径[J]. 企业经济，2020，39（7）：131-137.
- 李鹏宇. 植万亿棵树：应对气候变化 贡献中国碳中和目标[J]. 可持续发展经济导刊，2020，19（10）：45-46.
- 刘解龙. 论绿色发展时代的精准扶贫精准脱贫[J]. 贵州省党校学报，2016（5）：92-98.
- 刘静. 中国特色社会主义生态文明建设研究[D]. 北京：中共中央党校，2011.
- 卢风. 论基于自然的解决方案（NbS）与生态文明[J]. 福建师范大学学报（哲学社会科学版），2020（5）：44-53，169.
- 王应临，赵智聪. 自然保护地与生态保护红线关系研究[J]. 中国园林，2020，36（8）：20-24.
- 习近平. 决胜全面建成小康社会，夺取新时代中国特色社会主义伟大胜利[N]. 人民日报，2017-10-28（001）.
- 恩格斯. 自然辩证法[M]. 中共中央马克思恩格斯列宁斯大林著作编译局，译. 北京：人民出版社，1971.
- 邹海贵. 代际正义与关注社会弱势群体利益——基于现代社会救助（保障）制度道德正当性的分析[J]. 中南林业科技大学学报（社会科学版），2012（1）：43-46.

附录1：IUCN基于自然的解决方案全球标准

本书案例筛选标准的重要基础是世界自然保护联盟于2020年7月发布的《IUCN基于自然的解决方案全球标准》（以下简称《全球标准》）。该标准旨在为使用者提供一个用于设计和验证NbS的稳健框架，以解决一个或多个社会挑战，从而产生预期的成效。世界自然保护联盟鼓励国家政府、城市和地方政府、规划者、企业、捐赠者、包括开发银行在内的金融机构和非营利组织使用这一标准。

在《全球标准》中，世界自然保护联盟提出了优秀的NbS应适用的8项准则与28项指标，见下表。

《全球标准》的8项准则与28条指标

	准则		指标
1	NbS应有效应对社会挑战	1.1	优先考虑对权利持有者和受益者而言最紧迫的社会问题
		1.2	清楚了解并记录所解决的社会问题
		1.3	确定产生的人类福祉、设定基准并定期评估
2	应根据尺度来设计NbS	2.1	NbS的设计应认识到经济、社会和生态系统之间的相互作用并作出响应
		2.2	NbS应与其他相关措施互补，并联合不同部门产生协同作用
		2.3	NbS的设计应纳入干预场地以外区域的风险识别和风险管理

（续表）

	准则		指标
3	NbS应带来生物多样性净增长和生态系统完整性	3.1	NbS行动必须对基于证据的评估作出直接响应，评估内容包括生态系统的现状、退化及丧失的主要驱动力
		3.2	识别、设立基准并阶段性评估清晰的、可测量的生物多样性保护成效
		3.3	监测并阶段性评估NbS可能对自然造成的不利影响
		3.4	识别加强生态系统整体性与连通性的机会并整合到NbS策略中
4	NbS应具有经济可行性	4.1	确认和记录NbS项目的直接和间接成本及效益，包括谁承担成本谁受益
		4.2	采用成本有效性研究支持NbS的决策，包括相关法规和补贴可能带来的影响
		4.3	NbS设计时应与备选方案比照其有效性，并充分考虑相关的外部效应
		4.4	NbS设计应考虑市场、公共、自愿承诺等多种资金来源并保证资金使用合规
5	NbS应基于包容、透明和赋权的治理过程	5.1	在实施NbS前，应与所有利益相关方商定和明确反馈与申诉机制
		5.2	保证NbS的参与过程基于相互尊重和平等，不分性别、年龄和社会地位，并维护“原住民的自由，事前和知情同意权”（FPIC）
		5.3	应识别NbS直接和间接影响的所有利益相关方，并保证其能够参与NbS干预措施的全部过程
		5.4	清楚记录决策过程，并对所有参与及受影响的利益相关方权益的诉求作出响应
		5.5	当NbS的范围超出管辖区域时，应建立利益相关方联合决策机制
6	NbS应在首要目标和其他多种效益间公正地权衡	6.1	明确NbS干预措施不同方案的权衡，以及潜在成本和效益，并告知相关的保障措施和改进措施
		6.2	承认和尊重利益相关方在土地及其他自然资源方面的权利与责任
		6.3	定期检查已建立的保障措施，以确保各方遵守商定的权衡界限，并且不破坏整个NbS的稳定性

（续表）

	准则		指标
7	NbS应基于证据进行适应性管理	7.1	制定NbS策略，并以此为基础开展定期监测和评估
		7.2	制定监测与评估方案，并应用于NbS干预措施全生命周期
		7.3	建立迭代学习框架，使适应性管理在NbS干预措施全生命周期中不断改进
8	NbS应具有可持续性并在适当的辖区内主流化	8.1	分享和交流NbS在实施、规划中的经验教训，以此带来更多积极的改变
		8.2	以NbS促进政策和法规的完善，有助于NbS的应用和主流化
		8.3	NbS有助于实现全球及国家层面在增进人类福祉、应对气候变化、保护生物多样性和保障人权等方面的目标，包括《联合国原住民权利宣言》（UNDRIP）

以上8项准则相互关联，共同对NbS在环境、经济与社会中发挥作用的综合能力和实施过程中的透明性、监督保障措施等方面给予明确的评价标准。

NbS的概念示意图
（来源：世界自然保护联盟）

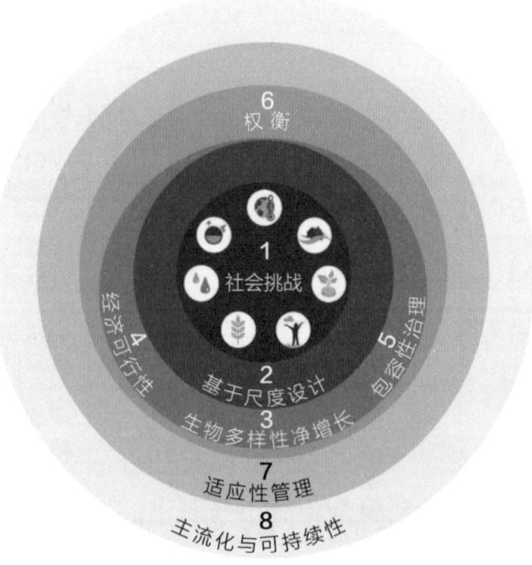

NbS《全球标准》8项准则的相互关系
（来源：世界自然保护联盟）

准则1：NbS应有效应对社会挑战

准则1的目的是确保NbS的设计能够应对有可能或即将受到直接影响的人群所识别出的社会挑战；所有利益相关方，尤其是权利拥有者和NbS的受益方，必须参与识别优先需要解决的社会挑战的决策过程（准则5）。

准则2：应根据尺度来设计NbS

准则2希望在设计NbS时，能将动态景观的复杂性和不确定性考虑进去，其范围不仅包括生物学、物理学或地理学，也包括经济制度、政策框架和文化方面。NbS将根据利益相关方对景观不同因素相互作用的情况来进行设计，在考虑景观的部分特征、景观本身情况和外部环境的基础上，采用三级尺度框架。例如，针对某地行政村内的几户居民来说，了解诸如文化价值、法律、土壤、森林和水等因素的相互作用非常重要，因为它们与评估不利变化的风险或创造有利变化的可能性有关。NbS的设计力求在保证生态系统生产能力的情况下，也能更好地维持人类福祉。

准则3：NbS应带来生物多样性净增长和生态系统完整性

NbS起源于生态系统的产品与服务，因而强烈依赖生态系统的健康程度。生物多样性丧失与生态系统的变化对该系统的功能与整体性具有显著影响。因此，NbS的设计与实施必须避免破坏系统的整体性，应有前瞻性地努力加强生态系统的功能性与连通性。这样做也能够确保NbS的长期恢复力与延续性。

准则4：NbS应具有经济可行性

NbS成功的关键取决于投资回报率、干预措施的效率和有效性，以及成本和效益分配的公平性。准则4要求在设计阶段和监测实施阶段充分考虑经济可行性。为了使NbS具有可持续性，必须认真考虑经济方面的问题，而且最需要考虑的是短期行动需要在长期（跨代）目标和计划的背景下制定。如果不能充分考虑经济可行性，NbS可能会成为短期项目。在项目完成后，其收益将不复存在，还可能使景观和社区的状况更加恶化。创新的基于证据的自然价值评估工具，以及NbS贡献于市场和就业的初衷，都鼓励NbS创新融资模式（包括混合融资），以增加其长期成功的可能性。

准则5：NbS应基于包容、透明和赋权的治理过程

准则5要求NbS承认、回应各利益相关方的关切，特别是权利所有者的关切，并主动让他们参与。实践证明，良好的治理安排不仅可以减少干预措施不可持续的风险，而且可以提高其社会“经营许可”。相反，本来计划周密的行动如果在治理方面考虑不周，则可能会对利益分配和成本分摊的正当性产生不利影响。作为最低要求，NbS要遵循现行法律法规，明确法律责任和义务。然而，与自然资源常见的情况一样，需要建立赋权当地社区和其他利益相关方并鼓励他们参与的辅助机制。

准则6：NbS应在首要目标和其他多种效益间公正地权衡

在土地和自然资源管理中，权衡是不可避免的。生态系统提供了丰富的不同效益，但并不是每个人都以同样的方式重视它们。虽然权衡不能避免，但可以有效和公平地加以管理。准则6要求NbS实施者承认这些权衡的必要性，并遵循公平、透明和包容的过程来协调和管理不同的时间和空间。这需要进行可信的评估、充分的披露，并在受影响最大的利益相关方之间就如何解决权衡问题达成共识。潜在受影响的利益相关方就任何对当地发展和生计产生损害的解决方案及补偿进行公平和透明的协商，是NbS取得长期成效的基础。至关重要的是，必须认识到权衡是有社会和生态限度的，超过这个限度，某些价值或利益可能会永远失

去，这意味着需要采取必要的保障措施，尤其需要确保生态系统的完整性和生态系统服务的稳定性不被破坏。

准则7：NbS应基于证据进行适应性管理

准则7要求NbS的实施计划应包含有关条款，以适应性管理应对不确定性，并成为有效利用生态系统恢复力的一种选择。由于生态系统的复杂、动态和自组织性质，在管理大多数生态系统时存在一定程度的不确定性。这意味着生态系统具有更大的恢复力，其为应对不可预计的社会、经济或气候事件提供了更广泛的选择。适应性管理的基础是利用基于证据的方法及传统知识，定期监测和评价所提供的证据和事实。通过主动采用适应性管理方法，NbS可以在整个项目周期中持续发挥作用，并将投资冗余和搁浅的风险降至最低。

准则8：NbS应具有可持续性并在适当的辖区内主流化

准则8要求NbS干预措施在制定和实施中有长期可持续发展的视野；同时，重视跨部门、国家和其他政策框架。NbS主流化有多种方法，然而所有方法都依赖于策略沟通和对外扩展。潜在的受众包括个人（如公众、学者）、机构（如国家政府、初创企业、企业和组织）和全球网络（如联合国可持续发展目标、《巴黎协定》）。

附录2：入选案例初级筛选指标得分情况

下表展示了在初级筛选阶段，8个类别中28个最终入选案例各个指标的表现情况（各指标详细说明与打分标准见3.2节）。值得注意的是，由于不同类型的NbS案例功能与特点各异，因此比较分析各指标的得分时仅限于同一案例类别内部的横向比较。

入选案例初级筛选指标得分

案例类别	案例名称	气候变化的减缓与适应得分	生物多样性保护得分	应对多重社会挑战得分	实现多重目标的协同得分	促进当地经济发展得分	关怀弱势群体得分	总分
林业类	哥伦比亚国家级森林保护计划	3	3	3	3	2	2	16
	尼泊尔社区林业运动	3	3	3	3	3	3	18
	“植万亿棵树领军者”倡议	3	3	1	3	1	1	12
草地类	非洲“绿色长城”计划	3	3	3	3	3	3	18
	中国毛乌素沙地治理	2	3	3	3	3	3	17
农业类	印度基于自然的“零预算农业”	3	1	3	3	3	3	16
	中国气候智慧型主要粮食作物生产	3	1	3	2	3	1	13
	中国杭州“三好农业”实践	1	1	3	3	3	1	12
湿地类	印度尼西亚泥炭地保护	3	1	2	2	2	1	11
	中国盐城黄海湿地生态修复	1	3	1	2	2	0	9
	中国东营湿地城市建设	1	2	2	2	2	0	9
	中国—东盟红树林保护与修复	2	2	3	3	2	2	14
	南太平洋小岛国海洋保护	1	1	2	2	3	1	10

（续表）

案例类别	案例名称	气候变化的减缓与适应得分	生物多样性保护得分	应对多重社会挑战得分	实现多重目标的协同得分	促进当地经济发展得分	关怀弱势群体得分	总分
城市类	意大利米兰基于自然的旧城改造	2	1	3	3	1	0	10
	英国伦敦可持续城市绿色建设	3	3	3	3	1	2	15
	中国成都公园城市建设	3	0	3	3	3	0	12
	荷兰鹿特丹气候适应性基础设施	3	1	3	3	0	0	10
	美国曼哈顿下城区气候适应性计划	3	1	3	3	0	0	10
	美国旧金山湾区“绿图计划”	3	1	3	3	0	0	10
国家类	哥斯达黎加国家级生态保护计划	3	3	2	3	2	1	14
	中国生态保护红线	3	2	2	3	1	1	12
	巴西保护水源地的“绿洲计划”	3	1	3	3	3	2	15
平台与倡议类	自然与气候联盟	3	2	1	3	1	1	11
	“倒计时”全球倡议	3	2	1	3	0	1	10
	C+NbS合作平台	3	3	3	2	1	1	13
企业类	宝洁公司利用NbS促进企业碳中和	3	2	2	3	3	2	15
	雀巢公司推广可再生农业公司与再造林以实现净零排放	3	1	2	3	3	3	15
	宝丰集团的农光互补与绿氢创新实践	3	1	2	3	3	3	15

附录3：入选案例高级评估维度具体表现

在初级筛选的基础上，本书基于治理模式、金融与市场机制、保障多元行为体及妇女参与3项高级评估维度（第3.3节），总结各案例的突出亮点并予以评分。下表展示了这一分析结果。

入选案例高级评估维度的亮点分析

案例类别	案例名称	治理模式	金融与市场机制	保障多元行为主体及妇女参与
林业类	哥伦比亚国家级森林保护计划	自上而下、政企合作	永久性项目融资、可撤销的过渡基金、生态系统付费、生态补偿机制、碳税项目	国家及地方政府、民间组织、企业、高校和社区的多方参与，关注气候公平和性别平等议题
	尼泊尔社区林业运动	自下而上的森林使用者小组	由政府部门、地方政府和非政府组织提供资金，森林使用者小组创造收入	保证妇女参与的选举机制，鼓励贫困人口和社区、土著、达利特斯人参与
	“植万亿棵树领军者”倡议	整合所有项目并提供资金和政治支持的平台	—	全球企业，国际组织，56个国家的政府、组织和个人参与
草地类	非洲“绿色长城”计划	自上而下和自下而上两种模式的有机结合	—	非洲国家、资助方（世界银行、欧盟、联合国等）、社区参与
	中国毛乌素沙地治理	自上而下、精准施策，创新防沙治沙模式，防用结合	—	中国中央政府，内蒙古自治区、陕西省和宁夏回族自治区政府与当地群众
农业类	印度基于自然的“零预算农业”	自上而下，推广可再生农业措施，开展农业大户培训和能力建设	—	政府、非政府组织、研究机构、当地农民、女性农民

（续表）

案例类别	案例名称	治理模式	金融与市场机制	保障多元行为主体及妇女参与
农业类	中国气候智慧型主要粮食作物生产	自上而下，对农业系统进行改造并重新确定发展方向	—	中国农业农村部、世界银行、全球环境基金合作
	中国杭州“三好农业”实践	从绿色消费推动流域生态产业发展，创新商业模式，探索长效机制	慈善信托增加了更加多元化且具有影响力的投资口径	政府、非政府组织、企业、农民
湿地类	印度尼西亚泥炭地保护	社区伙伴关系、统一预警体系	—	利用社区伙伴关系发动社区力量
	中国盐城黄海湿地生态修复	自上而下、统筹协调	—	—
	中国东营湿地城市建设	通过建立完善的指标体系规划发展	—	—
	中国—东盟红树林保护与修复	社区协议保护机制，统一修复指南	—	社区协议保护机制保证政府、民间社会组织和企业等的共同参与
	南太平洋小岛国海洋保护	设立执行委员会协调运营	—	非政府组织、当地政府与社区合作
城市类	意大利米兰基于自然的旧城改造	—	—	“领养一片绿地”等社区参与项目
	英国伦敦可持续城市绿色建设	自上而下与自下而上两种模式的深度融合	开发商融资+共同融资	社区技能培训
	中国成都公园城市建设	—	设施租赁、社会投资	居民参与“碳惠天府”
	荷兰鹿特丹气候适应性基础设施	政府针对城市进行适应性管理	—	—
	美国曼哈顿下城区气候适应性计划	—	政府融资	—
	美国旧金山湾区“绿图计划”	以非政府组织为主导，成立专门委员会	非政府组织筹资+农田信托	—

（续表）

案例类别	案例名称	治理模式	金融与市场机制	保障多元行为主体及妇女参与
国家类	哥斯达黎加国家级生态保护计划	政企合作	信托基金、生态系统服务付费	明确绩效目标的拨付
	中国生态保护红线	因地制宜、大数据使用	统筹协调、自上而下	—
	巴西保护水源地的“绿洲计划”	—	生态系统服务付费	—
平台与倡议类	自然与气候联盟	高层沟通机制	自然与气候联盟战略平台	强调青年动员
	“倒计时”全球倡议	各地特色讲座与研讨会	依托TED平台	讲座及研讨的网络开放
	C+NbS合作平台	跨界月度工作坊	依托清华大学资源	论坛在网络上同步直播
企业类	宝洁公司利用NbS促进企业碳中和	圆桌探讨、非政府组织携手	产业链全覆盖	牵头鼓励多方主体参与
	雀巢公司推广可再生农业与再造林以实现净零排放	外部机构评估、全生产链协作	对可再生农业溢价、增加采购数量和共同投资	鼓励农户改革种植方法
	宝丰集团的农光互补与绿氢创新实践	技术支撑氢能发展	—	就业扶贫

附录4：缩略语表

缩略语

缩写	中文全称	英文全称
AU	非洲联盟	African Union
BSDC	商业与可持续发展委员会	Business and Sustainable Development Commission
CBD	《生物多样性公约》	*Convention on Biological Diversity*
CI	保护国际基金会	Conservation International
COP	缔约方大会	Conference of Parties
FAO	联合国粮食及农业组织	Food and Agriculture Organization of the United Nations
FOLU	粮食与土地利用联盟	Food and Land Use Coalition
GCAI	气候变化全球行动	Global Climate Action Initiative
GEF	全球环境基金	Global Environment Fund
GEI	永续全球环境研究所	Global Environmental Institute
GIS	地理信息系统	Geographic Information System
IADB	美洲开发银行	Inter-American Development Bank
ICCSD	清华大学气候变化与可持续发展研究院	Institute of Climate Change and Sustainable Development, Tsinghua University
IUCN	世界自然保护联盟	International Union for Conservation of Nature
N4C	自然与气候联盟	Nature4Climate
NbS	基于自然的解决方案	Nature-based Solutions
NDCs	国家自主贡献	Nationally Determined Contributions
NGO	非政府组织	Non-Governmental Organization
PES	生态系统服务付费	Payment for Ecosystem Services
PFP	永久性项目融资	Project Finance for Permanence

（续表）

缩写	中文全称	英文全称
RAMSAR	关于特别是作为水禽栖息地的国际重要湿地公约	Convention on Wetlands of International Importance Especially as Waterfowl Habitat
SBTi	科学碳目标倡议	Science Based Targets Initiative
SDGs	可持续发展目标	Sustainable Development Goals
TNC	大自然保护协会	The Nature Conservancy
UNCCD	《联合国防治荒漠化公约》	*United Nations Convention to Combat Desertification*
UNDP	联合国开发计划署	United Nations Development Programme
UNEP	联合国环境规划署	United Nations Environment Programme
UNESCO	联合国教科文组织	United Nations Educational, Scientific and Cultural Organization
UNFCCC	《联合国气候变化框架公约》	*United Nations Framework Convention on Climate Change*
UNOPS	联合国项目事务署	United Nations Office for Project Services
WB	世界银行	World Bank
WCC	世界自然保护大会	World Conservation Congress
WCS	国际野生生物保护学会	Wildlife Conservation Society
WEF	世界经济论坛	World Economic Forum
WRI	世界资源研究所	World Resources Institute
WWF	世界自然基金会	World Wildlife Fund
ZBNF	基于自然的“零预算农业”	Zero Budget Natural Farming